Operator Theory
Advances and Applications
Vol. 60

Editor
I. Gohberg

Floquet Theory for Partial Differential Equations

Peter Kuchment

Springer Basel AG

Author

Peter Kuchment
Department of Mathematics and Statistics
Wichita State University
Wichita, KS 67208-0033
USA

A CIP catalogue record for this book is available from the Library of Congress, Washington D.C., USA

Deutsche Bibliothek Cataloging-in-Publication Data
Kuchment, Peter:
Floquet theory for partial differential equations / Peter
Kuchment. – Basel ; Boston ; Berlin : Birkhäuser, 1993
(Operator theory ; Vol. 60)
ISBN 978-3-0348-9686-3 ISBN 978-3-0348-8573-7 (eBook)
DOI 10.1007/978-3-0348-8573-7
NE: GT

Originally published by Birkhäuser Verlag in 1993
Softcover reprint of the hardcover 1st edition 1993
Printed on acid-free paper produced from chlorine-free pulp
Cover design: Heinz Hiltbrunner, Basel

ISBN 978-3-0348-9686-3

To my Teacher, Professor Selim Krein

TABLE OF CONTENTS

Introduction

Linear differential equations with periodic coefficients constitute a well developed part of the theory of ordinary differential equations [17, 94, 156, 177, 178, 272, 389]. They arise in many physical and technical applications [177, 178, 272]. A new wave of interest in this subject has been stimulated during the last two decades by the development of the inverse scattering method for integration of nonlinear differential equations. This has led to significant progress in this traditional area [27, 71, 72, 111-119, 250, 276, 277, 284, 286, 287, 312, 313, 337, 349, 354, 392, 393, 403, 404]. At the same time, many theoretical and applied problems lead to periodic partial differential equations. We can mention, for instance, quantum mechanics [14, 18, 40, 54, 60, 91, 92, 107, 123, 157-160, 192, 193, 204, 315, 367, 412, 414, 415, 417], hydrodynamics [179, 180], elasticity theory [395], the theory of guided waves [87-89, 208, 300], homogenization theory [29, 41, 348], direct and inverse scattering [175, 206, 216, 314, 388, 406-408], parametric resonance theory [122, 178], and spectral theory and spectral geometry [103-105, 381, 382, 389].

There is a significant distinction between the cases of ordinary and partial differential periodic equations. The main tool of the theory of periodic ordinary differential equations is the so-called Floquet theory [17, 94, 120, 156, 177, 267, 272, 389]. Its central result is the following theorem (sometimes called Floquet-Lyapunov theorem) [120, 267]. Consider a system

$$\frac{du}{dt} = A(t)u \tag{1}$$

with a periodic matrix coefficient $A(t)$. (We shall assume that, modulo a change of the variable, the period is 1.) Then there is a substitution $v = B(t)u$ with an invertible 1-periodic matrix $B(t)$ that converts (1) into a system with a constant matrix coefficient A_0. By the famous L. Euler's theorem [106], any solution of such a system is a linear combination of the exponential-polynomial solutions:

$$v(t) = e^{i\lambda t} \left(\sum_{k=0}^{n} a_k t^k \right).$$

Returning to the original unknown function, we obtain an expansion of all solutions of (1) into the solutions of the form

$$(2) \qquad u(t) = e^{i\lambda t} \left(\sum_{k=0}^{n} a_k(t)\, t^k \right).$$

Here the vector valued functions $a_k(t)$ are 1-periodic. Solutions (2) are usually called Floquet solutions. The non-zero complex number $z = exp\ i\lambda$ is the Floquet exponent (or Floquet multiplier) of the solution. The complex number λ is said to be the quasi-momentum of the solution. (This name has come from quantum theory of solids [18, 414].) Classical results [17, 94, 177, 178, 272, 389] show that various properties of equation (1) can be deduced from the distribution of Floquet exponents: stability, solvability of non-homogeneous equations, exponential dichotomy, spectral theory, etc.

There are two standard ways of proving the Floquet theorem. One of them is to find the aforementioned periodic substitution $B(t)$, and to apply Euler's theorem to the resulting constant coefficient equation. Another approach is to consider the monodromy operator, i.e. the shift along the trajectories of the system onto the period. The Floquet expansion of solutions corresponds to the Jordan representation of this operator. If one tries to carry over these two methods to periodic partial differential equations, one will face major difficulties. First of all, it is impossible in general to convert a partial differential equation that is periodic with respect to several variables into an equation with constant coefficients. Secondly, the monodromy operator can be correctly defined only for time-periodic evolution equations under the assumption that the corresponding Cauchy problem is correctly posed. This approach cannot be used, for instance, in the case of a stationary Schrödinger operator with a potential that is periodic in several variables. Even in the case of time-periodic evolution equations the approaches analogous to those applied to ordinary differential equations do not always work. In order to develop the Floquet theory for periodic partial differential equations, we need a new point of view; some significant progress has been made during the last decade. My main goal is to give an up to date account of this theory.

My approach to the Floquet theory is analogous to the one adopted in [99, 273, 274, 324] for the case of constant coefficient partial differential equations. The results and technique of V. Palamodov [321-326, 421] play an especially important role.

Let me describe briefly the contents of the book.

Chapter 1 contains some preliminary information on functional analysis, operator theory and complex analysis. The main results that will play the major role in the following chapters are concentrated in sections 1.7 and 1.8. They are rather technical, and I recommend to the reader who is interested mainly in differential equations to skip chapter 1 on the first reading. The statements of the main results of chapters 2-6 can be understood without referring to sections 1.7 and 1.8. The proofs, however, depend strongly on chapter 1.

Chapter 2 provides the definitions and properties of transforms that play the role of the Fourier transform.

Chapter 3 is devoted to the principal results of the Floquet theory for hypoelliptic equations.

In chapter 4 it is shown how the Floquet theory is related to other properties of periodic equations (for instance, to spectral theory).

Chapter 5 treats in more detail the case of evolution equations of hypoelliptic (in particular, parabolic) type. Writing this chapter, I found out that it was impossible to cover in one chapter even the case of time-periodic parabolic equations. I plan to give a more complete account of this topic in another publication.

Chapter 6 contains a brief consideration of related problems: equations with deviating arguments, more general groups of periods, etc.

Every chapter has a final section devoted to additional comments and references.

I would like to mention that the main emphasis of this book is on the Floquet theory, so many related topics (spectral problems, positive solutions, non-commutative groups of periods, etc.) are just touched on. Some of them (like non-commutative Floquet theory) are still in the beginning stages of development. My presentation of these subjects is very sketchy, so the reader who is interested in these areas is advised to refer to the corresponding literature. For instance, there are several books devoted completely or partially to the spectral theory of periodic operators [94, 341, 372, 389], so I do not go into the details of this deep theory, and present only some basic results, sometimes without complete proofs.

The numeration of theorems, formulas, etc. reflects the numbers of the chapter, the section and the item inside the section. For instance, the first theorem in section 1.1 will be referred to as theorem 1 inside the section, and as theorem 1.1.1 from other sections.

I am indebted to many people whose help enabled me to write this book. I would like to mention, first of all, the role of three people: Professor S. Krein, whose guidance and help has played a very important role in my life, and in this research in particular; Professor L. Zelenko, who made the suggestion to start this research, and with whom we obtained first results; Professor V. Palamodov, whose results and ideas were on many occasions my major sources of information and inspiration. I am thankful to Professors C. Berenstein, G. DaPrato, N. Firsova, B. Gramsch, S. Helgason, W. Kaballo, Yu. Korobeinik, E. Landis, J. Leiterer, V. Lin, A. Lunardi, F. Mantlik, A. Miloslavskii, A. Pankov, V. Papanicolaou, I. Pinchover, L. Ronkin, M. Shubin, B.A. Taylor, M. Zaidenberg, and V. Zhikov for valuable discussions and information. I am indebted to Professors A. Elcrat, I. Gohberg, K. Lancaster, K. Miller, L. Mogilevskaya, P. Parker and V. Papanicolaou and to my graduate student S. Lissianoi for their helpful comments on parts of the manuscript. I would like to thank Mrs. P. Altum, Mrs. C. MacDonald, Dr. L. Mogilevskaya and Ms. A. Olson for typing the text. I appreciate the help and understanding that I have had from my family. The work on some results of sections 1.5, 5.2 and 5.3 was partially supported by the NSF Grant No. DMS-9102111, and I would like to thank the National Science Foundation for this support.

CHAPTER 1

HOLOMORPHIC FREDHOLM OPERATOR FUNCTIONS

The purpose of this chapter is to set down some facts from operator theory, functional analysis and complex analysis that will be frequently used in the main body of the book. We do not intend to present here the whole theory of Fredholm operator functions, because it would take a book itself. The reader can find some references on this subject in the final section.

Sections 1.1-1.6 provide an account of some tools from functional and complex analysis that will be frequently used.

The main results of the chapter are contained in sections 1.7 and 1.8.

The present chapter terminates in 1.9 with some comments and references.

Some parts of this chapter are rather technical and involve, for instance, a lot of complex analysis in several variables. This might cause some difficulties for a reader specializing in PDEs. All main results of the book are formulated, however, in terms that are standard for the theory of partial differential equations and can be understood without reading the first chapter (except some necessary notations). In order to understand the proofs of the results, the reader should refer from time to time to the first chapter.

1.1. Lifting and open mapping theorems

We begin with some notations. Let E and F be Banach spaces. We denote by $L(E,F)$ the space of all continuous linear operators $A:E\to F$ equipped with the operator norm and set $L(E) = L(E,E)$. By $S(E)$ we denote the unit sphere $\{x\in E \mid \|x\| = 1\} \subset E$. The identity mapping on a set $\mathfrak{X}$ is denoted by id_X or I_X (in particular, $I_E = id_E$ is the unit linear operator in E). We use the notations id or I when the domain of the mapping is clear from the context. For a mapping $f\colon \mathfrak{X}\to Z$ we denote by $f|_J$ its restriction to a subset $J\subset \mathfrak{X}$.

Let E be a Banach space, $N \subset E$ – a closed subspace, and E/N – the corresponding factor-space. It is well known that the canonical projection operator $\Pi: E \to E/N$ does not possess in general any continuous linear right inverse operator $T: E/N \to E$, $\Pi \circ T = id$. Existence of such right inverse operator is obviously equivalent to the existence of a continuous linear projection $P: E \to N$ (just define $P = I - T \circ \Pi$). The known result of [261] says, however, that for every E which is not isomorphic to Hilbert space, there exists a closed subspace $N \subset E$ that does not possess any continuous linear projections.

The following beautiful result shows that omitting the linearity (and even less, additivity) condition we gain the existence of a non-linear right inverse operator.

Theorem 1.1.1 (Bartle and Graves [36]). *For every $\lambda > 1$ there exists a continuous homogeneous map $\rho: E/N \to E$ such that $\Pi \circ \rho = id_{E/N}$, and*

$$\| \rho(x) \|_E \leq \lambda \| x \|_{E/N}, \; \forall x \in E/N.$$

Proof. It suffices to construct a continuous map ψ of the unit sphere $S(E/N)$ into E such that $\Pi \psi(x) = x$ and $\| \psi(x) \| \leq \lambda \| x \|$. Then the map $h(x) = \| x \| \psi(x / \| x \|)$ (and $h(0) = 0$) has all the required properties, except that it is only positively homogeneous. A homogeneous map ρ can now be defined by

$$\rho(x) = \frac{1}{2} (h(x) - h(-x))$$

in the real case, and by

$$\rho(x) = (2\pi)^{-1} \int_0^{2\pi} e^{-i\phi} h(e^{i\phi} x) d\phi$$

in the complex one.

Let us denote by $\Pi^{-1}(x)_\lambda$ $(x \in S(E/N))$ the intersection of the preimage $\Pi^{-1}(x)$ with the ball $\{x: x \in E, \; \| x \| \leq \lambda\}$. This set is closed and convex. We are looking for a continuous map $\psi: S(E/N) \to E$ such that $\psi(x) \in \Pi^{-1}(x)_\lambda$ for all $x \in S(E/N)$. We construct this mapping as the limit of a sequence of maps ψ_i for which

$$\psi_i(x) \in U_{2^{-i}}(\Pi^{-1}(x)_\lambda).$$

Here $U_\epsilon(A)$ denotes the open ϵ-neighborhood of a set A.

We start with constructing ψ_1. For $x_0 \in S(E/N)$ we choose an element $\hat{x}_0 \in \Pi^{-1}(x_0)_\lambda$ and consider the map

$$\psi_{1,x_0}(x) \equiv x_0.$$

On the sphere $S(E/N)$ there is a neighborhood V_{x_0} of x_0 such that for all points x of it we have

$$\psi_{1,x_0}(x) = \hat{x}_0 \in U_{2^{-1}}(\Pi^{-1}(x)_\lambda) .$$

Covering $S(E/N)$ by the neighborhoods V_x , and choosing a locally-finite subcover V_{x_α} and a partition of unity $\eta_\alpha(x)$, we can define

$$\psi_1(x) = \sum_\alpha \eta_\alpha(x)\, \psi_{1,x_\alpha}(x).$$

Since $U_{2^{-1}}(\Pi^{-1}(x)_\lambda)$ is convex, the map is as required.

We now construct inductively a sequence of maps $\psi_j(x)$ such that

$$\psi_j(x) \in U_{2^{-j}}(\Pi^{-1}(x)_\lambda)$$

and

$$\|\psi_{j+1}(x) - \psi_j(x)\| \le 2^{-(j-1)}.$$

By the inductive hypothesis, the set $U_{2^{-j}}(\psi_j(x) \cap \Pi^{-1}(x)_\lambda)$ is non-empty and convex for each $x \in S(E/N)$. We need only to ensure that

$$\psi_{j+1}(x) \in U_{2^{-j-1}}(U_{2^{-j}}(\psi_j(x)) \cap \Pi^{-1}(x)_\lambda) .$$

We can achieve this in a neighborhood of each point x_0 by means of a constant map ψ_{j+1,x_0}; we then "paste together" by means of a partition of unity, as this was done before. It is clear that the sequence ψ_j converges to a mapping ψ that satisfies our requirements. The proof is complete.

Let $\mathfrak{X}$ be a topological space. We write $C(\mathfrak{X},E)$ for the Banach space of all bounded continuous functions $f\colon \mathfrak{X}\to E$ with the norm $\|f\|_C = \sup_{\tau\in\mathfrak{X}} \|f(\tau)\|_E$. The map $\Pi\colon E\to E/N$ induces a map $\Pi_*\colon C(\mathfrak{X},E)\to C(\mathfrak{X},E/N)$.

The following result about lifting of continuous functions is an immediate consequence of the theorem above.

Theorem 1.1.2. (on lifting) *The map Π_* is an epimorphism of $C(\mathfrak{X},E)$ onto $C(\mathfrak{X},E/N)$.*

Proof. ρ_* is a right inverse operator to Π_*.

In fact, if a lifting of our mapping into E/N is given over some "good" subspace of $\mathfrak{X}$, it can be extended to a lifting of the whole mapping. The next result makes this statement precise.

Corollary 1.1.3. *Let $\mathfrak{X}$ be a CW-complex, $Y\subset\mathfrak{X}$ a sub-complex and $f\colon \mathfrak{X}\to E/N$ and $f_1\colon Y\to E$ continuous maps such that $\Pi\circ f_1 = f|_Y$. Then there is a continuous map $\tilde f\colon \mathfrak{X}\to E$ such that $\tilde f|_Y = f_1$ and $\Pi\circ\tilde f = f$.*

To prove this we construct in accordance with the preceding theorem a map $f_2\colon \mathfrak{X}\to E$ such that $\Pi\circ f_2 = f$. Consider the map $(f_1-f_2)|_Y : Y\to N$. Since N is contractible, we can extend this map to a map $f_3\colon \mathfrak{X}\to N$ (extending by skeleton). The map $\tilde f = f_2 + f_3$ satisfies all the requirements of the corollary.

We shall use theorem 1 in the following equivalent form:

Corollary 1.1.4. *For any surjective operator $A\in L(E,F)$ there exists a continuous homogeneous bounded right inverse mapping $\rho\colon F\to E$ (i.e. $A\circ\rho = I$).*

We shall need a stronger version of theorem 1. Some refinement of the proof leads to the analogous result for the case of Fréchet spaces.

Theorem 1.1.5. *Let E and F be Fréchet spaces and $A\colon E\to F$ be a continuous linear surjective mapping. Then it possesses a homogeneous continuous right inverse mapping*

$\rho : F \to E$.

The proof of the last theorem and some related discussion can be found in the papers [293, 294, 242].

We shall need also some open mapping theorems. The first one is the famous Banach's theorem [30].

Theorem 1.1.6. *Every continuous surjective linear mapping of Fréchet spaces is open.*

This result follows, for instance, from the stronger assertion contained in theorem 1.1.5.

A locally convex vector space is called a *space of the type* (β), if it is the inductive limit of a family of Banach spaces. Inductive limits of sequences of Frechet spaces are of the type (β). The class of spaces of the type (β) is stable under countable inductive limits. A locally convex vector space is called an $\mathfrak{UF}$-space, if it is the union of a countable family of its Frechet subspaces. The next theorem was proved by A.Grothendieck [152].

Theorem 1.1.7. *Every continuous linear surjection of an* $\mathfrak{UF}$*-space onto a space of the type* (β) *is open.*

1.2. Some classes of linear operators

Let E be a Banach space. We denote by $GL(E) \subset L(E)$ the group of all invertible operators in E. $C(E,F) \subset L(E,F)$ denotes the set of compact linear operators from E into F, and $C(E) = C(E,E)$. $GL_C(E)$ is the subgroup of $GL(E)$ consisting of operators $I + A$ where $A \subset C(E)$. For any operator $A \in L(E,F)$ we denote by *Im A, Ker A* and *Coker A* the image (=range), the kernel and the cokernel of A correspondingly:

$$Im\, A = \{y \in F \mid\ y = Ax,\ \exists x \in E\} \subset F$$

$$Ker\, A = \{x \in E \mid Ax = 0\} \subset E$$

$$Coker\, A = \{f \in F^* \mid f(Ax) = 0,\ \forall x \in E\} \subset F^*$$

We employ standard notations $\sigma(A)$, $\sigma_d(A)$, $\sigma_c(A)$, $\sigma_{ac}(A)$ for the spectrum, discrete spectrum, continuous spectrum and absolutely continuous spectrum of A correspondingly.

The main class of operators that we are going to deal with is the one of Fredholm operators.

Definition 1.2.1. An operator $A \in L(E,F)$ is said to be *Fredholm* if

i) $dim\ Ker\ A < \infty$;

ii) $Im\ A$ is closed in F;

iii) $dim\ Coker\ A < \infty$;

In this case $ind\ A = dim\ Ker\ A - dim\ Coker\ A$ is called the *index* of the operator A.

We denote by $\Phi(E,F)$ ($\subset L(E,F)$) the set all Fredholm operators from E into F, and $\Phi(E) = \Phi(E,E)$.

We shall also use the notations

$S_p\Phi(E) = \{A \in \Phi \mid dim\ Ker\ A \geq p\}$,

$CS_p\Phi(E) = \{A \in \Phi \mid dim\ Coker\ A = codim\ Im\ A \geq p\}$.

The semi-Fredholm operators form wider classes of operators than Fredholm ones. The class $\Phi_+(E)$ (correspondingly $\Phi_-(E)$) consists of all continuous linear operators in E with closed range and finite-dimensional kernel (correspondingly, cokernel). If we add the condition that the range (correspondingly, the kernel) is a complemented subspace, then we obtain the class $\Phi_l(E)$ (respectively, $\Phi_r(E)$). In the Hilbert case the class $\Phi_-(E)$ coincides with $\Phi_r(E)$, and $\Phi_+(E)$ coincides with $\Phi_l(E)$. We shall sometimes use the classes $GL_r(E)$ and $GL_l(E)$ of all right- and left- invertible operators from $L(E)$.

Let us describe briefly the common properties of Fredholm operators.The reader can find missing details in the references mentioned in section 1.9.

First of all, we mention the following fundamental fact.

Theorem 1.2.2. *If* $A,B \in \Phi(E)$, *then* $AB \in \Phi(E)$, *and*

$$ind(AB) = ind\ A + ind\ B. \tag{1.2.1}$$

We leave to the reader the proof of Fredholm property, and just check the last identity.

Let $n_A = \dim Ker\ A$, $m_A = Codim\ Im\ A$, and n_B, m_B be defined analogously. Then $ind\ A = n_A - m_A$, $ind\ B = n_B - m_B$. Let us denote by ν the dimension of the intersection $Ker\ A \cap Im\ B$. Then

$$\dim Ker\ AB = n_B + \nu,$$

$$Codim\ Im\ AB = m_A + m_B + \nu - n_A,$$

so $ind\ AB = (n_B + \nu) - (m_A + m_B + \nu - n_A) = ind\ A + ind\ B$.

The theorem says that $\Phi(E)$ is a semigroup of operators, and *ind* is a semigroup homomorphism from $\Phi(E)$ into $\mathbb{Z}$.

The proof of the following simple fact we also leave to the reader.

Theorem 1.2.3. *$A \in \Phi(E)$ if and only if $A^* \in \Phi(E^*)$. Moreover,*

$$ind\ A = -\ ind\ A^*.$$

Let us denote by $\Phi^{(n)}$ the subset of Φ consisting of all Fredholm operators of index $n \in \mathbb{Z}$. Then $\Phi(E,F) = \cup_{n \in \mathbb{Z}} \Phi^{(n)}(E,F)$.

Theorem 1.2.4. *The set $\Phi^{(0)}(E,F)$ consists of all sums $A + B$ where $A \in L(E,F)$ is an invertible operator, and $B \in C(E,F)$.*

Proof. First of all, let $\Phi^{(0)}(E,F) \neq \emptyset$. Each operator $A \in \Phi^{(0)}(E,F)$ provides an isomorphism between any closed direct complement $M \subset E$ to $Ker\ A$, and the range $Im\ A$. The condition $ind\ A = 0$ implies that the codimensions of M and of $Im\ A$ are equal, so A can be easily extended to an isomorphism between E and F. Conversely, if there is an invertible operator $A \in L(E,F)$, then $A \in \Phi^{(0)}(E,F)$, and so $\Phi^{(0)}(E,F) \neq \emptyset$. This evidently implies that we may assume $E = F$.

Let $A \in \Phi^{(0)}(E)$. Denote, as above, by M any closed direct complementary subspace to $Ker\ A$:

$$Ker\ A \oplus M = E.$$

We denote by N any (finite dimensional, and so closed) complement subspace to $Im\ A$ in E. The condition $ind\ A = 0$ implies that $dim\ Ker\ A = dim\ N(<\infty)$. Consider any isomorphism $A': Ker\ A \to N$ and define the linear operator

$$\tilde{A}x = \begin{cases} Ax & \text{for } x \in M \\ A'x & \text{for } x \in Ker\ A. \end{cases}$$

Then obviously $\tilde{A} \in GL(E)$, and $\tilde{A} = A + A'P$ where P is the (finite-dimensional) projection onto $Ker\ A$ parallel to M. We get now: $A = -\tilde{A} + A'P$, and $A \in GL(E)$, $A'P \in C(E)$.

Suppose now that $A \in GL(E)$, $B \in C(E)$, and prove that $T = A + B \in \Phi^{(0)}(E)$. First of all, multiplying by A^{-1}, we obviously reduce the problem to the case when $A = I$, so $T = I + B$. It is easy to see that $Ker\ T$ is finite dimensional, because on this subspace $B = -I$, and B is compact. Now let us prove that $Im\ T$ is closed. Considering any closed complement M to $Ker\ T$, we see that $Im\ T = Im\ (T|_M)$. Let us show that there exists a constant $c>0$ such that $\|x\| \le C\|Tx\|$ for all $x \in M$. If this is not true, then there exists a sequence of points $x_n \in M$ such that $\|x_n\| = 1$, and $\|Tx_n\| \to 0$. Due to compactness of B, choosing a subsequence, we may assume that Bx_n converges to some $x \in E$. Then $x_n = Tx_n - Bx_n$ converges to the point $(-x) = y$. This point must belong to M, and $Ty = \lim_{n\to\infty} Tx_n = 0$, so $y \in Ker\ T$. On the other hand, $Ker\ T \cap M = 0$ by construction, so $y = 0$. This contradicts, however, the condition $\|x_n\| = 1$.

The inequality $\|x\| \le C\|Tx\|$ $(x \in M)$ implies that $Im\ T = Im\ (T|_M)$ is closed. In fact, if $y_n = Tx_n$ converges to some point y, then by the inequality the sequence x_n is fundamental in M, so $x_n \to x \in M$, and $y = Tx$. Since $(Im\ T)^{\perp} = Ker\ T^*$, and also due to the fact that $T^* = I + B^*$ has the same structure as T, we see that $Codim\ Im\ T<\infty$, so $T \in \Phi(E)$. We have only to prove that $ind\ T = 0$. It is sufficient to prove that $ind\ T \le 0$, because the opposite inequality $ind\ T \ge 0$ is equivalent to $ind\ T^* \le 0$, and $T^* = I + B^*$ has the same structure as $T = I + B$. Now, suppose that $ind\ T > 0$. Then $ind\ T^k = k\ ind\ T \underset{k\to\infty}{\to} \infty$. This means that

$$dim\ Ker\ T^k = ind\ T^k + dim\ Im\ T^k \to \infty \text{ as } k\to\infty.$$

Let us denote by N_k the space $Ker\ T^k$. Then $dim\ N_k < \infty$ for each $k \in \mathbf{Z}_+$, $N_k \subset N_{k+1}$, and $N_k \ne N_{k+1}$ for any k. Indeed, if $N_k = N_{k+1}$ for some k, then we would have $N_k = N_m$ for all $m \ge k$, so $dim\ N_k$ would not diverge. Let us choose elements $x_k \in N_k$ such that $\| x_k \| = 1$, and $dist\ (x_k, N_{k-1}) \ge 1/2$. Due to the compactness of B, the sequence $\{Bx_k\}$ must be compact. On the other hand,

$$\| Bx_m - Bx_k \| = \| Tx_m - Tx_k + x_k - x_m \|$$

Let us note that $Tx_m - Tx_k + x_k \in N_{m-1}$ for $m > k$. In fact,

$$T^{m-1}(Tx_m - Tx_k + x_k) = T^m x_m - T^m x_k + T^{m-1} x_k = 0.$$

This means that $\| Tx_m - Tx_k + x_k - x_m \| \ge dist\ (x_m, N_{m-1}) \ge 1/2$, so the sequence Bx_k is noncompact. This contradiction finishes the proof.

The reader can notice that the operator B in the theorem can be assumed to be even finite-dimensional, not only compact.

Corollary 1.2.5. *If $A \in \Phi(E)$, then $(A + K) \in \Phi(E)$ for any $K \in C(E)$.*

When $ind\ A = 0$, this statement follows immediately from the theorem. The general case can be easily reduced to this one.

Corollary 1.2.6. *The set $\Phi^{(0)}(E)$ is open in $L(E)$.*

Corollary 1.2.7. *For every $n \in \mathbf{Z}$ the set $\Phi^{(n)}(E,F)$ is open in $\Phi(E,F)$. In particular, the set $\Phi(E,F)$ is open.*

The case of general $n \in \mathbf{Z}$ can be easily reduced to $n = 0$. Namely, let $n > 0$. Consider the space $F_1 = F \oplus \mathbf{C}^n$. Then for each $A \in \Phi^{(n)}(E,F)$ the operator $A_1 = A \oplus 0$

(where 0 is zero operator from E into $\mathbf{C}^n$) belongs to $\Phi^{(0)}(E,F_1)$, and we can apply the last corollary. If $n<0$, we consider $E_1 = E \oplus \mathbf{C}^n$, and $A_1 = A \oplus O_1$ where O_1 is zero operator from $\mathbf{C}^n$ into F.

Theorem 1.2.8. *An operator $A \in L(E,F)$ belongs to $\Phi(E,F)$ if and only if there exists an operator $B \in L(F, E)$ ("regularizer" of A) such that the operators $AB - I$ and $BA - I$ are compact.*

Proof. Let $A \in \Phi(E,F)$. Assume first that $A \in \Phi^{(0)}$. Then, due to theorem 4, $A = T + K$, where T is invertible, and K is compact. In this case the operator $B = T^{-1}$ is obviously the regularizer we need. The case of nonzero *ind* A can be easily reduced to *ind* $A = 0$ by the trick we applied above.

Another way to find a regularizer (for any value of *ind* A) is to consider a closed complement subspace M to *Ker* A, and to define $B = (A \mid_M)^{-1} P$ where P is any projector onto *Im* A. It can be easily checked that in this case the operators $AB - I$ and $BA - I$ are even finite dimensional.

Suppose now that there exists a regularizer B, and prove the Fredholm property of A. We see that *Ker* $A \subset$ *Ker* BA , so *Ker* A is finite dimensional (because $BA = I + K$ for some compact operator K). In the same way we get the finite dimensionality of *Ker* $A^* \subset$ *Ker* $B^*A^* =$ *Ker*$(AB)^*$. We must prove only that the image of A is closed. The operator AB is Fredholm, because it can be represented as $I + K$ with a compact K. Therefore, *Im* AB is closed and of finite codimension. On the other hand, *Im* $A \supset$ *Im* AB. Now the reader can easily prove that any subspace containing a closed subspace of finite codimension, is closed. The theorem is proved.

It can be easily shown that A belongs to $\Phi_l(E)$ ($\Phi_r(E)$) if and only if it has a *left* (respectively, *right*) regularizer B such that $BA - I$ (respectively, $AB - I$) is compact.

If there exists an *invertible* regularizer B, we call it an *equivalent regularizer*. The next result also can be proved easily by the reader.

Theorem 1.2.9. *An operator $A \in L(E,F)$ has an equivalent regularizer if and only if $A \in \Phi^{(0)}(E,F)$.*

Let us restrict ourself to the case $E = F$. The subspace $C(E) \subset L(E)$ consisting of all compact operators is a two-sided ideal in the algebra $L(E)$ (i.e. $A \in C(E)$, $B \in L(E)$

imply AB, $BA \in C(E)$). Then we can define the following factor-algebra $Cal(E) = L(E)/C(E)$, so called *Calkin algebra* of the space E. We denote by π: $L(E) \to Cal(E)$ the natural projection. Then the result of theorem 8 can be reformulated in the following way:

Theorem 1.2.10. *An operator $A \in L(E)$ is Fredholm if and only if $\pi(A)$ is an invertible element of Calkin algebra.*

The Bartle-Graves theorem 1.1.1, applied to the projection $L(E) \xrightarrow{\pi} Cal(E)$, shows that topologically this is a trivial bundle with the (contractible) fiber $C(E)$. This means, in particular, that π is a homotopic equivalence between the set of Fredholm operators and the set of invertible elements of Calkin algebra. The reader can also show that an operator $A \in L(E)$ belongs to $\Phi_l(E)$ ($\Phi_r(E)$) if and only if its image in Calkin algebra is left (right) invertible.

We shall use the following lemma.

Lemma 1.2.11. *Let $A_0 \in \Phi(E,F)$, and M be a closed subspace of finite codimension in E such that $M \cap Ker\ A_0 = \{0\}$. Then for all operators A in a neighborhood U of A_0 we have $M \cap Ker\ A = \{0\}$.*

Proof. Let M_1 be a finite dimensional subspace of E such that $M_1 \oplus M = E$, and $M_1 \supset Ker\ A_0$. Define P as the projection onto M parallel to M_1. Consider also the subspace $N = Im\ (A_0|_M) \subset F$. Due to the Fredholm property of A_0, N is closed in F and of finite codimension. Denote by Q any continuous projection operator from F onto N. Let $dim\ M_1 = m$, and B: $M_1 \to \mathbf{C}^m$ be any isomorphism. For any $A \in L(E)$ consider the operator $\tilde{A} = QAP + B(I-P)$: $E \to N \oplus \mathbf{C}^n$. We notice that 1) $\tilde{A}_0$ is an isomorphism, 2) if A is close to A_0 then $\tilde{A}$ is close to $\tilde{A}_0$. This means that in a neighborhood of A_0 all operators $\tilde{A}$ are isomorphisms. The statement of the lemma now follows immediately because if $x \in M$ and $Ax = 0$, then $\tilde{A}x = 0$. The lemma is proved.

Remarks. 1. Let M be a subspace of finite codimension such that $M \cap Ker\ A = \{0\}$ for some Fredholm operator A. Let also P be a projector onto M. Then it is easily verified that $ind\ A = ind\ AP$. In fact, corollary 13 below implies that the condition $M \cap Ker\ A = \{0\}$ is not necessary here. 2. Using the lemma, we can obviously find a finite-codimensional subspace M such that $M \cap Ker\ A = \{0\}$ for all operators A from a compact subset in $\Phi(E)$. 3. The statement of the lemma holds also for semi-Fredholm operators from $\Phi_l(E)$.

Corollary 1.2.12. *The function ind A is continuous (and hence locally constant) on* $\Phi(E)$.

Proof. Let A_0 be a Fredholm operator, and M be a subspace satisfying the assumptions of lemma 11. Due to the lemma, $M \cap Ker\ A = \{0\}$ in a neighborhood of A_0. Let P, Q be continuous projections onto M and $Im\ (A_0|_M)$ correspondingly. Then the remark shows that $ind\ A = ind\ AP$ for all A from the neighborhood. We can observe that $Ker\ AP$ is constant in this neighborhood. Consider the operator family $QA : M \to Im\ A_0$. This family of operators is continuous with respect to the norm of the space $L(M, Im\ A_0)$, and its value is invertible when $A = A_0$. This means that it is invertible also for all A from some neighborhood of A_0. In this neighborhood Q is an isomorphism of $Im\ (A|_M)$ onto $Im\ (A_0|_M)$. In particular,

$$codim\ Im\ (A|_M) = codim\ Im\ (A_0|_M)\ .$$

This implies that $ind\ A = ind\ A_0$ in this neighborhood, and the statement is proved.

Corollary 1.2.13. *If* $A \in \Phi(E)$, *and* $K \in C(E)$, *then* $ind\ A = ind\ (A + K)$.

Proof. Consider the continuous family of operators $A + tK$, where $t \in [0,1]$, and apply corollaries 5 and 12.

Corollary 1.2.14. *The function dim Ker A is upper-continuous on the space* $\Phi(E,F)$. *Due to the continuity of ind A, the same is true for codim Im A.*

We shall see later that if $dim\ Ker\ A$ remains constant under some continuous deformation in the Fredholm class, then the kernel itself behaves "nicely" (i.e. continuously).

We know already that every $\Phi^{(0)}$ - operator (i.e. Fredholm operator of zero index) can be perturbed by a compact (or even finite-dimensional) operator to an invertible one. In some applications (for instance, in chapter 4) we need to choose the perturbation from a special subspace in the space of all bounded operators. The following result provides sufficient condition for the existence of the perturbation.

Theorem 1.2.15. *Let* $\mathfrak{A} \subset L(E,F)$ *be a linear subspace, and* $A \in \Phi^{(0)}(E,F)$. *Assume* *that for every pair* $x\ (\neq 0) \in Ker\ A$, $f\ (\neq 0) \in Coker\ A$ *there exists an operator* $B \in \mathfrak{A}$ *such that*

$$(f, Bx) \neq 0. \tag{1.2.1}$$

Then there exists an operator $T \in \mathfrak{A}$ *such that* $A + \epsilon T$ *is invertible for all sufficiently small values of* $\epsilon \neq 0$. *If, in particular, we can satisfy* (1) *for all* $x\ (\neq 0) \in E$, $f\ (\neq 0) \in F^*$, *then the conclusion holds for all* $A \in \Phi^{(0)}(E,F)$.

Proof. Because $\Phi^{(0)}(E,F) \neq \emptyset$ (there is nothing to prove otherwise) we can assume that $E = F$. Given any $B \in \mathfrak{A}$, consider the family $A + \epsilon B$. If it is invertible for small $\epsilon \neq 0$, then the statement is proved. Assume now that $A + \epsilon B$ is non-invertible for an arbitrarily small nonzero ϵ. This means that there exists a sequence $\epsilon_n \to 0$ and a sequence of vectors $x_n \in E$ such that $\| x_n \| = 1$ and $(A + \epsilon_n B)x_n = 0$. Let P be any continuous projection onto $Ker\ A$ and $x_n = Px_n + (1-P)x_n = y_n + z_n$. Here $y_n \in Ker\ A$, $z_n \in Ker\ P$. Then we have

$$Az_n = -\epsilon_n Bx_n \to 0\ .$$

Due to Fredholm property, A is a topological isomorphism of $Ker\ P$ onto $Im\ A$, so $lim\ z_n = 0$. We can conclude now that $lim \| y_n \| = 1$. Since $y_n \in Ker\ A$, and this space is finite dimensional, there is a subsequence y_{n_m} that converges to a vector $x \in Ker\ A$, $\| x \| = 1$. Moreover, since $Bx_n = -\epsilon_n^{-1} Ax_n \in Im\ A$, and $Im\ A$ is closed, $Bx \in Im\ A$. We conclude therefore that non-invertibility of $A + \epsilon B$ for an arbitrarily small $\epsilon \neq 0$ implies the existence of an element $x \neq 0$ from $Ker\ A$ such that $Bx \in Im\ A$, and hence $(f, Bx) = 0$ for all elements f from $Coker\ A$.

Let us consider a basis $x_1, \dots, x_n$ in $Ker\ A$ and a basis $f_1, \dots, f_n$ in $Coker\ A$. (Recall that $ind\ A = 0$). Suppose that the conclusion of the theorem is false. Then the above considerations imply that for each $B \in \mathfrak{A}$ there is a nonzero linear combination $x = \sum a_i x_i$ such that $(f_j, \sum a_i Bx_i) = 0$, $j = 1, \dots, n$. Thus for all $B \in \mathfrak{A}$ the $n \times n$ matrix

$$M = (f_j,\ Bx_i)$$

is singular. Consider the space $\mathcal{M}$ of all matrices M arising from $B \in \mathfrak{A}$. This is a linear space of singular $n \times n$ matrices. Let us prove now the following algebraic statement.

Lemma 1.2.16. *Let $\mathcal{M}$ be a linear set of singular $n \times n$ matrices. Then there is a sequence of elementary row and column operations which, when applied to the matrices of $\mathcal{M}$, yields a set of matrices all of which have entry zero in the same position.*

Remark. The lemma shows that there exist nonzero $n-$vectors a and b such that $(Ma, b) = 0$ for all $M \in \mathcal{M}$.

Proof can be obtained by induction. The statement holds for $n = 1$ by obvious reasons. Suppose that it holds for all orders less than n. Consider now our space $\mathcal{M}$ of singular $n \times n$ matrices. We can assume that $\mathcal{M} \neq \{0\}$, otherwise there is nothing to prove. Choose any matrix $C \in \mathcal{M} \backslash \{0\}$. We can bring it to the form

$$J = \begin{bmatrix} 1 & 0 \\ 0 & 0 \end{bmatrix}$$

by elementary operations , so we can assume that $J \in \mathcal{M}$.

Let us write any $M \in \mathcal{M}$ in the block form

$$M = \begin{bmatrix} M_{11} & M_{12} \\ M_{21} & M_{22} \end{bmatrix}$$

corresponding to the block structure of J. Our assumption guarantees that $det\ (M + tJ) = 0$ for all complex t. A simple conclusion is that $det\ M_{22} = 0$. This shows that the (linear) space of these blocks M_{22} contains only singular matrices (of order less then n). The induction hypothesis finishes the proof of the lemma.

Let us return to the proof of the theorem now. Consider the space of matrices $\mathcal{M}$ arising from $\mathfrak{A}$. The last lemma shows that there exist nonzero elements $x \in Ker\ A$, $f \in Coker\ A$ such that $(f, Bx) = 0$ for all $B \in \mathfrak{A}$ which contradicts the assumptions. The theorem is proved.

We shall return to the consideration of Fredholm operators (depending on some parameters) later in this chapter. At this moment we shall consider other classes of operators we need.

From now on, until the end of this section we shall deal with operators in a separable Hilbert space H. We shall introduce some important operator ideals in $L(H)$ (so called Shatten-von Neumann classes).

Let $A \in L(H)$ be a compact linear operator. If A is positive, we denote by $\lambda_j(A)$ $(j \geq 1)$ its nonzero eigenvalues numerated in descending order, where each eigenvalue is repeated as many times as the order of its multiplicity.

Definition 1.2.17. The $j-th$ *singular value* (or $j-th$ $s-number$) of A is

$$s_j(A) = \lambda_j((A^*A)^{1/2}) .$$

It can be proved that $s_j(A)$ allows also the following descriptions:

$$\text{(1.2.2)} \quad \text{i) } s_j(A) = \min_{\dim M = j-1} \; \max_{0 \neq x \perp M} (\| Ax \| / \| x \|), \; j \geq 1 .$$

$$\text{(1.2.3)} \quad \text{i i) } s_j(A) = \min \{ \| A - K \| \mid K \in L(H), \; rank\, K \leq j-1 \},$$

where $rank\, K = \dim Im\, K$.

Singular values play a very important role in the operator theory, and their properties have been studied in many works. The reader can find the pertinent references in the last section of this chapter.

We denote by $\mathfrak{I}_p$ $(1 \leq p < \infty)$ the collection of all compact operators A such that

$$\sum_{j=1}^{\infty} s_j^p(A) < \infty \quad (1 \leq p < \infty).$$

This class is a Banach space with respect to the norm

$$\| A \|_p = (\sum s_j^p(A))^{1/p}.$$

When $p=\infty$ we denote by $\mathfrak{S}_\infty$ the space $C(E)$ of all compact operators in H equipped with the operator norm. We also consider the Matsaev's ideal $\mathfrak{S}_\omega$ of compact operators A such that

$$\sum_{j=1}^{\infty} s_j(A)\,(2j-1)^{-1}<\infty.$$

Let us mention some known properties of these operator classes. The space $\mathfrak{S}_p$ is a two-sided ideal in the algebra $L(H)$ in the following sense: if $A\in\mathfrak{S}_p$, $B,C\in L(H)$, then $BAC\in\mathfrak{S}_p$ and $\|BAC\|_p\le\|B\|\,\|A\|_p\|C\|$. The ideal $\mathfrak{S}_1$ consists of trace class operators and $\mathfrak{S}_2$ consists of Hilbert-Schmidt operators. If $A\in\mathfrak{S}_p$, then $A^r\in\mathfrak{S}_{p/r}$, and $\|A^r\|_{p/r}\le\|A\|_p{}^r$.

Let us consider some standard examples of operators of $\mathfrak{S}_p$ class. We shall need these examples later.

Consider the torus $\mathbf{T}^n=\mathbf{R}^n/\mathbf{Z}^n$ and Sobolev space $H^s(\mathbf{T}^n)$ that consists of all functions

$$f(x)=\sum_{\gamma\in\mathbf{Z}^n} f_\gamma e^{2\pi i\gamma\cdot x}\quad\text{with}\quad \|f\|_s^2=\sum(1+|\gamma|^{2s})\,|f_\gamma|^2<\infty.$$

Here $|\gamma|=\max_{i=1}|\gamma_i|$. This space is obviously continuously embedded into $L_2(\mathbf{T}^n)$. Let us denote by Λ: $H^s(\mathbf{T}^n)\to L_2(\mathbf{T}^n)$ the corresponding embedding operator.

Lemma 1.2.18. *The operator Λ belongs to the class $\mathfrak{S}_p$ for $p>n/s$.*

Proof. Consider the operator $\Sigma=(1-\Delta)^{-s/2}$ where $\Delta=\sum\partial_{x_j}^2$ is the standard Laplacian. Then it obviously is an isomorphism of $L_2(\mathbf{T}^n)$ onto $H^s(\mathbf{T}^n)$. If we prove that the operator $\Lambda\Sigma$ belongs to some class $\mathfrak{S}_p$, as an operator in $L_2(\mathbf{T}^n)$, then the operator $\Lambda=(\Lambda\Sigma)\Sigma^{-1}$ will also belong to the same class , due to the boundedness of $\Sigma^{-1}: H^s(\mathbf{T}^n)\to L_2(\mathbf{T}^n)$. The operator $A=\Lambda\Sigma$ in $L_2(\mathbf{T}^n)$ is self-adjoint, positive-definite, and diagonal with respect to the Fourier series expansion:

$$A\Big(\sum_{\gamma\in\mathbf{Z}^n} f_\gamma e^{2\pi i\gamma\cdot x}\Big)=\sum_{\gamma\in\mathbf{Z}^n}\frac{1}{(|2\pi\gamma|^2+1)^{s/2}}f_\gamma\, e^{2\pi i\gamma\cdot x}.$$

The compactness of A follows from this equality immediately (because A is the operator norm limit of finite dimensional operators), and the eigenvalues of A are equal to $\lambda_\gamma = (|2\pi\gamma|^2 + 1)^{-s/2}$. Then

$$\sum \lambda_\gamma^p < \infty$$

for all $p > n/s$, so A belongs to the corresponding class $\mathfrak{I}_p$. The lemma is proved.

We shall need also an analogous result for the case of anisotropic Sobolev spaces. Let us denote by $H^{1,s}(\mathbf{T}^n)$ the space of functions f on $\mathbf{T}^n$ such that

$$\|f\|_{1,s}^2 = \sum_{\gamma \in \mathbf{Z}^n} (1 + |\gamma_1| + |\gamma'|^s)^2 |f_\gamma|^2 < \infty$$

where $\sum f_\gamma e^{2\pi i \gamma \cdot x}$ is the Fourier series of f. Here $\gamma' = (\gamma_2, \dots \gamma_n) \in \mathbf{Z}^{n-1}$. Consider now the embedding operator Λ: $H^{1,s} \to L_2(\mathbf{T}^n)$.

Lemma 1.2.19. *The operator Λ belongs to the class $\mathfrak{I}_p$ for $p > (n-1)/s + 1$.*

Proof. A construction analogous to the one used in the previous proof leads to the operator in $L_2(\mathbf{T}^n)$:

$$A\Big(\sum_{\gamma \in \mathbf{Z}^n} f_\gamma e^{2\pi i \gamma \cdot x}\Big) = \sum_{\gamma \in \mathbf{Z}^n} \frac{1}{1 + |\gamma_1|^2 + |\gamma'|^s} f_\gamma \, e^{2\pi i \gamma \cdot x} .$$

(we just use an appropriate operator Σ instead of $(-\Delta + 1)^{-s/2}$). A simple estimate shows now that $\sum s_j(A)^p < \infty$ for $p > (n-1)/s + 1$. Lemma is proved.

We introduce now one of the most important notions of operator theory, the one of regularized determinant.

Let $A \in \mathfrak{I}_p(H)$, and p be a positive integer. We define the regularized determinant of $I - A$ as

$$(1.2.4) \qquad det_p(I - A) = \prod_j (1 - \lambda_j(A)) \, exp \sum_{k=1}^{p-1} \frac{1}{k} \lambda_j^k(A).$$

The reader should notice that the definition depends on p. Let us mention also that if $A \in \mathcal{S}_r(H)$ with some non-integer $r > 0$, then we can define the determinant $det_p(1 - A)$ for any integer $p > r$.

The next theorem collects the main properties of regularized determinants.

Theorem 1.2.20.

i) *If $r \leq p$, then the determinant $det_p(I - A)$ is analytic as a complex-valued function on the Banach space $\mathcal{S}_r(H)$.*

ii) *If $p - 1 \leq r \leq p$, then*

$$| \, det_p(I - A) \, | \leq C \, exp \, (C \, \| A \|_r^r),$$

where the constant C depends only on p and r.

iii) *If $p \geq 2$, $r \leq p - 1$, then for $A \in \mathcal{S}_r(H)$*

$$det_p(1 - A) = exp(\frac{1}{p-1} \, tr(A^{p-1})) \, det_{p-1}(1 - A).$$

iv) *If $p - 1 \leq r \leq p$, then the operator function $det_p(1 - A)(1 - A)^{-1}$ (a priori defined only for $1 \notin \sigma(A)$) is analytic on the whole Banach space $\mathcal{S}_r(H)$, and satisfies the estimate*

$$\| \, det_p(1 - A)(1 - A)^{-1} \| \leq exp(C(\| A \|_r^r + 1))$$

for some constant C depending on p and r.

We shall not prove this theorem. The reader can find the pertinent references in the final section of the chapter.

Remarks. 1. If $A_1, A_2 \in \mathcal{S}_p$, and $\sigma(A_1) = \sigma(A_2)$, then $det_p(1 - A_1) = det_p(1 - A_2)$.
2. The resolvent $(I - A)^{-1}$ for $A \in \mathcal{S}_p(H)$ (if $1 \notin \sigma(A)$, so the operator is invertible) can be represented in the following way:

$$(1-A)^{-1} = det_p(1-A)(1-A)^{-1}/det_p(1-A) = B_p(A)/\Delta_p(A).$$

Here the operator-valued function $B_p(A)$: $\mathcal{S}_p(H) \to L(H)$ and the scalar function $\Delta_p(A)$: $\mathcal{S}_p(H) \to \mathbb{C}$ are analytic *on the whole* space $\mathcal{S}_p(H)$ and satisfy the estimates

$$\| B_p(A) \|, \ |\Delta_p(A)| \le exp(C(\| A \|_p^p + 1)).$$

We can claim also that $\Delta_p(A) = 0$ if and only if the operator $1 - A$ is not invertible.
3. Consider the following operator valued polynomial function on $\mathbb{C}^n$:

$$\mathcal{L}(\lambda) = I + \sum_{|\alpha| \le m} \lambda^\alpha B_\alpha S^{|\alpha|},$$

where $B_\alpha \in L(H)$, and $S \in \mathcal{S}_p(H)$. We would like to represent the resolvent $\mathcal{L}^{-1}(\lambda)$ of this operator pencil as the ratio of two entire functions of a finite order.

Lemma 1.2.21. $\mathcal{L}^{-1}(\lambda) = B(\lambda)/\Delta(\lambda)$ *for all* $\lambda \in \mathbb{C}^n$ *such that* $\mathcal{L}(\lambda)$ *is invertible. Here* B: $\mathbb{C}^n \to L(H)$, Δ: $\mathbb{C}^n \to \mathbb{C}$ *are entire functions satisfying the estimates*

$$\| B(\lambda) \|, \ |\Delta(\lambda)| \le C \ exp \ C|\lambda|^p.$$

Proof. Let us denote by $\alpha = (\alpha_1, \ldots, \alpha_n) \in \mathbb{Z}_+^n$ a multi-index and consider a vector

$$\{(\psi_\alpha) \ | \ |\alpha| \le m-1, \ \psi_\alpha \in H\}.$$

For any $\alpha \ne (0,\ldots,0)$ we denote by $j(\alpha)$ the first number j such that $\alpha_{j(\alpha)} = max\ \alpha_i$. We also denote by α' the multi-index such that $\alpha_j' = \alpha_j \ \ \forall j \ne j(\alpha)$, and $\alpha'_{j(\alpha)} = \alpha_{j(\alpha)} - 1$. Then it is obvious that the equation $\mathcal{L}(\lambda)\psi = \phi$ in H is equivalent to the following system of equations (where $\psi = \psi_0 = \psi(0,\ldots 0)$):

(1.2.5)
$$\begin{cases} \psi_0 + \sum_{0<|\alpha|\le n} \lambda_{j(\alpha)} B_\alpha S\psi_{\alpha'} = \phi \\ \psi_\alpha - \lambda_{j(\alpha)} S\psi_{\alpha'} = 0, \qquad \alpha \neq (0,\dots,0). \end{cases}$$

This is an equation in a finite direct sum $\mathcal{H}$ of copies of the space H. The operator matrix of this system has the form

$$\mathcal{A}(\lambda) = I + \sum_{j=1}^{n} \lambda_j \tilde{S}_j$$

where the matrices $\tilde{S}_j$ have entries belonging to $\mathcal{S}_p(H)$. Therefore, all $\tilde{S}_j$ also belong to the class $\mathcal{S}_p$. We can notice also that $\mathcal{A}(0) = I$, and so $\mathcal{A}(\lambda)$ is invertible for some $\lambda \in \mathbf{C}^n$ (and hence almost everywhere in $\mathbf{C}^n$, as it will follow from section 1.6). Due to the last theorem and remarks that follow it, we can represent the resolvent $\mathcal{A}(\lambda)^{-1}$ in the form

$$\mathcal{A}(\lambda)^{-1} = B_p(\sum \lambda_j \tilde{S}_j) / \Delta_p(\sum \lambda_j \tilde{S}_j) ,$$

where $B\colon \mathcal{S}_p(\mathcal{H}) \to L(\mathcal{H})$, $\Delta_p\colon \mathcal{S}_p(\mathcal{H}) \to \mathbf{C}$ are entire functions of order p, and Δ_p has zeros only when $\mathcal{A}(\lambda)$ is not invertible. Denoting by $\tilde{B}(\lambda)$ and $\Delta(\lambda)$ the functions $B_p(\sum \lambda_j \tilde{S}_j)$ and $\Delta_p(\sum \lambda_j \tilde{S}_j)$ correspondingly, we get the representation

$$\mathcal{A}(\lambda)^{-1} = \tilde{B}(\lambda)/\Delta(\lambda),$$

and $\Delta(\lambda)$ is zero only when $\mathcal{A}(\lambda)$ is not invertible. Considering now the operators

$$P\colon H \to \mathcal{H}, \quad P(\psi) = (\psi, 0, 0, \dots, 0)$$

$$Q\colon \mathcal{H} \to H \quad Q(\{\psi_\alpha\}) = \psi_{(0,0,\dots,0)} ,$$

we can represent $\mathcal{L}^{-1}(\lambda)$ as $Q\mathcal{A}(\lambda)^{-1}P = Q\tilde{B}(\lambda)P/\Delta(\lambda)$. Denoting by $B(\lambda)$ the operator $Q\tilde{B}(\lambda)P$, we prove the statement of the lemma.

1.3. Banach vector bundles

We assume that the reader is familiar with the notion of a vector bundle. In this section we shall consider briefly some basic definitions and facts about an infinite dimensional version of vector bundles – Banach vector bundles.

Let $\mathfrak{X}$ and $\mathcal{E}$ be topological spaces and let $p\colon \mathcal{E}\to\mathfrak{X}$ be a continuous surjection (*projection*). We assume that the *fiber* $p^{-1}(\tau)=\mathcal{E}(\tau)$ over each point $\tau\in\mathfrak{X}$ is equipped with a Banach space structure such that the corresponding strong topology coincides with the one induced from $\mathcal{E}$.

An open cover $\{\mathfrak{U}_\alpha\}$ of $\mathfrak{X}$ is called *trivializing* for p if for each α there exists a Banach space E_α and a homeomorphism $\phi_\alpha\colon p^{-1}(\mathfrak{U}_\alpha)\to\mathfrak{U}_\alpha\times E_\alpha$ having the properties

i) the diagram

$$\begin{array}{ccc} p^{-1}(\mathfrak{U}_\alpha) & \overset{\phi_\alpha}{\to} & \mathfrak{U}_\alpha\times E_\alpha \\ & {\scriptstyle p}\searrow \quad \swarrow{\scriptstyle \nu} & \\ & \mathfrak{U}_\alpha & \end{array}$$

is commutative, where ν is the natural projection on the first factor, and for each $\tau\in\mathfrak{U}_\alpha$ the map ϕ_α induces on the fiber $\mathcal{E}_\tau$ an isomorphism $\phi_{\alpha\tau}$ of the Banach spaces $\mathcal{E}_\tau$ and E_α.

ii) If $\mathfrak{U}_\alpha$ and $\mathfrak{U}_\beta$ are two elements of the cover that have non-empty intersection then the map of $\mathfrak{U}_\alpha\cap\mathfrak{U}_\beta$ to $L(E_\alpha, E_\beta)$ given by

$$\tau\to(\phi_\beta\circ\phi_\alpha{}^{-1})|_{\nu^{-1}(\tau)}$$

is continuous in the operator norm.

The maps ϕ_α are called *trivializing* and the maps

$$\tau\to(\phi_\beta\circ\phi_\alpha{}^{-1})|_{\nu^{-1}(\tau)}=\phi_{\alpha,\beta}(\tau)$$

are the *transition functions.*

Two trivializing covers are said to be *equivalent* if their union satisfies i) and ii). A class of equivalent trivializing covers gives the structure of a (locally trivial) *Banach vector bundle* on $(\mathfrak{X},\mathcal{E},p)$. Here $\mathcal{E}$ is the *bundle space,* and $\mathfrak{X}$ is its *base.* A bundle is of *countable* (*finite*) type if it has a countable (finite) trivializing cover.

If $\mathfrak{X}$ is connected, then all the fibers E_α are necessarily isomorphic, so they can be identified with a particular Banach space E. In this case we say that $\mathcal{E}$ is a Banach vector bundle with *fiber E.*

By a *morphism* $A: \mathcal{E}_1 \to \mathcal{E}_2$ of Banach vector bundles $\mathcal{E}_i \xrightarrow{p_i} \mathfrak{X}$ $(i = 1,2)$ we mean a continuous map such that 1) $p_2 A = p_1$ (i.e. A preserves fibers), 2) at each point $\tau \in \mathfrak{X}$ the induced map $A_\tau: \mathcal{E}_{1\tau} \to \mathcal{E}_{2\tau}$ of fibers is linear and continuous and 3) there are trivializing maps

$$\phi^{(i)} : p_i^{-1}(\mathfrak{U}) \to \mathfrak{U} \times E^i \quad (i = 1,2) \quad (\tau \in \mathfrak{U})$$

for which the map from $\mathfrak{U}$ into $L(E_1, E_2)$ given by $\tau \to \phi_\tau^{(2)} \circ A_\tau \circ (\phi_\tau^{(1)})^{-1}$ is continuous in the operator norm.

A *Fredholm morphism* is a morphism A such that every restriction A_τ onto a fiber is a Fredholm operator.

An *isomorphism of* $\mathcal{E}_1$ onto $\mathcal{E}_2$ is a morphism which is a one-to-one map (or A_τ is an isomorphism of Banach spaces for every $\tau \in \mathfrak{X}$). In this case we say that our bundles are *isomorphic.*

A bundle isomorphic to the $\mathfrak{X} \times E \xrightarrow{p} \mathfrak{X}$ (where p is just the projection onto the first factor) is said to be *trivial.*

An important example of a trivial bundle is generated by the standard projection map $E \xrightarrow{\pi} E/N$ where E is a Banach space and N is its closed subspace. To find the Banach vector bundle structure we use the map ρ of Theorem 1.1.1 to construct a map ϕ: $E \to E/N \times N$ by the formula $\phi(x) = (\pi(x), x - \rho \circ \pi(x))$. This is an isomorphism. The whole base $U = E/N$ is a trivializing cover and the map ϕ defines on $(E/N,E,\pi)$ the structure of a trivial Banach vector bundle with the fiber N and base E/N.

A section of a bundle $\mathcal{E} \xrightarrow{p} \mathfrak{X}$ is a continuous right inverse map of p, i.e. δ: $\mathfrak{X} \to \mathcal{E}$, $p \circ \delta = id_{\mathfrak{X}}$. We shall say, for brevity, "the Banach bundle $\mathcal{E}$", or even "the bundle $\mathcal{E}$", instead of "the Banach vector bundle $\mathcal{E}$".

Let $\mathcal{E} \xrightarrow{p} \mathfrak{X}$ be a bundle with a fiber E and let $\mathcal{F}$ be a subspace of $\mathcal{E}$ such that for each $\tau \in \mathfrak{X}$ the intersection $\mathcal{E}_\tau \cap \mathcal{F}$ is a closed linear subspace of $\mathcal{E}_\tau$. We write q for the restriction of p on $\mathcal{F}$ and assume that $(\mathfrak{X}, \mathcal{F}, q)$ is also a Banach vector bundle. This bundle is said to be a subbundle of $\mathcal{E}$ if for each point $\tau \in \mathfrak{X}$ there are trivializations ϕ: $p^{-1}(\mathcal{U}) \to \mathcal{U} \times E$ and ψ: $q^{-1}(\mathcal{U}) \to \mathcal{U} \times F$ $(\tau \in \mathcal{U})$ such that the map $\tau \to \phi_\tau \circ (\psi_\tau)^{-1}$ is continuous on $\mathcal{U}$.

A subbundle $\mathcal{F} \subset \mathcal{E}$ is *direct* (or *complemented*) if its fibers are complemented subspaces of the corresponding fibers of $\mathcal{E}$. (For instance, all subbundles of Hilbert bundles are direct.)

The following construction provides an important (and in fact generic) example of a direct subbundle. Suppose that on a Banach space E we are given a family of bounded projection operators $P(\tau)$ depending continuously (with respect to the operator norm) on the point τ of a topological space $\mathfrak{X}$. We consider the trivial bundle $\mathcal{E} = \mathfrak{X} \times E$ over $\mathfrak{X}$. Then the subspace $\mathcal{F} \subset \mathcal{E}$ consisting of all pairs (τ, x) for which $P(\tau)x = x$ is a direct subbundle in $\mathcal{E}$. As follows later (theorem 4), practically every Banach bundle can be obtained by this construction.

The subbundle $\mathcal{F}$ constructed above is trivial if and only if there exists an operator following the evolution of the image *Im* $P(\tau)$, in other words, an operator $U(t,\tau)$ on E, depending continuously on the parameters $t, \tau \in \mathfrak{X}$, that maps isomorphically *Im* $P(t)$ onto *Im* $P(\tau)$.

It can be easily proved that, for a family of operators $T(\tau)$ $(\tau \in \mathfrak{X})$ having zero kernel and complemented images and depending continuously on $\tau \in \mathfrak{X}$, there is a continuous family of projections onto *Im* $T(\tau)$. Therefore, the subspace of the bundle $\mathcal{E} = \mathfrak{X} \times E$ consisting of the pairs $(\tau, T(\tau)\, x)$ $(\tau \in \mathfrak{X}, x \in E)$ forms a direct subbundle of it, which we denote by *Im* T. The same happens if the kernels of $T(\tau)$ are identical.

Similarly, if all the operators $T(\tau)$ are surjective and have complemented kernels, then the kernels form a direct subbundle $Ker\ T = \bigcup_{\tau \in \mathfrak{X}} Ker\ T(\tau)$ in $\mathcal{E} = \mathfrak{X} \times E$. Both these facts hold for paracompact bases, and may be deduced from theorem 1.4.6.

Let $\mathcal{E}$ be a Banach vector bundle with fiber E, and let G be a Banach space for which there is a continuous homomorphism θ of the group $GL(E)$ into the group $GL(G)$. The bundle *associated* with $\mathcal{E}$ is the bundle with the same base and trivializing cover, with the fiber G and transition functions $\theta\phi_{\alpha\beta}$. Two important examples of this construction are:

a) Let $G = L(E)$ and for every $g \in GL(E)$ we define $\theta g \in GL(L(E))$ as

$$(\theta g)A = gAg^{-1} \quad , \quad A \in L(E).$$

The corresponding bundle is denoted by $L(\mathcal{E})$. Sections of it may be identified with endomorphisms of the bundle $\mathcal{E}$.

b) Let $G = E^*$ and $\theta g = (g^*)^{-1}$ for $g \in GL(E)$. The corresponding associated bundle is denoted by $\mathcal{E}^*$ and called the *adjoint* bundle to $\mathcal{E}$. The reader can easily check that the standard duality between E and E^* defines a duality between the fibers E_τ and $E_\tau{}^*$ of $\mathcal{E}$ and $\mathcal{E}^*$ over the same point τ.

The operation of direct sum of two Banach bundles (over the same base) can be defined exactly as for vector bundles.

We present now without proof some simple basic facts about Banach bundles.

Theorem 1.3.1. *If $\mathcal{F}$ is a direct subbundle of a bundle $\mathcal{E}$ over a paracompact base, then there is a projection from $\mathcal{E}$ onto $\mathcal{F}$, that is an endomorphism of $\mathcal{E}$ such that on each fiber it is a bounded projection onto the corresponding fiber of $\mathcal{F}$.*

A simple usage of unity partition gives the proof of this theorem.

Under the conditions of the theorem the bundle $\mathcal{E}$ decomposes into the direct sum of the subbundle $\mathcal{F}$ and of a complementary subbundle $\mathcal{J}$ (the kernel of the projection). In this case we say that $\mathcal{F}$ is complemented.

The next theorem contains a standard result of the theory of fiber bundles.

Theorem 1.3.2. *If the base of the Banach bundle $\mathcal{E}$ with fiber E is a CW-complex and if $GL(E)$ is contractible in the uniform operator topology, then $\mathcal{E}$ is trivial.*

The group $GL(E)$ is not contractible when $\dim E < \infty$. The contractibility of the general linear group of an infinite dimensional Banach space E is, however, rather common situation. The first theorem of this kind was proved by N. Kuiper [240].

Theorem 1.3.3. (N. Kuiper). *If H is an infinite dimensional Hilbert space, then the group $GL(H)$ is contractible.*

The last two theorems show that infinite dimensional Hilbert bundles are in general trivial .

We write $l_1(E)$ for the Banach space of sequences $\{x_i\}_{i=1}^{\infty}$, $x_i \in E$ with the norm

$$\| \{x_i\} \|_{l_1(E)} = \sum \| x_i \|_E.$$

Theorem 1.3.4. *Every bundle $\mathcal{E}$ of countable type with a fiber E and paracompact base $\mathfrak{X}$ is isomorphic to a complemented subbundle of the trivial bundle $\mathcal{L}_1(E) = \mathfrak{X} \times l_1(E)$. Moreover, $\mathcal{E} \oplus \mathcal{L}_1(E)$ is isomorphic to $\mathcal{L}_1(E)$.*

In this theorem the space $l_1(E)$ can be replaced by a number of other spaces. If, for example, $E \oplus E$ is isomorphic to a complemented subspace of E, then we can take E in place of $l_1(E)$ in the first statement of the theorem.

Let us consider now the relations between Banach vector bundles. We shall see that every infinite dimensional Banach bundle is sufficiently rich to "accommodate" all vector bundles. First of all, we need a new definition.

Let E be a Banach space. We write $G_n(E)$ for the Grassmanian of the n-dimensional subspaces of E with the gap metric, i.e. if E_1, E_2 are two n-dimensional subspaces of E, then

$$\rho(E_1, E_2) = max(dist(S(E_1), E_2), dist(S(E_2), E_1)).$$

Here, as before, $S(E)$ is the unit sphere of E. The group $GL(E)$ acts continuously on $G_n(E)$ in the natural way. Therefore, for each bundle $\mathcal{E}$ with fiber E we can consider the associated bundle $G_n(\mathcal{E})$ with fiber $G_n(E)$. (We have to mention that $G_n(\mathcal{E})$ is not a Banach vector bundle, but it is a locally trivial bundle in the topological sense.) It is easy to understand that there is a one-to-one correspondence between all n-dimensional vector subbundles of $\mathcal{E}$ and continuous sections of $G_n(\mathcal{E})$.

Theorem 1.3.5. *If $\mathcal{E}$ is a Banach bundle with paracompact base space $\mathfrak{X}$ and infinite dimensional fiber E, then there is a section δ: $\mathfrak{X} \to \mathcal{E}$ that vanishes nowhere. Moreover, given an open set $\mathcal{U} \subset \mathfrak{X}$, a non-vanishing section δ over $\mathcal{U}$ and a closed subset $D \subset \mathcal{U}$, there exists a non-vanishing section ψ over $\mathfrak{X}$ such that $\psi|_D = \delta|_D$. Every two such sections are homotopic (in the class of non-vanishing sections).*

This theorem actually means that every infinite-dimensional Banach bundle (over a paracompact base) contains a trivial one-dimensional subbundle. (This is definitely not true in the finite dimensional case [176].) Taking the complement and iterating the procedure, we can get n-dimensional trivial subbundles, and then (by stable triviality) any vector bundle of finite type.

Theorem 1.3.6. *Let $\mathcal{E}$ be an infinite dimensional Banach bundle over a paracompact base $\mathfrak{X}$ and $\mathcal{K}$ be a finite dimensional vector bundle of finite type over $\mathfrak{X}$. Then $\mathcal{E}$ contains a subbundle $\mathcal{F}$ isomorphic to $\mathcal{K}$. If $\mathcal{F}_1$, $\mathcal{F}_2$ are two isomorphic n-dimensional subbundles of finite type in $\mathcal{E}$, then the corresponding sections of $G_n(\mathcal{E})$ are homotopic. In particular, complements to $\mathcal{F}_i$ $(i = 1,2)$ are also isomorphic.*

Corollary 1.3.7. *For any infinite dimensional Banach bundle $\mathcal{E}$ over a compact space $\mathfrak{X}$ there exists a one-to-one correspondence between the set $Vect_n(\mathfrak{X})$ of isomorphic classes of n-dimensional vector bundles over $\mathfrak{X}$ and the set of homotopy classes of continuous sections of $G_n(\mathcal{E})$.*

Theorem 1.3.8. *Let $\mathcal{E}$ be a Banach bundle over a paracompact base $\mathfrak{X}$, that splits over an open set $\mathcal{U} \subset \mathfrak{X}$ into the direct sum of subbundles: $\mathcal{E}|_{\mathcal{U}} = \mathcal{F}_1 \oplus \mathcal{F}_2$,* where *$\mathcal{F}_1$ is finite dimensional and trivial. Then for each closed set $D \subset \mathcal{U}$ there is a decomposition $\mathcal{E} = \mathcal{E}_1 \oplus \mathcal{E}_2$, where $\mathcal{E}_1$ is trivial and finite dimensional, and $\mathcal{E}_i|_D = \mathcal{F}_i|_D$ $(i = 1,2)$.*

We are able now to prove an important generalization of lemma 1.2.11.

Theorem 1.3.9. *Let A: $\mathcal{E} \to \mathcal{F}$ be a morphism of Banach bundles over a compact space $\mathfrak{X}$ whose kernel is finite-dimensional in every fiber and whose image is complemented. Then $\mathcal{E}$ contains a subbundle $\mathcal{M}$ of finite codimension on which A is a monomorphism. Moreover, the image of the morphism A restricted to $\mathcal{M}$ is a subbundle* *of* $\mathcal{F}$.

Proof. First of all, let us mention that the statement is non-trivial only in the case of infinite-dimensional fibers (otherwise $\mathcal{M}$ can be chosen zero-dimensional).

Let us consider a finite collection of open sets $\mathcal{U}_i^1$, $\mathcal{U}_i^2$ ($i=1, ..., n$) such that 1) $\overline{\mathcal{U}_i^2} \subset \mathcal{U}_i^1$, 2) $\bigcup_i \mathcal{U}_i^2 = \mathfrak{X}$, 3) the restrictions of the bundles $\mathcal{E}$ and $\mathcal{F}$ onto $\overline{\mathcal{U}_i^1}$ are trivial. We are going to construct inductively a sequence $\mathcal{M}_1 \supset \mathcal{M}_2 \supset \dots \supset \mathcal{M}_n$ of subbundles of finite codimension such that the restriction of A onto $\mathcal{M}_k$ is injective over $\bigcup_{i=1}^{k} \mathcal{U}_i^2$. Then $\mathcal{M} = \mathcal{M}_n$ will satisfy the required condition.

Let us start with $\mathcal{M}_1$. After trivialization of $\mathcal{E}$ and $\mathcal{F}$ over $\overline{\mathcal{U}_1^2}$, we are in the situation of the remark to lemma 1.2.11, so we can choose a subbundle $\tilde{\mathcal{M}}_1$ of a finite codimension m in $\mathcal{E}|_{\overline{\mathcal{U}_1^2}}$ such that $A|_{\tilde{\mathcal{M}}_1}$ has zero kernel, and $\tilde{\mathcal{M}}_1$ has trivial complement. Due to the theorem 8, there is an expansion $\mathcal{E} = \mathcal{M}_1 + \mathcal{N}_1$ where $\mathcal{N}_1$ is a finite dimensional trivial bundle and $\mathcal{M}_1|_{\overline{\mathcal{U}_1^2}} = \tilde{\mathcal{M}}_1|_{\overline{\mathcal{U}_1^2}}$. A simple manipulation with the corollary 7 shows that $\mathcal{M}_1|_{\overline{\mathcal{U}_2^2}}$ is trivial. Then we can repeat the previous step, considering $\mathcal{M}_1$ instead of $\mathcal{E}$, and $\mathcal{U}_2^2 \subset \mathcal{U}_2^1$ instead of $\mathcal{U}_1^2 \subset \mathcal{U}_1^1$. This leads to a subbundle $\mathcal{M}_2 \subset \mathcal{M}_1$ (of finite codimension) such that A has zero kernel on $\mathcal{M}_2|_{\overline{\mathcal{U}_2^2}}$. Because A already has zero kernel on $\mathcal{M}_1|_{\overline{\mathcal{U}_1^2}}$, we conclude that it also has zero kernel on $\mathcal{M}_2|_{\overline{\mathcal{U}_1^2 \cup \mathcal{U}_2^2}}$. Repeating this procedure, we finish the proof of the first statement. The last statement of the theorem can be proved like in theorem 1.4.10.

Remarks. 1. The compactness condition of the last theorem is essential, because in the case of a non-compact base $\mathfrak{X}$ the dimension of the kernel of A can be unbounded over the whole base, so we can not cut it out by subbundles of a finite codimension. **2.** Up to this point we have not imposed any restriction on the fiber E of the Banach bundle $\mathcal{E}$. A number of the assertions above can be strengthened considerably if the space E has a symmetric basis, for instance in Hilbert or l_p cases.

In this case we can drop in some results the condition that our bundles are of finite type. An unconditional basis $\{e_i\}$ in a Banach space E is said to be *symmetric* if it satisfies the following two conditions:

a) if the series $\sum \alpha_i e_i$ converges and $|\beta_i| \leq |\alpha_i|$ for all i, then the series $\sum \beta_i e_i$ also converges and

$$\| \sum \beta_i e_i \| \leq \| \sum \alpha_i e_i \| .$$

b) for every permutation σ of the natural numbers the convergence of the series $\sum \alpha_i e_i$ implies that the series $\sum \alpha_i e_{\sigma(i)}$ also converges and

$$\| \sum \alpha_i e_i \| = \| \sum \alpha_i e_{\sigma(i)} \| .$$

One can prove also the following result.

Theorem 1.3.10. *Suppose that E has a complemented subspace with a symmetric basis, and let $\mathcal{E}$ be a trivial bundle with the fiber E over a paracompact base $\mathfrak{X}$. Then $\mathcal{E} \oplus \mathcal{K}$ is isomorphic to $\mathcal{E}$ for any finite-dimensional bundle $\mathcal{K}$ of finite type over $\mathfrak{X}$.*

1.4. Fredholm operators that depend continuously on a parameter

Let $\mathfrak{X}$ be a compact topological space, and $T\colon \mathfrak{X} \to \Phi(E)$ be a continuous Fredholm operator function. We can introduce a notion of index of such operator family that generalizes the notion of the index of a single Fredholm operator. The values of this index are not numbers anymore. Let *Vect* $(\mathfrak{X})$ be the semigroup (with respect to the direct sum) of isomorphism classes of finite dimensional vector bundles over $\mathfrak{X}$. We denote by $K(\mathfrak{X})$ the corresponding Grothendieck group (see[19], for instance). If $\mathfrak{P}$ is a vector bundle, we denote by $[\mathfrak{P}]$ its equivalence classes in $K(\mathfrak{X})$. In this case $K(\mathfrak{X})$ consists of differences $[\mathfrak{P}] - [\mathfrak{Q}]$.

Consider again an operator family $T\colon \mathfrak{X} \to \Phi(E)$. Due to the remark after lemma 1.2.11, we can find a closed subspace $M \subset E$ of finite codimension such that $M \cap Ker\ T(\tau) = \{0\}$ for all $\tau \in \mathfrak{X}$. In this case $Im\ (T|_M)$ is a subbundle $\mathcal{E}$ of finite codimension in the trivial bundle $\mathfrak{X} \times E$ over $\mathfrak{X}$. Consider a complement vector subbundle $Q \subset \mathfrak{X} \times E$, i.e. $Q \oplus \mathcal{E} = \mathfrak{X} \times E$. Consider also the "kernel subbundle" $\mathfrak{P} = \mathfrak{X} \times E/M$ over $\mathfrak{X}$.

Definition 1.4.1. The *index* of the family T is the element of the group $K(\mathfrak{X})$ given by the formula

$$\text{index } T = [\mathfrak{P}] - [\mathfrak{Q}]. \tag{1.4.1}$$

The correctness (i.e. the independence on the choice of M) of the definition can be easily checked. It can be also shown that *index* $(T_1 T_2) = \text{index}(T_1) + \text{index}(T_2)$, so *index* is a semigroup homomorphism.

We introduce the following notation: given topological spaces $\mathfrak{U}$ and $\mathfrak{V}$ we write $[\mathfrak{U}, \mathfrak{V}]$ for the set of homotopy classes of continuous maps from $\mathfrak{U}$ to $\mathfrak{V}$.

Theorem 1.4.2. *Let $\mathfrak{X}$ be a compact topological space. Then the sequence of group homomorphisms*

$$[\mathfrak{X}, GL_c(E)] \xrightarrow{i} [\mathfrak{X}, GL(E)] \xrightarrow{j} [\mathfrak{X}, \Phi(E)] \xrightarrow{\text{index}} K(\mathfrak{X})$$

is exact.

Here the homomorphisms i and j are induced by the inclusions $GL_c(E) \subset GL(E) \subset \Phi(E)$.

This theorem can be sharpened by imposing some conditions on the Banach space E.

Theorem 1.4.3. *Suppose that a Banach space E contains a complemented infinite-dimensional subspace E with a symmetric basis. Then for every compact set $\mathfrak{X}$ there is an exact sequence of group homomorphisms*

$$0 \to [\mathfrak{X}, GL(E)] \xrightarrow{j} [\mathfrak{X}, \Phi(E)] \xrightarrow{\text{index}} K(\mathfrak{X}) \to 0$$

The group $GL(E)$ has been proved to be contractible for a wide class of Banach spaces (see theorem 1.3.3 for the Hilbert space case).

Corollary 1.4.4. *If $GL(E)$ is contractible (for instance, if E is a Hilbert space), then the index mapping generates the group isomorphism*

$$0 \to [\mathfrak{X}, \Phi(E)] \overset{index}{\rightleftarrows} K(\mathfrak{X}) \to 0 .$$

The significance of the index mapping can be emphasized also by the following statement.

Theorem 1.4.5. *Let $T_i : \mathfrak{X} \to \Phi(E)$ $(i = 0, 1)$ be continuous maps from a compact space $\mathfrak{X}$ and let index $T_0 =$ index T_1. Then there are continuous maps $\mathcal{K}_i : \mathfrak{X} \to C(E)$ $(i = 0, 1)$ and $G: \mathfrak{X} \to GL(E)$ such that the operators $\mathcal{K}_i(\tau)$ are finite-dimensional and*

$$G(\tau)(T_1(\tau) + \mathcal{K}_1(\tau)) = T_0(\tau) + \mathcal{K}_0(\tau), \ \tau \in \mathfrak{X}.$$

We shall consider now the question about the existence of left, right and two-sided inverses and regularizers.

Theorem 1.4.6. *Let A be a Banach algebra with unit e, $\mathfrak{X}$ a topological space, and $T: \mathfrak{X} \to A$ a continuous map. If $T(\tau)$ is right (left, two-sided) invertible in A for all $\tau \in \mathfrak{X}$, then there exists a right (left, two-sided) inverse $T(\tau)$ depending continuously on $\tau \in \mathfrak{X}$.*

Proof. Since A is a metric space, we can assume that $\mathfrak{X}$ is paracompact. We give the proof for right inverse elements. First we construct $T'(\tau)$ locally. Let $\tau_0 \in \mathfrak{X}$ and let S_0 be some right inverse of $T(\tau_0)$, i.e. $T(\tau_0) S_0 = e$. There is a neighborhood U_0 of τ_0 such that $\| T(\tau)S_0 - e \|_A < 1$. Then in this neighborhood $T(\tau)S_0$ is invertible, and $T_0'(\tau) = S_0(T(\tau) S_0)^{-1}$ is a continuously depending on $\tau \in U_0$ right inverse to $T(\tau)$. Let us mention now that the set of all right inverse elements to a fixed $T \in A$ is convex. Now the local construction together with a unity decomposition leads to a global right inverse. This finishes the proof.

Theorem 1.4.7. *Let $\mathfrak{X}$ be a paracompact topological space and $T\colon \mathfrak{X} \to \Phi_r(E)$ $(\Phi_l(E),\ \Phi(E))$ be a continuous map. Then there exists a continuous map $T' : \mathfrak{X} \to \Phi_l(E)$ $(\Phi_r(E),\ \Phi(E))$ such that $T'(\tau)$ is a right (left, two-sided) regularizer for $T(\tau)$, $\tau \in \mathfrak{X}$.*

Proof. Consider the natural projection $\pi\colon L(E) \to Cal(E)$ onto the Calkin algebra. Due to theorem 1.2.10, $\Phi_r(E)$ $(\Phi_l(E),\ \Phi(E))$ consists of pre-images under π of all right (left, two-sided) invertible elements of $Cal(E)$. Consider now $t(\tau) = \pi(T(\tau))$. According to theorem 6, we can find corresponding continuous one-sided or two-sided inverse $t'(\tau)$ in $Cal(E)$. Choosing a Bartle-Graves map $\rho : Cal(E) \to L(E)$ (see theorem 1.1.1), we get a regularizer $T'(\tau) = \rho(t'(\tau))$. The proof is complete.

Here we present the conditions under which $T(\tau)$ has an equivalent regularizer that depends continuously on $\tau \in \mathfrak{X}$.

Theorem 1.4.8. *Let $\mathfrak{X}$ be a topological space and $T\colon \mathfrak{X} \to \Phi(E)$ be a continuous map. A global left (right, two-sided) equivalent continuous regularizer $T'(\tau)$ exists if and only if the equivalence class of T in $[\mathfrak{X}, \Phi(E)]$ contains a mapping*

$$T_1 : \mathfrak{X} \to \Phi(E) \cap GL_r(E) \quad (\Phi(E) \cap GL_r(E),\ GL(E)).$$

Corollary 1.4.9. *Let $\mathfrak{X}$ be a compact space and $T : \mathfrak{X} \to \Phi(E)$ a continuous map. There is a global continuous left (right, two-sided) equivalent regularizer for T if and only if $index\ T = [P]$ $(index\ T = -[P]$, $index\ T = 0)$, where $P \in Vect(\mathfrak{X})$.*

Let us consider now the case when the kernel (or, equivalently, cokernel) of a continuous Fredholm operator function is of a fixed dimension. The result that we are interested in can be stated in some more general setting - for Fredholm morphisms of Banach bundles.

Theorem 1.4.10. *Let $\mathcal{E}$ and $\mathcal{F}$ be Banach bundles over some topological space $\mathfrak{X}$ and $A : \mathcal{E} \to \mathcal{F}$ be a Fredholm morphism. If the dimension of the kernel (or cokernel) of A on each fiber is the same, then*

$$\bigcup_{\tau \in \mathfrak{X}} Ker\ A(\tau)$$

is a subbundle of $\mathcal{E}$, *and*

$$\bigcup_{\tau \in \mathfrak{S}} Im\ A(\tau)$$

is a subbundle of $\mathcal{F}$.

This theorem shows that the condition $dim\ Ker\ A(\tau) = const$ implies that the kernel $Ker\ A(\tau)$ and the range $Im\ A(\tau)$ of $A(\tau)$ depend continuously on the parameter $\tau \in \mathfrak{S}$.

Proof. Since the statement is local, and our bundles are locally trivial, we may assume that we are dealing with a continuous family $A(\tau) \in \Phi(E,F)$ of Fredholm operators. The standard trick allows us to assume that $E = F$. Let $\tau_0 \in \mathfrak{S}$ and $M \subset E$ be a closed complement to $Ker\ A(\tau_0)$. Lemma 1.2.11 shows then that $M \cap Ker\ A(\tau) = \{0\}$ for all τ in some neighborhood U of τ_0. Since $dim\ Ker\ A(\tau) = const$, we conclude that $M \oplus Ker\ A(\tau) = E$ for all $\tau \in U$. This means that $Im\ A(\tau) = Im\ (A(\tau)\,|\,_M)$ for all $\tau \in U$. Let us restrict $A(\tau)$ to M. Then we get a family $B(\tau)$ of left invertible operators from M to E. Theorem 6 shows now that there is a left inverse operator $C(\tau) \in L(E,M)$ that depends continuously on the parameter $\tau \in U$. In other words, $C(\tau)B(\tau)\ x = x$ for all $x \in M$, $\tau \in U$. Consider now the operator $P(\tau) = B(\tau)C(\tau)$. This is obviously a continuous family of projectors onto $Im\ A(\tau)\ (= Im(A(\tau)\,|\,_M))$, so $\cup\, Im\ A(\tau)$ forms a subbundle. We can trivialize this subbundle locally. Then we may assume that $A(\tau)$ is a continuous family of *surjective* Fredholm operators. Let $D(\tau)$ be a continuous family of right inverse operators to $A(\tau)$ (see theorem 6). Then $I - D(\tau)A(\tau) = Q(\tau)$ is a continuous family of projections onto $Ker\ A(\tau)$, so $\cup\, Ker\ A(\tau)$ is a subbundle. The theorem is proved.

1.5. Some information from complex analysis

Let us introduce first some functional spaces that we shall need throughout the book. Let U be a complex manifold (for instance, a region in $\mathbf{C}^n$) and H a Hilbert space. We denote by $\mathcal{H}(U,H)$ the space of all holomorphic functions $U \to H$. The space is equipped with the topology of uniform convergence on compacts in U. This is a Frechet space, and the topology is defined by semi-norms

$$\rho_{\mathcal{K}}(\varphi) = \max_{z \in \mathcal{K}} \|\varphi(z)\|$$

where $\mathcal{K} \subset\subset U$ is an arbitrary compact (here and further $\subset\subset$ denotes compact embedding).

When H=C, we use the notation $\mathcal{H}(U):=\mathcal{H}(U,\mathbf{C})$. In the case when $U=\mathbf{C}^n$ we use the notations $A(\mathbf{C}^n,H) := \mathcal{H}(\mathbf{C}^n,H)$, $A(\mathbf{C}^n) := \mathcal{H}(\mathbf{C}^n)$. By $\widetilde{A}(\mathbf{C}^n,H)$ and $\widetilde{A}(\mathbf{C}^n)$ we denote the subspaces in $A(\mathbf{C}^n,H)$ and $A(\mathbf{C}^n)$ correspondingly, consisting of functions that are $2\pi i$-periodic in every z_j $(j=1, ..., n)$.

We denote by $A_p(\mathbf{C}^n, H)$ the space of H-valued entire functions in $\mathbf{C}^n$ of order p and of minimal type. This means that an entire H-valued function $\varphi(z)$ belongs to the space if

$$q_a(\varphi) = \sup_{z \in \mathbf{C}^n} \|\varphi(z)\| \exp(-a|z|^p) < \infty$$

for all $a>0$. This is also a Frechet space with respect to semi-norms q_a $(a>0)$.

By $\widetilde{A}_p(\mathbf{C}^n,H)$ $(\widetilde{A}_p(\mathbf{C}^n))$ we denote the subspace of $A_p(\mathbf{C}^n,H)$ $(A_p(\mathbf{C}^n))$ consisting of functions that are $2\pi i$-periodic with respect to z_j $(j=1, ..., u)$.

Let now $p_0<p$, $q \in (p_0, p)$, and $a>0$. We denote by $B^{(q)}_{(a)}$ the following Banach space:

$$B^{(q)}_{(a)} = \{f \in A(\mathbf{C}^n,H) \mid \|f\|_{a,q} = \sup \|f(z)\|_H \exp(-a|z|^q) < \infty\}.$$

Then $B^{(q)} = \lim_{a \to 0} proj\, B^{(q)}_{(a)}$ is a Frechet space. We introduce the space

$$B_{p,p_0}(\mathbf{C}^n, H) = \lim_{q \to p} ind\, B^{(q)}.$$

We denote, as before, by $\widetilde{B}_{p,p_0}(\mathbf{C}^n,H)$ the corresponding subspace of $2\pi i$-periodic functions (with the induced topology on it).

Consider the torus

$$\mathcal{T}^n = \{z \in \mathbf{C}^n \mid |z_j| = 1, j=1, ..., n\} \subset \mathbf{C}^n.$$

This torus is the skeleton (Shilov boundary) of the unit polydisk in $\mathbf{C}^n$. Let E be a

Banach space. We denote by $W(\mathcal{T}^n, E)$ the space of continuous functions $f\colon \mathcal{T}^n \to E$ that have absolutely convergent Fourier (=Taylor) series

$$f(z) = \sum_{k \in \mathbf{Z}^n} f_k z^k.$$

This is a Banach space with respect to the norm

$$\|f\|_w = \sum_{k \in \mathbf{Z}^n} \|f_k\|_E < \infty.$$

We abbreviate $W(\mathcal{T}^n, \mathbf{C})$ as $W(\mathcal{T}^n)$. We can observe that any E-valued function, analytic in a neighborhood of $\mathcal{T}^n$, belongs to $W(\mathcal{T}^n, E)$. If $A(z)$ is an analytic operator function in a neighborhood of $\mathcal{T}^n$ with values in $L(E, F)$, then multiplication by $A(z)$ is a bounded operator from $W(\mathcal{T}^n, E)$ into $W(\mathcal{T}^n, F)$. These statements follow easily from standard estimates of Laurent's series coefficients.

In some cases analogs to the spaces described above can be introduced for sections of Banach bundles. Let, for instance, $\mathcal{E}$ be a subbundle of the trivial bundle $\mathbf{C}^n \times H$ over $\mathbf{C}^n$. We denote by $A_p(\mathbf{C}^n, \mathcal{E})$ the subspace of $A_p(\mathbf{C}^n, H)$ consisting of sections of the bundle $\mathcal{E}$. Analogously, if $\mathcal{E}$ is a subbundle of $\mathcal{T}^n \times \mathcal{E}$ over $\mathcal{T}^n$, we denote by $W(\mathcal{T}^n, \mathcal{E})$ the subspace of $W(\mathcal{T}^n, E)$, consisting of sections of $\mathcal{E}$. In both cases the subspaces are obviously closed.

A. Interpolation of entire functions of finite order.

We shall discuss in this subsection the following problem. Let $F(\lambda)$ $(\lambda \in \mathbf{C})$ be an entire function of a finite order. Consider its zero set $\{\lambda_k\} \subset \mathbf{C}^n$, where λ_k is a root of $F(\lambda)$ of multiplicity p_k. What are the conditions on complex numbers $\{a_{k,l}\}$ such that the interpolation problem

$$\text{(1.5.1)} \qquad \left.\frac{d^l f}{d\lambda^l}\right|_{\lambda_k} = a_{k,l}\,,\ k=1, 2, \ldots, l= 0, \ldots, p_k - 1$$

is solvable in the same class of entire functions of finite order?

We shall actually deal with case of $2\pi i$-periodic functions. The next two results show that if $F(z)$ and the data $\{a_{k,l}\}$ are $2\pi i$-periodic, then the interpolation problem is solvable in periodic functions $f(\lambda)$ if and only if it is solvable without periodicity condition. This means that we can disregard the condition of periodicity of $f(\lambda)$ when we discuss solvability of (1).

Let $F(z) \equiv 0$ be an entire function in $\mathbf{C}^n$ *of order p>1* and of *normal type*, i.e.

$$| F(z) | \leq C \ exp \ a|z|^p.$$

Suppose that $F(z)$ is $2\pi i$-periodic with respect to every variable z_j $(j=1, \ldots, n)$. We denote by χ_F the analytic subset

$$\chi_F = \{z \in \mathbf{C}^n \mid F(z) = 0\}.$$

Let us denote by χ_F^* the subset of all regular points of χ_F, i.e. the points in whose neighborhood χ_F is an analytic submanifold in $\mathbf{C}^n$ [154, 307]. (Let us note that when $n=1$ we have $\chi_F^* = \chi_F$.)

For every point $z_o \in \chi_F^*$ we can choose some coordinates $(\zeta_1, \ldots, \zeta_n)$ such that the ζ_n-axis is transversal to χ_F^* in a neighborhood of z_0. This enable us to find a locally finite cover of χ_F^* by open subsets with corresponding coordinate systems of described type. We can define then at $z_0 \in \chi_F^*$ the multiplicity of zero for $F(z)$ as

$$\gamma(z_0) = max \ \{k \in \mathbf{Z}_+ \mid \ \frac{\partial^q F}{\partial \zeta_n^q} |_{\chi_F^* \cap U} = 0 \ \forall q \leq k-1\}.$$

Here U is the element of the cover containing z_0 and $\{\zeta_j\}$ are corresponding coordinates.

Due to the periodicity of $F(z)$ we may assume the cover (with the related coordinate systems) invariant with respect to the $2\pi i$-translations in each variable z_j $(j=1, \ldots, n)$. It is obvious that the multiplicity $\gamma(z_0)$ is also periodic.

Let now H be a separable Hilbert space.

Theorem 1.5.1. *Let* $f: \mathbf{C}^n \to H$ *be an entire function of order p and of normal type* (*i.e.* $\| f(z) \| \leq C_f \ exp \ a_f \ |z|^p$). *Suppose that*

$$\frac{\partial^k f(z+2\pi i m)}{\partial \zeta_n^k} = \frac{\partial^k f(z)}{\partial \zeta_n^k} \tag{1.5.2}$$

for all $z \in \chi_F^*$, $m \in \mathbb{Z}^n$, $k=0, 1, \ldots, \gamma(z)-1$. *Then there exists an entire function* $\widetilde{f}: \mathbb{C}^n \to H$ *of order p and of normal type that is 2πi-periodic with respect to* z_j ($j=1, \ldots, n$) *and such that*

$$\frac{\partial^k \widetilde{f}(z)}{\partial \zeta_n^k} = \frac{\partial^k f(z)}{\partial \zeta_n^k}$$

for all $z \in \chi_F^*$, $k = 0, 1, \ldots, \gamma(z)-1$.

Remark. The theorem obviously can be reformulated in the following way:

If $f(z)$ is an entire function of order p and of normal type such that

$$(f(z+2\pi i m) - f(z)) \,/\, F(z)$$

is entire for all $m \in \mathbb{Z}^n$, then there exists a $2\pi i$-periodic function $\widetilde{f}(z)$ of the same class such that

$$(\widetilde{f}(z) - f(z)) \,/\, F(z)$$

is also entire.

Proof. We split $\mathbb{C}^n$ into the union of tubular domains

$$\mathfrak{T}_k = \{z=x+iy \mid 2\pi k_j \le y_j \le 2\pi(k_j+1),\ j=1, \ldots, n\}$$

Here $k \in \mathbb{Z}^n$. Let U_k be the open ϵ-neighborhood of $\mathfrak{T}_k$ for some $\epsilon<\pi$. Consider the functions $f_k(z) = f(z-2\pi i k)$. The functions f_k satisfy for $z \in U_k$ the following estimate:

$$\|f_k(z)\| = \|f_k(x+iy)\| \le C \, exp \, a|x|^p \tag{1.5.3}$$

with constants that do not depend on k. For $k=0$ and $z \in U_0$ it is just the exponential estimate for $f(z)$ in $\mathbb{C}^n$. For $k \neq 0$ we have

$$||f_k(z)|| = ||f_k(x+i(y_0 + 2\pi k))|| = || f_0(x+iy_0)|| \leq C \; exp \; a|x|^p ,$$

where $z=x+iy_0 + 2\pi ik \in U_k$.

The condition (2) implies that for every $k \in \mathbb{Z}^n$ the difference $f_k(z) - f(z)$ can be represented as $F(z)g(z)$ with some entire H-valued function g. Namely, if we define $g = (f_k - f)/F$, then, due to (2), g is analytic everywhere in $\mathbb{C}^n$ except for the set $N = \chi_F \backslash \chi_F^*$. Because N is an analytic subsets of codimension ≥ 2, g is an entire function (see [154]).

Now we define the functions

$$g_{k,l}(z) = (f_k(z) - f_l(z))/F(z) \; .$$

The previous consideration shows that $g_{k,l}$ is an entire function for all $k,l \in \mathbb{Z}^n$. In fact, these functions are entire functions or order p and of normal type. This is the result of some routine considerations in complex analysis. Namely, $f_k - f_l$ is obviously an entire function of the discussed class, and $(f_k - f_l)/F$ is holomorphic in the whole $\mathbb{C}^n$. We can assume that $F(0) \neq 0$ (otherwise we shift the coordinate system). Consider any unit vector $w \in \mathbb{C}^n$, and restrict $g_{k,l}$ to the complex line $z=\lambda\omega$ ($\lambda \in \mathbb{C}$). The function $F_\omega(\lambda) = F(\lambda\omega)$ is entire of order p and of normal type (uniformly with respect to ω). Besides, $F_\omega(0) \neq 0$ does not depend on ω. Standard estimates of entire functions (of one variable) from below (see theorem 6) show that for every ω there exists a system of circles in the λ-plane whose radii tend to ∞ such that for λ on the circles we have the estimate

$$|F^{-1}(\lambda\omega)| \leq C exp \; a|\lambda|^p = C \; exp \; a \; |z|^p \tag{1.5.4}$$

uniformly with respect to ω. Moreover, for $R>0$ large enough we can choose one of these circles in the annulus $R<|z|<CR$ ($C=const$). The estimate (4) together with the maximum principle applied to the entire function $g_{k,l}= (f_k - f_l) \; /F$ shows that $g_{k,l}$ has the order p and normal type.

We are interested in the functions $g_{k,l}$ restricted to the domains $U_{k,l} = U_k \cap U_l$. Since $U_{k,l} \neq \emptyset$ only when $\max_j \; |k_j - l_j| \leq 1$, we shall consider only the functions $g_{k,l}$ for such values of k and l. For these values of k and l the simple calculation

$$g_{k,l}(x+iy)=g_{k-l,0}(x+i(y-2\pi l))$$

shows that in $U_{k,l}$ the function $g_{k,l}$ can be estimated uniformly with respect to k,l (where $max\ |k_j-l_j|\le 1$) as

(1.5.5.) $$\|g_{k,l}(x+iy)\| \le C\ exp\ a|x|^p.$$

Due to the analyticity of $g_{k,l}$ analogous estimates hold for all derivatives of $g_{k,l}$ with respect to x and y. In fact, $\partial g_{k,l}/\partial\bar{z}=0$, and $\partial g_{k,l}\ /\ \partial z$ can be estimated by using the standard Cauchy inequalities for a circle of fixed radius.

Consider the intersections U_k^1 of the regions U_k with the "imaginary" subspace $i\mathbf{R}^n$ in $\mathbf{C}^n$. Let $\varphi_k(y)$ be a C^∞-partition of unity in $i\mathbf{R}^n$, subordinated to the cover $\{U_k^1\}$. It is obvious (follows the standard construction of partitions) that we can choose a partition satisfying $\varphi_k(y)=\varphi_0(y-2\pi k)$ for all $k\in\mathbf{Z}^n$.

Consider now the following functions g_k in U_k.

(1.5.6) $$g_k(x+iy)=\sum_{l\neq k}\varphi_l(y)\ g_{l,k}(x+iy).$$

It follows from the construction that the functions $g_k(x+iy)$ and their x,y-derivatives of arbitrary order can be estimated in U_k by an expression $C\ exp\ a|z|^p$ with constants that might depend on the order of the derivative. In fact, we have the estimates (5) for $g_{k,l}$ and its derivatives, the functions $\varphi_l(y)$ and their derivatives are bounded uniformly with respect to l, and the number of terms in (6) is fixed.

We shall need also the equality

(1.5.7) $$g_{k,m}-g_{l,m}=g_{k,l}$$

that follows immediately from the definition of $g_{k,l}$. This equality together with (6) leads to

(1.5.8) $$g_k - g_l = g_{k,l} \quad \text{in} \quad U_{k,l}.$$

Consider the H-valued (0, 1)-form in $\mathbf{C}^n$

$$h = \bar\partial g_k \text{ on } U_k.$$

Due to the holomorphity of $g_{k,l}$ and to (8), we see that h is correctly defined on the whole $\mathbf{C}^n$:

$$\bar\partial(g_k - g_l) = \bar\partial g_{k,l} = 0 \quad \text{on} \quad U_{k,l} = U_k \cap U_l.$$

Let $h = \sum h_j d\bar z_j$. Our estimates on g_k and their derivatives show that the coefficients $h_j(x+iy)$ and their x,y-derivatives can be estimated from above by expressions

$$C\ exp\ a\ |x|^p$$

with constants C, a that might depend only on the order of the derivative. Let us note also that

$$g_k(x+iy) = \sum_{l \neq k} \varphi_l(y) g_{l,k}(x+iy)) = \sum_{l \neq k} \varphi_0(y - 2\pi l)\ g_{l\text{-}k,0}(x+i(y-2\pi k)) =$$
$$= \sum_{l\text{-}k \neq 0} \varphi_{l\text{-}k}(y-2\pi k)\ g_{l\text{-}k,0}(x+i(y-2\pi k)) = g_0\ (x+i(y-2\pi k)).$$

This means that $h(z) = h(z - 2\pi ik)$ for all $k \in \mathbf{Z}^n$, so the form h is $2\pi i$-periodic in every variable z_j.

Our next goal is to find an H-valued function g which obeys the same type of estimates as h, which is also $2\pi i$-periodic, and such that

(1.5.9) $$\bar\partial g = h \quad \text{in} \quad \mathbf{C}^n.$$

Taking into account the periodicity of the coefficient h_j, we can expand it into the Fourier series

$$h_j(x+iy) = \sum_{k \in \mathbf{Z}^n} e^{ik \cdot y} h_{j,k}(x).$$

Consider h_j as a function on the "cylinder" $\mathbf{R}^n_x \times (i\mathbf{R}^n_y / 2\pi i \mathbf{Z}^n) = \mathbf{R}^n \times \mathbf{T}^n$. Let $H^s(\mathbf{T}^n, H)$ be the Sobolev space of H-valued functions $t(y) = \sum_{k \in \mathbf{Z}^n} e^{ik \cdot y} t_k$ on the torus $\mathbf{T}^n$ with the norm

$$\Big(\sum_{k \in \mathbf{Z}^n} \| t_k \|^2_H (|k| + 1)^{2s} \Big)^{1/2}.$$

Due to the estimates on the derivatives of h_j, we have for any $s \geq 0$:

$$\sum_{k \in \mathbf{Z}^n} \|h_{j,k}(x)\|^2 (|k| + 1)^{2s} \leq C_s \; exp \; a_s |x|^p. \tag{1.5.10}$$

We have the same type of estimates for any derivative $D^\alpha_x h_{j,k}(x)$ instead of $h_{j,k}$.

We shall look for the function g also expanded into the Fourier series:

$$g(x+iy) = \sum e^{iky} g^{(k)}(x).$$

The $\overline{\partial}$-equation (9) can be written as

$$\overline{\partial} g(z) = \sum_{j=1}^n \Big(\sum_{k \in \mathbf{Z}^n} \big(\frac{\partial g^{(k)}}{\partial x_j} - k_j g^{(k)} \big) e^{iky} \Big) d\overline{z}_j$$

$$= \sum_{j=1}^n \Big(\sum_k e^{iky} h_{j,k} \Big) d\overline{z}_j.$$

Comparing the coefficients before $e^{iky} d\overline{z}_j$, we get the system

$$\frac{\partial g^{(k)}(x)}{\partial x_j} - k_j g^{(k)}(x) = h_{j,k}(x), \; j = 1, \dots, n, \;\; k \in \mathbf{Z}^n. \tag{1.5.11}$$

Let $k \in \mathbf{Z}^n$ and $x_a(k)$ be point in $\mathbf{R}^n$ of the maximum

$$\max_{x \in \mathbf{R}^n} \exp\,(kx - 2a|x|^p)$$

(such point exists due to $p>1$). We are able now to suggest an explicit solution to (11):

$$g^{(k)}(x) = \int_{x_a(k)}^{x} \sum_j e^{kx-k\eta} h_{j,k}(\eta) d\eta_j,$$

where the path of the integration is the interval between $x_a(k)$ and x. We still preserve the freedom of choice of an appropriate value of $a>0$. Let us estimate $|g^{(k)}(x)|$. Due to convexity of $|x|^p$, the minimal value of the function

$$\exp\,(k\eta - 2a|\eta|^p)$$

over the path of integration is attained at the point x. We can estimate now:

$$\exp\,(kx - 2a|x|^p) \leq \exp\,(k\eta - 2a|\eta|^p),$$

or

$$\exp\,(kx - k\eta) \leq \exp\,(2a|x|^p - 2a|\eta|^p).$$

This enables us to estimate

$$\| g^{(k)}(x) \| \leq \sum_j \int_{x_{a(k)}}^{x} \exp(2a|x|^p - 2a|\eta|^p) \, \| h_{j,k}(\eta) \| d\eta_j \tag{1.5.12}$$

$$\leq \sum_j \exp 2a|x|^p \int_{x_{a(k)}}^{x} \exp(-a|\eta|^p) \, \| h_{j,k}(\eta) \| \frac{d\eta_j}{\exp a|\eta|^p}$$

$$\leq \exp 2a|x|^p \sup_{j,\eta} \frac{\| h_{j,k}(\eta) \|}{\exp a|\eta|^p} \sum_j \int_{x_a(k)}^{x} \exp\,(-a|\eta|^p) d\eta_j.$$

The last integral is bounded by a constant (depending on a).

The estimate (10) gives us

$$(|k|+1)^s \|h_{j,k}(x)\| \leq C_s \; exp \; a_s|x|^p,$$

or

$$\sup_{j,\eta} \frac{\|h_{j,k}(\eta)\|}{exp \; a_s|\eta|^p} \leq (|k|+1)^{-s} C_s$$

for all $s \geq 0$. This and (12) show that for every $s \geq 0$ there exist $a>0$, $C>0$ such that

$$\sup_x \frac{\|g^{(k)}(x)\|}{exp \; 2a|x|^p} \leq C(|k|+1)^{-s} \tag{1.5.13}$$

For $a>0$ sufficiently large we can guarantee estimates of (13) type for any given number of derivatives of $g^{(k)}(x)$. This is obvious from the equation (11) written in the form

$$\frac{\partial g^{(k)}(x)}{\partial x_j} = k_j g^{(k)}(x) + h_{j,k}(x).$$

We can estimate now the first derivatives if the function itself has been estimated already. Differentiation of both sides enables us to estimate derivatives of arbitrary order. In other words, given $s \geq 0$ we can find a $2\pi i$-periodic solution

$$g = \sum g^{(k)}(x) \; exp(iky)$$

to the equation (9) such that

$$g \in H^s_{loc}(\mathbb{C}^n, H) := H^s_{loc}(\mathbb{R}^{2n}, H)$$

and

$$\| g(x+iy) \|_{H^s(\mathbb{T}^n_y, H)} \leq C \; exp \; a|x|^p.$$

Due to the embedding theorems, for sufficiently large s we can estimate in the $C(\mathbb{T}^n)$ norm instead of the Sobolev one, i.e.

$$||g(x+iy)|| \leq C\ exp\ a|x|^p$$

We solved the $\overline{\partial}$-problem in the class of periodic functions with the required estimates. The rest is simple. Consider in U_k the function

$$\tilde{f}_k(z) = f_k(z) - F(z)\,(g_k(z) - g(z)).$$

Due to the equality $\overline{\partial} g_k = h = \overline{\partial} g$, the function $\tilde{f}_k(z)$ is analytic. Moreover, it obeys the required exponential estimates. It follows from the construction that $\tilde{f}_k(z) = \tilde{f}_0(z - 2\pi ik)$. Let us note also that $\tilde{f}_k = \tilde{f}_l$ in $U_{k,l}$. In fact,

$$\tilde{f}_k - \tilde{f}_l = f_k - F(g_k - g) - f_l + F(g_l - g) = (f_k - f_l) - F(g_k - g_l)$$

$$= f_{k,l} - F g_{k,l} = f_{k,l} - f_{k,l} = 0.$$

So, an entire function $\tilde{f}\,|_{U_k} = \tilde{f}_k$ is defined. Due to the construction, it is $2\pi i$-periodic and of the required exponential type. The difference $f - \tilde{f}$ is analytically divisible by F, because both $f - f_k$ (the theorem's condition) and $f_k - \tilde{f}$ (the definition of $\tilde{f}$) are. This finishes the proof of the theorem.

Let us recall the notion of an entire function $f(z)$ of order p and of minimal type. It means that $f(z)$ can be estimated as

$$||f(x)|| \leq C_\epsilon\ exp\ \epsilon|z|^p$$

for all $\epsilon > 0$.

Theorem 1.5.2. *The statement of theorem 1.5.1 also holds if we assume all functions F, f and $\tilde{f}$ be of minimal (instead of normal) type.*

Proof follows the scheme of the last theorem's one. We have to make only two changes. The first of them is that the estimate (4) can be made with arbitrarily small constant $a>0$ provided F has the minimal type. This follows from theorem 6. The next is in the solution of $\overline{\partial}$-problem. We describe the change in the solution of $\overline{\partial}$-problem for $n=1$, though it can be done for any n. (We shall need later only the one-dimensional result.) First of all, the functions $g_{k,l}$ are of order p and of minimal type. This follows from the representation $g_{k,l}=(f_{k,l}-f)F^{-1}$ and the version of (4) that we mentioned above.

This means that

$$||g_{k,l}(z)|| \leq C \; exp \; \varepsilon(|z|) \; |z|^p,$$

where $\varepsilon(r)$ monotonically tends to zero when $r \to \infty$. Let us define now $m(r)$ as $lnM(r)$, where

$$M(r) = \max_{|z|=r} ||g_{k,l}(z)||$$

The Hadamard's three circles theorem [390] shows that $m(r)$ is a convex function. Besides, the estimates on $||g_{k,l}||$ show that

$$\lim_{r\to\infty} \frac{m(r)}{r^p} = 0.$$

We define now the function

$$\tilde{m}(r)=m(r) + \sum_{j=1}^{\infty} j^{-2} \, 2^{-jp} m(2^j r).$$

It is easy to verify convergence of the series and the following properties of the function $\tilde{m}(r)$:

i) $\tilde{m}(r)$ is convex;

ii) $\tilde{m}(2r) \leq C\tilde{m}(r)$;

iii) $\lim_{r\to\infty} r^{-p} \, \tilde{m}(r) = 0.$

iv) $||g_{k,l}(z)|| \leq C \; exp \; \tilde{m}(|z|)$ on $U_{k,l}$.

Consider now the function

$$p(z) = p(x+iy) = \tilde{m}(|x|) + |x|^{p-\delta},$$

where $0<\delta<p-1$. Then $p= p(|x|)$ and p is convex. In U_0 we can estimate

$$||g_{k,l}(z)|| \leq C \; exp \; C \; p(z).$$

Indeed, we can apply the estimate *iv*) above:

$$||g_{k,l}(z)|| \leq C \; exp \; C \; \tilde{m}(z).$$

Taking into account that on U_0 we have $y \in [0, 2\pi]$ and choosing $|x| > 7$, we get

$$\tilde{m}(|z|) = \tilde{m}\left(\sqrt{|x|^2+|y|^2}\right) \leq \tilde{m}(\sqrt{2}|x|) \leq$$

$$\leq \tilde{m}(2|x|) \leq C\tilde{m}(|x|) \leq Cp(z).$$

We can achieve analogous estimates also for every derivative of $g_{k,l}$. This leads to the estimate of the function h:

$$||h(z)|| \leq C \; exp \; C \; p\,(|x|)$$

(and analogous estimates for the derivatives).

The last step is to solve $\bar{\partial}$-problem $\bar{\partial}g=h$. We can repeat here completely the corresponding part of the last theorem's proof changing $exp\ 2a|x|^p$ onto $exp\ 2ap(|x|)$ everywhere. All the details can be preserved due to the convexity of $p(t)$ and due to the equality

$$\lim_{t\to\infty} p(|t|)t^{-1} = \infty.$$

This leads to a function $\tilde{f}\,(z)$ with the estimate

$$||\tilde{f}\,(z)|| \leq C \; exp \; Cp\,(|x|).$$

Due to $\lim_{t\to\infty} t^{-p}\, p(t) = 0$, we conclude that $\tilde{f}$ has order p and minimal type. This proves the theorem.

Remark. On the last step, solution of the $\overline{\partial}$-problem, we could also use the known L. Hormander's results on L_2-estimates in $\overline{\partial}$-problem [169].

Let us return to solution of the mentioned earlier multiple interpolation problem.

Let $\{\lambda_n\}$ be all zeros of a function $F \in A_{p_0}(\mathbb{C})$. We denote by p_n the multiplicities of these zeros. Choose a number $\delta \in (0.5, 1)$ and denote by U_n the disk in $\mathbb{C}$ with center at λ_n and radius $\delta|\lambda_n|$. Suppose we have a collection $\{a_{m,l}\}$ of complex numbers, where $\lambda_m \in U_n$, $0 \leq l \leq p_m - 1$. Consider the Newton's interpolation polynomial Q_n for the following interpolation problem in U_n:

$$\frac{\partial^l f}{\partial \lambda^l}\Big|_{\lambda = \lambda_m} = a_{m,l} \quad (\lambda_m \in U_n, 0 \leq l \leq p_m - 1).$$

Assuming that m runs from 1 to q, we can write Q_n down as

$$Q_n(\lambda) = \sum_{i=1}^{q} \sum_{j=1}^{p_i} A_{i,j}^{(n)} \prod_{m=1}^{i-1} \left(1 - \frac{\lambda}{\lambda_m}\right)^{p_m} \left(1 - \frac{\lambda}{\lambda_i}\right)^{j-1}.$$

We renumerate the coefficients $A_{i,j}^{(n)}$ somehow as $A_i^{(n)}$ (choosing, for instance, as new lower index i the number $\sum_{m=1} p_m + i - 1$). The coefficients $A_i^{(n)}$ depend linearly on $\{a_{m,l}\}$, i.e.

$$A_i^{(n)} = \sum_{k,m} \beta_{i,k,n,m}\, a_{m,k}\,.$$

Here $\lambda_m \in U_n$ and $0 \leq k \leq p_m - 1$.

Let p be greater than 1. We shall define now some Frechet space $\Gamma_p(\{\lambda_n, p_n\}, H)$, where H is a Hilbert space. The elements of the space are collections $A = \{a_{n,l}\}$ of elements of H, defined for $0 \leq l \leq p_n - 1$, such that the semi-norm

$$p_b(A) = \sup_n \left(\max_i \left\| \sum_{k,m} \beta_{i,k,n,m} a_{m,k} \right\| \exp\left(-\, b|\lambda_n|^p\right)\right)$$

is finite for every $b>0$. These semi-norms define the structure of a Frechet space. If all the data (i.e. $\lambda_n, p_n, a_{m,k}$) are $2\pi i$-periodic, we get the space $\widetilde{\Gamma}_p(\{\lambda_n, p_n\}, H)$.

Now let us assume that $p > p_0$.

Theorem 1.5.3. *For solvability of the interpolation problem*

$$\frac{\partial^k f}{\partial \lambda^k}\Big|_{\lambda=\lambda_n} = a_{n,k}$$

in the space $A_p(\mathbb{C}, H)$, *it is necessary and sufficient that*

$$\{a_{n,k}\} \in \Gamma_p(\{\lambda_n, p_n\}, H) \ (p > p_0).$$

Moreover, the map from $A_p(\mathbb{C}, H)$ *into* $\Gamma_p(\{\lambda_n, p_n\}, H)$

$$T: f \to \left\{ \frac{\partial^k f}{\partial \lambda^k}\Big|_{\lambda=\lambda_n} \right\}$$

is continuous.

Proof. We shall describe only the necessity part to check the continuity of the map T. The corresponding result (in slightly different terms) was proved in [57, 58] for the case $H = \mathbb{C}$. The case of general H does not cause any additional problems.

Let $f \in A_p(\mathbb{C}, H)$. The construction of the last theorem's proof leads to a subharmonic function $\tilde{p}(|z|)$ such that

i) $\|f(z)\| \leq C \, exp \, \tilde{p}(|z|)$;
ii) $\tilde{p}(2t) \leq A\tilde{p}(t) + B$;
iii) $\lim_{t\to\infty} t^{-p}\tilde{p}(t) = 0$.
iv) $\lim_{t\to\infty} t^{-1}p(t) = \infty$

We can also assume that

$$|F(z)| \leq C \, exp \, \tilde{p}(|z|).$$

Let k_n be the number of points $\lambda_m \in U_n$ (taking into account the multiplicities). Then we have the standard estimate [46]:

$$|k_n| \leq C \; exp\, \widetilde{p}\,(|z|) \;\; (z \in U_n).$$

Consider the disks U_n^1 that are $7\delta|\lambda_n|$-neighborhoods of U_n. Then, due to the properties of the function p,

$$||f(z)|| \leq C \; exp\; C\widetilde{p}\;\; (|z|),\; z \in U_n^1.$$

We denote by M_h the number

$$M_h = \max_{z \in U_n^1} \;\; |(f(z),\, h)| \;,\; h \in H,$$

then

$$M_h \leq C \; exp \; C\, \widetilde{p} \;\; (|\lambda_n|) \; ||h||. \tag{1.5.14}$$

Let us denote the points $\lambda_m \in U_n$ (counted with their multiplicities) by ζ_j. Then the numbers

$$(A_j^{(n)}, h) \prod_{j=1}^{i} \zeta_j^{-1}$$

are the coefficients of the Newton's interpolation polynomial for the function $f_h(z) = \big(f(z),\, h\big)$ at these points, so these numbers are equal to divided differences of $f_h(z)$ over the points ζ_j. Because every point ζ_j belongs to U_n^1 with the disk of radius $7\delta|\lambda_n|$, we have the standard estimate [46]:

$$|(A_i^{(n)}, h)| \prod_{j=1}^{i} |\zeta_j|^{-1} \leq 2^{-1}(7\delta|\lambda_n|)^{-i} M_h.$$

This gives us the estimate

$$||\; A_i^{(n)} \;|| \; \leq C\, exp\; C\widetilde{p}\,(|\lambda_n|).$$

Taking into account the property *iii)* of $\widetilde{p}$ and the definition of the space Γ_p, we conclude that the collection

$$\{a_{n,k}\} = \left\{\frac{\partial^k f}{\partial \lambda^k}\Big|_{\lambda = \lambda_n}\right\}$$

belongs to $\Gamma_p(\{\lambda_n, p_n\}\ , H)$. The continuity follows easily from the previous arguments. This finishes the proof of the necessity part.

Corollary 1.5.4. *If the function $F \neq 0$ is $2\pi i$-periodic, and $p>p_0$, then the sequence of continuous linear mappings*

(1.5.15) $$0 \to \widetilde{A}_p\,(\mathbb{C}, H) \xrightarrow{F} \widetilde{A}_p(\mathbb{C}, H) \xrightarrow{T} \Gamma_p\,(\{\lambda_n, p_n\},\ H) \to 0$$

is exact, where $(Fg)(z) = F(z)\ g(z)$ and

$$(Tf)_{n,k} = \frac{\partial^k f}{\partial z^k}\Big|_{z = \lambda_n}.$$

Proof. The exactness on the left is obvious because $F(z)$ is not identically equal to zero. The equality $Ker\ T = Im\ F$ is proved in the following way. If $f(z) = F(z)g(z)$, then $f(z)$ has all zeros of $F(z)$ with at least the same multiplicities, i.e. Tf=0. If now Tf=0, then the function $g(z)$=$f(z)/F(z)$ is an entire one. The standard estimates from below show that $g \in A_p(\mathbb{C}, H)$ (see theorem 6), and due to the obvious periodicity, $g \in \widetilde{A}_p(\mathbb{C}, H)$. This proves the exactness in the middle term. Now let us assume that a=$\{a_{n,k}\}\ \in \widetilde{\Gamma}_p(\{\lambda_n, p_n\}, H)$. Due to the preceding theorem, a=Tf for some $f \in A_p(\mathbb{C}, H)$. Theorem 2 shows that there exists a function $\widetilde{f} \in \widetilde{A}_p(\mathbb{C}, H)$ such that $T\widetilde{f} = a$. This proves the exactness on the right. The proof is now complete.

We shall present now a slightly different version of the interpolation theorem 3. In order to do this we have to introduce another space of interpolation data.

Let again $1<p_0<p$, $F \in A_{p_0}(\mathbb{C})$ be not identically zero and $\{\lambda_n, p_n\}$ be zeros and their multiplicities of the function F. Proposition 4 of [46] shows that there exist A>0, ϵ>0 such that all connected components U_α of the set

$$S(F,\epsilon, A) = \{z|\ |F(z)| < \epsilon\ exp\ (-A|z|^{p_0})\}$$

are relatively compact in $\mathbb{C}$, and for some B>0 (that does not depend on α) we have

$$|z|^{p_0} \le B|\zeta|^{p_0}+B$$

for all $z,\zeta \in U_\alpha$.

Let us denote by d_α the diameter of U_α and by $\zeta_1, \ldots, \zeta_{n_\alpha}$ the points $\lambda_m \in U_\alpha$, counted with their multiplicities. Introduce the polynomials

$$P_j(\lambda)=\prod_{i=1}^{j}(\lambda-\zeta_i).$$

Let $Q_\alpha(\lambda) = \sum \Delta^{(j)} p_j(\lambda)$ be the Newton's interpolation polynomial. We denote by $\gamma_{i,k,\alpha,m}$ the coefficients of the linear expansion of $\Delta^{(j)}$ into $a_{m,k}$ ($\lambda_m \in U_\alpha$, $0 \le k \le p_m - 1$).

We define now the Frechet space $\sqcap^{(q)}(\{\lambda_n, p_n\},H)$ (where $q \in (p_0, p)$) as the space of all collections $\{a_{m,k}\}$ of vectors in H such that for every $a>0$ the norm

$$\sup_\alpha \left(\max_i \| \sum_{k,m} \gamma_{i,k,\alpha,m}\, a_{m,k} \| \, d_\alpha^{j} \; \exp(- a|z_\alpha|^q)\right)$$

is finite, where $z_\alpha \in U_\alpha$ is arbitrary. The space $\sqcap_{p,p_0}(\{\lambda_n, p_n\},H)$ is the inductive limit when $q \to p$ of the Frechet spaces $\sqcap^{(q)}$. The definition of this space does not depend on the choice of $z_\alpha \in U_\alpha$. If the collection $\{\lambda_n, p_n\}$ is $2\pi i$-periodic (i.e. $\lambda_n+2\pi i=\lambda_{n'}$ and $p_n=p_{n'}$), and we add the periodicity condition

$$a_{n,k}=a_{n',k} \,,$$

we get the corresponding space $\widetilde{\sqcap}_{p,p_o}(\{\lambda_n, p_n\}H$.

Theorem 1.5.5. *If $F \in \widetilde{A}_{p_0}$ (C) is not identically zero, then the sequences of linear mappings*

(1.5.16) $$0 \to B_{p,p_0}(\mathbb{C},H) \xrightarrow{F} B_{p,p_0}(\mathbb{C},H) \xrightarrow{T} \sqcap_{p,p_0}(\{\lambda_n, p_n\},H) \to 0$$

(1.5.17) $$0 \to \widetilde{B}_{p,p_0}(\mathbb{C},H) \xrightarrow{F} \widetilde{B}_{p,p_0}(\mathbb{C},H) \xrightarrow{T} \widetilde{\sqcap}_{p,p_0}(\{\lambda_n, p_n\},H) \to 0$$

are exact, and all mappings are continuous.

Proof. The continuity of the mapping T can be proved by exactly the same arguments as in theorem 3. Let us prove the exactness of (16). The exactness on the left is trivial. The exactness in the middle term can be proved like in the last theorem. Let us prove the exactness on the right (i.e. surjectivity of T). Let $\{a_{m,k}\} \in \sqcap^{(q)}$. Then the Newton's interpolation polynomials $Q_\alpha(z)$ in U_α, constructed for the data $\lambda_n, p_n, a_{n,k}$ (where $\lambda_n \in U_\alpha$) can be estimated in U_α uniformly with respect to α as

$$\|Q_\alpha(z)\| \leq C\,exp\ c\ |z|^q. \tag{1.5.18}$$

This is just a consequence of the definition of the space $\sqcap^{(q)}$, because

$$\|\Delta^{(j)} d_\alpha^j\| \leq C\ exp\ c|z|^q$$

by the definition, and

$$\|Q_\alpha(z)\| \leq C\ exp\ c|z|^q \sum_{j=0}^{n_\alpha-1} d_\alpha^{j}|(z-\zeta_1)\ \dots\ (z-\zeta_j)| \ \leq n_\alpha c\ exp\ c|z|^q.$$

Lemma 4 of [46] shows, however, that

$$n_\alpha \leq C\ \exp\ c|z|^q.$$

This finishes the proof of (18). The collection $Q_\alpha(z)$ solves the interpolation problem

$$\frac{\partial^k f}{\partial\lambda^k}|_{\lambda=\lambda_n} = a_{n,k}$$

in $S(F,\epsilon,A)$ with the bounds that we need. The last step now is to solve this problem in the whole plane $\mathbb{C}$. Choosing $0<\epsilon_1<\epsilon$ and $A_1>A$, we can conclude that the distance from $z \in S(F,\epsilon_1,A_1)$ to the complement to $S(F,\epsilon,A)$ is at least $exp\ (-D|z|^{p_0} - E)$. Let us choose a function $\chi \in C^\infty(\mathbb{C})$ such that

a) $0 \leq \chi \leq 1$
b) $\chi=1$ on $S(F,\epsilon_1,A_1)$
c) $supp\ \chi \subset S(F,\epsilon,A)$

d) $\left|\frac{\partial\chi}{\partial\bar{\lambda}}\right| \le C \exp C|\lambda|^{p_0}$.

Such functions can be easily chosen (see [46]). Let us define now $\widetilde{Q}(z)|_{U_\alpha} = Q_\alpha(z)$. Then the functions $\chi\widetilde{Q}$ and $v = -F^{-1}\partial(\chi\widetilde{Q})/\partial\bar{\lambda}$ belong to $C^\infty(\mathbf{C})$, because in a neighborhood of zeros of F we have $\partial(\chi\widetilde{Q})/\partial\bar{\lambda}=0$. Moreover, $\|v(z)\| \le C\ exp\ c|z|^q$. Theorem 1 from [46] shows that there exists a function $u \in (\mathbf{C}, H)$ such that $\partial u/\partial\bar{\lambda} = v$ and

$$\int_{\mathbf{C}} \|u(\lambda)\|^2 e^{-c|\lambda|^q}(1+|\lambda|^2)^{-2} d\lambda \wedge d\bar{\lambda} < \infty.$$

Now the function $Q=\chi\widetilde{Q}+uF$ is square integrable with the weight $exp\,(-C|\lambda|^q)$ and an entire one. This means that $Q \in B_{p,p_0}$. Besides, we obviously have that $TQ=a$.

We have to prove the exactness on the right in (17). Let $a=\{a_{n,k}\} \in \widetilde{\Pi}_{p,p_0}$. Due to exactness of (16), there exists $f \in B_{p,p_0}$ such that $Tf=a$. Theorem 2 shows that there exists also a function $\widetilde{f} \in \widetilde{B}_{p,p_0}$ such that $T\widetilde{f} = Tf = a$. This finishes the proof.

We also present here one estimate from below for entire functions. This estimate follows from the following well known theorem (see section 8, ch. 1 of [256], or [55]).

Theorem 1.5.6. *Let $f(z)$ be holomorphic in the circle $|z| \le 2eR$ $(R>0)$ with $f(0)=1$ and let η be an arbitrary positive number not exceeding $3e/2$. Then inside the circle $|z| \le R$, but outside of a family of excluded circles the sum of whose radii is not greater than $4\eta R$, we have*

$$\ln|f(z)| > -H(\eta)\ \ln M_f(2eR) \tag{1.5.18}$$

for

$$H(\eta)=2+\ln\frac{3e}{2\eta}$$

and

$$M_f(r) = \max_{|z|=r} |f(z)|.$$

In section 6.3 we shall require the following version of this theorem.

Corollary 1.5.7. *Let $f(z)$ be an entire function with $f(0)=1$ and with the estimate*

$$|f(z)| \le C\ exp\ c|z|^p. \tag{1.5.19}$$

Then for every $p_1>p$ and $\gamma \in (0,\ 1)$ there exists a system of circles such that every point $z \in \mathbb{C}$ belongs to a circle U of the system with radius $r \le D_1(1+|z|)^\gamma$ and

$$|f(w)| \ge D_2\ exp\ (-D_3|z|^{p_1}) \tag{1.5.20}$$

for all $w \in \partial U$. The constants D_j $(j=1,2,3)$ depend only on C, p and p_1.

Proof. Let us choose numbers $\gamma_1 \in (0,\gamma)$, $p_2 \in (p, p_1)$, and set $\eta = R^{\gamma_1 - 1}$. For R big enough this number η satisfies the conditions of theorem 6. Then, according to the theorem we can find a number $\widetilde{R}$ such that $|\widetilde{R} - R| < 4\eta R^{\gamma_1}$, and on the boundary of the circle $B_{\widetilde{R}} = \{|z| \le \widetilde{R}\}$ we have the estimate

$$ln|f(z)| > -const\ (1+lnR)|R|^p > -const\ (1+|R|^{p_2}),$$

or

$$|f(z)| \ge C_1 exp\ (-C_2|z|^{p_2}) \quad (z \in \partial B_{\widetilde{R}}). \tag{1.5.21}$$

Let us consider any point $z_0 \in \partial B_{\widetilde{R}}$ (i.e. $|z_0| = \widetilde{R} \sim R$) and apply the same theorem to the function $\varphi_{z_0}(z) = f(z)/f(z_0)$ in the ball $B_{R_1, z_0} = \{z|\ |z - z_0| \le R_1 = R^\gamma\}$. First of all, due to (21) we have

$$|\varphi_{z_0}(z)| \le C_3\ exp\ C_3|z|^{p_2} \tag{1.5.22}$$

in this ball. Due to the theorem, applied for $\eta=0.5$, there exists a ball B_{R_2, z_0} of a radius $R_2 \in (R_1/2, R_1)$ (and so $R_2 \sim R^\gamma$) such that on $\partial B_{R_2, z_0}$ we have the estimate

$$ln\ |\varphi_{z_0}(z)| \ \ge -\ const \max_{|w| = 2eR_2} |\varphi_{z_0}(z_0 + w)| \ \ge -const\ (|z_0| + 2eR_2)^{p_2}$$

$$\geq -const(R^{p_2}).$$

Since for every point $z \in B_{R_2, z_0}$ we have $|z| \sim R$, we obtain the estimate

$$|\varphi_{z_0}(w)| \geq C_4 \; exp\,(-C_4|z|^{p_1}) \qquad (1.5.23)$$

for every point in B_{R_2, z_0} , and besides $R_2 \sim (1+|z|)^{\gamma}$. Combining (23) with (21), we get the system of circles we need, except that they cover only a neighborhood of infinity. However, the theorem 6 allows us to add one big circle at the origin that completes the system. This finishes the proof.

B. Some information from the complex analysis in several variables.

We assume that the reader is familiar with basic notions of several complex variables theory. It is sufficient to read the chapter II in Hörmander's book [169], and some parts of the chapters V and VII concerning Stein spaces, sheaves and cohomologies; cf. also [154, 167].

We recall briefly some of the notions that we are going to use most often.

A complex analytic manifold X that is countable at infinity is called a *Stein manifold* if

i) it is holomorphically convex (the holomorphic convex hull of a compact set is compact),

ii) it is holomorphically separable (for any two distinct points $x, y \in X$ there is a holomorphic function f on X with $f(x) \neq f(y)$),

iii) for any point $x \in X$ there is a set of holomorphic functions on X forming local coordinates in a neighborhood of x.

Every direct product of one-dimensional regions $U_1 \times \ldots \times U_n$ ($U_j \subset \mathbf{C}$ is open) is a Stein manifold.

Every analytic submanifold in $\mathbf{C}^n$ is a Stein manifold (moreover, analytic submanifolds of a Stein manifold are Stein manifolds). In fact, every Stein manifold is an analytic submanifold of some $\mathbf{C}^n$.

We assume that the basic notions of the sheaf theory are known (see 7.1 in

[169]). If X is an analytic manifold, we denote by O_X the sheaf of germs of analytic functions on X. When the manifold X can be determined from the context, we drop the subscript X and denote the sheaf by O. If $\mathcal{F}$ is a sheaf over X, then we denote by $\Gamma(X,\mathcal{F})$ the corresponding space of sections, and by $H^p(X,\mathcal{F})$ the p-th cohomology group with coefficients in X.

A sheaf $\mathcal{F}$ of O_X-modules over a complex manifold is called an *analytic sheaf*.

An analytic sheaf $\mathcal{F}$ over X is said to be *locally finitely generated* if for every point $x \in X$ there exists a neighborhood ω of x and a finite number of sections $f_1, \ldots f_q \in \Gamma(\omega,\mathcal{F})$ such that every stalk $\mathcal{F}_z$ $(z \in \omega)$ is generated by the germs of $f_1,\ldots, f_q$ at z.

Let us have sections $f_1,\ldots, f_1 \in \Gamma(\omega,\mathcal{F})$ over an open subset $\omega \subset X$. Then we can define a sub-sheaf $R(f_1,\ldots, f_q) \subset O^q$, the *sheaf of relations*, as follows: $(g_1, \ldots, g_1) \in O_z^q$ belongs to the stalk of $R(f_1, \ldots, f_1)$ over $z \in X$ if $\sum g_j f_j = 0$.

An analytic sheaf is *coherent* if

i) $\mathcal{F}$ is locally finitely generated,

ii) if $\omega \subset X$ is an open subset and $f_1,\ldots, f_q \in \Gamma(\omega,\mathcal{F})$ then the sheaf of relations $R(f_1,\ldots, f_q)$ is locally finitely generated.

We shall describe some examples of coherent analytic sheaves in the next theorem.

Let $A(z)$ be an analytic $m\times n$-matrix valued function on X. It defines in the natural way morphism A^0 of sheaves:

$$A^0\colon O^n \longrightarrow O^m.$$

Theorem 1.5.8. *Ker A^0, Im A^0 and Coker A^0 are coherent analytic sub-sheaves in O^n and O^m correspondingly.*

The reader can find many important properties of coherent sheaves in the chapter VII of [169]. We collect some of them in the following theorem.

Theorem 1.5.9. *Let $\mathcal{F}$ be a coherent analytic sheaf over a Stein manifold X. Then:*

i) If K is an $\mathcal{H}(X)$-convex compact and f is a section of $\mathcal{F}$ over a neighborhood *of K, then there exists a sequence $f_j \in \Gamma(X, \mathfrak{F})$ such that*

$$\|f - f_j\|_K \longrightarrow 0 \ \ for \ \ j \longrightarrow \infty.$$

ii) For every point $z \in X$ the stalk $\mathfrak{F}_z$ is generated by germs at z of global sections $g \in \Gamma(X, \mathfrak{F})$.

iii) $H^p(X, \mathfrak{F}) = 0$ for all $p>0$.

A vector bundle $\mathcal{E}$ whose base is an analytic manifold X is said to be *holomorphic* if it has a trivializing cover with holomorphic transition functions.

For such a bundle $\mathcal{E}$ we can naturally define the notions of holomorphic morphisms, holomorphic sections, and the corresponding sheaf of germs $O^{\mathcal{E}}$.

Theorem 1.5.10. *Let $\mathcal{E}$ be a holomorphic vector bundle over a Stein base X. Then*

i) $\mathcal{E}$ is holomophically trivial if and only if it is trivial in the category of (continuous) vector bundles.

ii) $H^p(X, O^{\mathcal{E}}) = 0$ for $p>0$.

An *analytic subset* is a subset that can be locally represented as the set of common zeros of a set of analytic functions. We shall present now a very important result of V. Palamodov that plays the central role in the Floquet theory. Let K be a field and A be a commutative K-algebra (in what follows we shall deal only with Noetherian algebras A (see [413])). A prime ideal $\mathfrak{P}$ of A is said to be *associated* to M if there exists an element $m \in M$ whose annulator ideal coincides with $\mathfrak{P}$. The module M is called $\mathfrak{P}$-*coprimary* if $\mathfrak{P}$ is the only ideal associated to M. Any A-module M of finite type over a Noetherian algebra A admits an *irreducible Lasker-Noether decomposition* :

$$0 = M_1 \cap \dots \cap M_m \tag{1.5.24}$$

(see [413]) where

i) $M_i \subset M$ and M/M_i is $\mathfrak{P}_i$ - coprimary $\{i=1, \dots, m\}$,

ii) no one of M_i in (24) can be omitted,

iii) all prime ideals $\mathfrak{P}_i$ are different.

The set *Assoc* $(M):=\{\mathfrak{P}_1, \dots, \mathfrak{P}_m\}$ is defined uniquely.

Definition 1.5.11. A K-linear mapping δ: $M \longrightarrow N$ of A-modules is called a *differential operator of order* $\leq d$, if

$$(ad\ b)^{d+1}\delta=0 \tag{1.5.25}$$

for any $b \in A$, where

$$(ad\ b)\ \gamma = \gamma b - b\gamma \tag{1.5.26}$$

Definition 1.5.12. Let $\mathfrak{P} \in Assoc\ (M)$. We call a map

$$\nu: M \longrightarrow [A\ /\mathfrak{P}]^r\ (r < \infty)$$

a $\mathfrak{P}$-*Noether operator* if

i) ν is a differential operator in A-modules,

ii) *Ker* ν is a submodule of M, and $\mathfrak{P} \notin Assoc\ (Ker\ \nu)$.

Definition 1.5.13. Let $\{\mathfrak{P}_1, \dots, \mathfrak{P}_m\} = Assoc\ (M)$. Then the operator

$$\nu = \sum \nu_j \colon M \longrightarrow \sum_j [A/\mathfrak{P}_j]^{r_j}$$

is said to be a *Noether operator* for the module M, where ν_j is a $\mathfrak{P}_j$-Noether operator.

It is easy to show that the kernel of ν is zero.

Now let X be a complex manifold, and O be the sheaf of germs of holomorphic functions on X. (The following considerations are valid in the more general situation of complex spaces [326], however we shall need only the case of regions in $\mathbb{C}^n$.) Consider two O-sheaves M and N over X. A morphism δ: $M \longrightarrow N$ is called a *differential operator*, if for any point $x \in X$ the fiber morphism δ_x: $M_x \longrightarrow N_x$ is a differential operator over $\mathbb{C}$-algebra O_x. We define $ord\ \delta$ as $\sup_x ord\ \delta_x$.

Let $O_x(\mathbb{C}^n)$ be the algebra of germs of holomorphic functions at some point

$x \in \mathbf{C}^n$. There is one-to-one correspondence between prime ideals of A and irreducible germs of analytic sets at x. Namely, every such ideal is equal to the ideal $I(G)$ of all germs vanishing on an irreducible germ G. We call such G *associated* to an A-module M, if $I(G) \subset Assoc\ (M)$. For any analytic germ Y in $\mathbf{C}^n$ the germs $G_1, ..., G_l$ associated to the A-module $O(Y)$ are exactly all the irreducible components of Y.

Theorem 1.5.14. [322, 323] *For any coherent sheaf M on X and any point $x \in X$ there exists a neighborhood U of x and a Noether operator $\nu = \sum \nu_j$,*

$$\nu_j\colon M \longrightarrow [O(Y_j)]^{r_j}\ (j=1, ..., m)\ ,$$

where $Y_1, ..., Y_m$ are some closed analytic subsets of U, whose germs at x are all germs associated to M_x.

We recall that a closed analytic subset Y in X is said to be *irreducible*, if it cannot be represented as the union of two proper analytic subsets.

Definition 1.5.15. Let M be a coherent analytic sheaf over X. An analytic subset $Y \subset X$ is called *associated to M*, if

i) Y is closed and irreducible,

ii) for any point $x \in Y$ its germ Y_x is an union of irreducible germs $G_1,...,G_l$ ($l>0$), any of which is associated to the O_x-module M_x.

We denote by $Assoc\ (M)$ the collection of analytic sets associated to M.

If for a given point $x \in X$ we have a germ $G \subset Assoc\ (M_x)$, then G is the germ of an analytic set $Y \in\ Assoc\ (M)$ ([365, 366], and theorem 14 for the local case).

Definition 1.5.16. If Y is an analytic set associated to M, we call *Y-Noether operator* for M any differential operator $\nu\colon M \longrightarrow \sum O(Y)$ where the direct sum is finite, such that for any point $x \in Y$ and any irreducible component G of Y_x the composition $\rho_G \nu$ is a G-Noether operator, where

$$\rho_G\colon \sum O(Y) \longrightarrow \sum O(G)$$

is the restriction mapping.

Theorem 1.5.17. *Let X be a Stein manifold, M be a coherent analytic sheaf on X, and $Y \in$ Assoc (M). Then there exists a Y-Noether operator*

$$\nu_Y : M \longrightarrow [O(Y)]^{r(Y)} \quad (r(Y) < \infty)$$

and a holomorphic function $s \neq 0$ on X such that for any element $a \in \Gamma(X, O(X))$ there exists an $O(Y)$-endomorphism b of $[O(Y)]^{r(Y)}$ with

$$s \cdot (ad\ a)\nu_Y = b\nu_Y. \tag{1.5.25}$$

Theorem 1.5.18. *Let X be a Stein manifold, M be a coherent sheaf on M. Let also for any $N \in$ Assoc (M) ν_N be a N-Noether operator for M, and $S(N)$ be a proper closed analytic subset of N containing the singular set sing$N$ of N. Then for any open set $U \subset X$ the linear operator*

$$\nu = \sqcap \nu_N : \Gamma(U, M) \longrightarrow \sqcap_N [\Gamma(U \cap N \setminus S(N))]^{r(N)} \tag{1.5.26}$$

is continuous and open. Here N runs over Assoc (M) and the second space in (26) *is endowed with the topology induced from the distribution space*

$$\sqcap_{N \in Assoc(M)} \mathcal{D}'(U \cap N \setminus S(N)).$$

We shall also need the following simple fact from complex analysis. Consider again the torus

$$\mathcal{T}^n = \{z \in \mathbf{C}^n |\ |z_j| = 1,\ j=1, \ldots, n\} \subset \mathbf{C}^n.$$

Theorem 1.5.19. *Let $N \subset \mathbf{C}^n \setminus 0$ be an analytic subset. If the n-dimensional Lebesgue measure of the intersection $N \cap \mathcal{T}^n$ is positive, then $N = \mathbf{C}^n \setminus 0$.*

Proof. Choosing appropriate local coordinates, we reduce the consideration to the case of an analytic subset $N \subset \mathbf{C}^n$ such that *mes* $(N \cap \mathbf{R}^n) > 0$. Then it is rather obvious that $N=\mathbf{C}^n$. For instance, dimension of $N\cap\mathbf{R}^n$ as of a real analytic set must be equal to n, so the complex dimension of N is n, and therefore $N=\mathbf{C}^n$ (the reader can find all pertinent properties of real and complex analytic sets in [307]).

C. Some problems of infinite-dimensional complex analysis.

The definitions in the theory of holomorphic functions with values in Banach spaces are analogous to those adopted in the theory of scalar holomorphic functions (see [73, 168]).

Definition 1.5.19. Let U be a region in $\mathbf{C}^n$ and $f:U\to E$ a function with values in a Banach space E. Then f is *holomorphic* in U if in a neighborhood of each point $z_0\in U$ it can be expanded into a power series

$$f(z)=\sum_{|\alpha|=0}^{\infty} a_\alpha(z-z_0)^\alpha,\ a_\alpha\in E.$$

Here $\alpha=(\alpha_1, ..., \alpha_n)$ is a multi-index, $|\alpha| = \alpha_1 + ... + \alpha_n$ and $z^\alpha = z_1^{\alpha_1} \dots z_n^{\alpha_n}$.

A function $f(z)$ is holomorphic in U if and only if it is *weakly holomorphic*, i.e. all the scalar functions $(u, f(z))$ are holomorphic for all functionals $u\in E^*$.

Since holomorphy is defined locally, it is easy to extend the definition to functions defined on a complex analytic manifold.

A function f defined on an analytic set $V\subset\mathbf{C}^n$ is said to be holomorphic if each point $z_0\in V$ has a neighborhood $U_{z_0}\subset\mathbf{C}^n$ such that $f|_{V\cap U_{z_0}}$ admits a holomorphic extension to U_{z_0}.

The maximum modulus principle obviously holds for holomorphic E-valued functions. A very useful tool is provided by the following theorem.

Theorem 1.5.20. *Let X_i ($i=1,2, ...$) be a sequence of analytic manifolds that are countable at infinity, and $f_i:X_i\to E$ a sequence of holomorphic functions. Then there exists a sequence of pairwise disjoint one-dimensional projections P_n such that each function f_i can be expanded as a series*

$$f_i=\sum_n P_n f_i=\sum f_i^{(n)} b_n \tag{1.5.27}$$

where $b_n \in E$, $P_n b_n = b_n$, $||b_n|| = 1$, and the $f_i^{(n)}$ are holomorphic scalar functions. For any compact set $K_i \subset X_i$ there are compact sets $K_i' \subset X_i$ and constants C_{K_i} independent of the sequence f_i, such that

$$\sum_{n=1}^{\infty} \sup_{z \in K_i} |f_i^{(n)}(z)| \leq C_{K_i} \sup_{z \in K_i'} ||f_i(z)||.$$

This theorem allows us to carry over many results of complex analysis to E-valued holomorphic functions.

Let us denote by O_X^E (or just O^E, if the context determines the manifold X uniquely) the sheaf of germs of holomorphic E-valued functions on X.

Theorem 1.5.21. *If X is a Stein manifold, then $H^p(X,O^E) = 0$ for $p \geq 1$.*

This theorem can be readily proved by expanding every co-chain into the series (27) and applying theorem 9.

One of standard conclusions from the last theorem is that any holomorphic function $f:Y \to E$ defined on an analytic subset Y of a Stein manifold X can be holomorphically extended onto the whole X.

Another example, where theorem 20 plays the crucial role is the following lifting theorem.

Theorem 1.5.22. *Let E be a Banach space, N a closed subspace, and $\pi: E \to E/N$ the canonical surjection. Let also X be a Stein manifold, and $Y \subset X$ an analytic subset. Suppose that $f:X \to E/N$, $f_1:Y \to E$ are holomorphic mappings such that $\pi \circ f_1 = f|_Y$. Then there exists a holomorphic map $f_2: X \to E$ such that $f_2|_Y = f_1$* and $\pi \circ f_2 = f$.

Proof. Due to theorem 20, we obtain an expansion

$$f(z) = \sum f^{(n)}(z) b_n \qquad (b_n \in E/N,\ ||b_n||=1).$$

In each coset $b_n \in E/N$ we can choose an element $e_n \in E$ such that $||e_n||_E \leq 2$. Then the series $\hat{f}(z) = \sum f^{(n)}(z) e_n$ is convergent for all $z \in X$, $\hat{f}(z)$ is holomorphic, and $\pi \circ \hat{f} = f$. Consider now the map $(\hat{f} - f_1)$: $Y \to N$. By the remark above, there exists a holomorphic extension $\hat{f}_1$: $X \to N$ of $(\hat{f} - f_1)$. Then the map $f_2 = f - f_1$ satisfies the requirements of the theorem.

Remark. When $Y=\emptyset$, we do not require the Stein property of X.

It is easy to prove (using the corresponding finite-dimensional results) that for any infinite-dimensional Banach space E and any Stein manifold X there exists an embedding $f\colon X \to E$. Moreover, the set of all such embeddings is dense (with respect to the compact-open topology) in the space of all holomorphic mappings $f\colon X \to E$.

A Banach vector bundle $\mathcal{E}$ whose base is an analytic manifold X is said to be *holomorphic* if it admits a trivializing cover with holomorphic transition functions.

A *holomorphic section* $\delta\colon X \to \mathcal{E}$ is defined in the natural way. We write $O_X^{\mathcal{E}}$ (or just $O^{\mathcal{E}}$) for the sheaf of germs of holomorphic sections.

The concepts of morphisms of holomorphic bundles and of holomorphic subbundles can be introduced in natural way. Namely, in the corresponding definitions of section 1.3 we require holomorphity of all mappings.

A holomorphic subbundle $\mathcal{F}$ of a holomorphic Banach bundle $\mathcal{E}$ and the quotient bundle $\mathcal{E}/\mathcal{F}$ are themselves holomorphic Banach bundles. The following result plays the central role in the theory of holomorphic Banach bundles.

Theorem 1.5.23. *Let X be a Stein manifold and $\mathcal{E}$ a Banach bundle over X. Then there exist a holomorphic Banach bundle which is isomorphic to $\mathcal{E}$ as a Banach bundle. Two holomorphic Banach bundles over X are isomorphic if and only if they are isomorphic as Banach bundles (i.e. in the continuous category).*

This nontrivial theorem is a reflection of the famous Oka's principle: everything that can be done on Stein manifolds "continuously," can be also done holomorphically.

Theorem 1.5.24. *Any holomorphic Banach bundle $\mathcal{E}$ with fiber E over a Stein base X can be embedded as a complemented subbundle in the trivial bundle $\mathcal{L}_1(E) = X \times l_1(E)$, and $\mathcal{E} \oplus \mathcal{L}_1(E)$ is isomorphic to $\mathcal{L}_1(E)$.*

This theorem, which follows directly from the preceding one and from theorem 1.3.4, often allows us to carry over results obtained for holomorphic functions to the case of holomorphic sections. One important example is the following result.

Theorem 1.5.25. *If X is a Stein manifold and $\mathcal{E}$ a holomorphic Banach bundle* *over X, then $H^p(X, O^{\mathcal{E}}) = 0$ for $p \geq 1$.*

Theorem 1.5.26. *A direct holomorphic subbundle $\mathcal{F}$ of a holomorphic Banach bundle $\mathcal{E}$ over a Stein manifold X is holomorphically complemented. In other words, there exists a holomorphic endomorphism of the bundle $\mathcal{E}$ which is a projection onto $\mathcal{F}$ (or, equivalently, there is a holomorphic complement subbundle $\mathcal{F}_1$ to $\mathcal{F}$ in $\mathcal{E}$).*

Proof. First of all, simple considerations show the local existence of holomorphic projections. Let now $U=\{U_\alpha\}$ be a cover of X by domains of holomorphy such that there are projections P_α mapping $\mathcal{E}|_{U_\alpha}$ onto $\mathcal{F}|_{U_\alpha}$. For the intersections $U_\alpha \cap U_\beta$ of two neighborhoods U_α and U_β the operators $P_\alpha - P_\beta = Q_{\alpha\beta}$ vanish on $\mathcal{F}|_{U_\alpha \cap U_\beta}$. If we found morphisms $G_\alpha : \mathcal{E}|_{U_\alpha} \longrightarrow \mathcal{F}|_{U_\alpha}$ such that $G_\alpha - G_\beta = Q_{\alpha\beta}$ and the G_α vanish on $\mathcal{F}|_{U_\alpha}$, then the morphisms $P_\alpha - G_\alpha$ would provide the required projection (since $P_\alpha - G_\alpha = P_\beta - G_\beta$ on $U_\alpha \cap U_\beta$). In order to find these G_α we consider the holomorphic subbundle N of the associated morphisms bundle $L(\mathcal{E})$ formed by the subspace of operators in $L(E)$ that annihilate F and carry E into F. The morphisms $Q_{\alpha\beta}$ form a one-dimensional cocycle. By theorem 25, $H^1(U, O^N)=0$, therefore, this cocycle is cohomologous to zero. This means that the required morphisms G_α exist. This proves the theorem.

The above theorems show that the sheaves O^E and $O^{\mathcal{E}}$ demonstrate some properties analogous to the properties of coherent analytic sheaves. They are not, however, coherent. The following infinite-dimensional generalization of coherence, introduced by J. Leiterer, plays an important role.

Definition 1.5.27. An analytic sheaf $\mathcal{F}$ over X is said to be an *analytic Fréchet sheaf* (AF-sheaf) if for every open set $U \subset X$ the $\mathcal{H}(U)$-module $\Gamma(U, \mathcal{F})$ carries a topology such that $\Gamma(U, \mathcal{F})$ becomes a Fréchet $\mathcal{H}(U)$-module (i.e. $\Gamma(U, \mathcal{F})$ is a Frechet space and the multiplication $\mathcal{H}(U) \times \Gamma(U, \mathcal{F}) \longrightarrow \Gamma(U, \mathcal{F})$ is continuous) and the restriction map

$$\rho_{VU} : \Gamma(U, \mathcal{F}) \longrightarrow \Gamma(V, \mathcal{F})$$

is continuous for every $V \subset U$.

An O_X-homomorphism of two AF-sheaves is said to be an AF-homomorphism if the induced mappings of spaces of sections are continuous.

Definition 1.5.28. An AF-sheaf $\mathcal{F}$ over X is called a *BCAF-sheaf* (*BC* stays for *"Banach coherent"*) if for every point $z \in X$ and every number n there is a neighborhood U of z and an exact sequence of AF-homomorphisms

$$O_U^{E_n} \longrightarrow O_U^{E_{n-1}} \longrightarrow \dots \longrightarrow O_U^{E_0} \longrightarrow \mathcal{F}|_U \longrightarrow 0$$

where $E_0, \dots, E_n$ are Banach spaces.

There are some simple properties of $BCAF$-sheaves that can be seen directly from the definition. For instance:

i) the property of being a $BCAF$-sheaf is a local one;

ii) for every Banach space E the sheaf O^E is a $BCAF$ sheaf;

iii) for every holomorphic Banach bundle $\mathcal{E}$ the sheaf $O^{\mathcal{E}}$ is a $BCAF$-sheaf.

The important property of $BCAF$-sheaves is described in the following theorem.

Theorem 1.5.29. *Let $\mathcal{F}$ be a BCAF-sheaf over a Stein space X, then*

i) $H^p(X, \mathcal{F}) = 0$ for all $p \geq 1$.

ii) If K is an $\mathcal{H}(X)$-convex compact and f is a section of $\mathcal{F}$ over a neighborhood of K, then there exists a sequence $f_j \in \Gamma(X, \mathcal{F})$ such that

$$||f - f_j||_K \longrightarrow 0 \quad \textit{for} \quad j \longrightarrow \infty.$$

We shall need sometimes the following simple lemma about polynomials.

Lemma 1.5.30. *Let $g_i(\lambda)$ $(i = 1, \dots, n)$ be polynomials of n complex variables. Suppose that*

$$g_i(\lambda + e_j) - g_i(\lambda) \equiv g_j(\lambda + e_i) - g_j(\lambda)$$

for all $i,j = 1, \dots, n$ and all $\lambda \in \mathbb{C}^n$, where e_i be standard basic vectors in $\mathbb{C}^n$. Then there exists a polynomial $g(\lambda)$ such that

$$g(\lambda + e_i) - g(\lambda) \equiv g_i(\lambda)$$

for all $i = 1, \dots, n$ *and all* $\lambda \in \mathbb{C}^n$. *The order of the polynomial* g *is equal to max ord* $g_i + 1$.

Proof. We shall prove the lemma by induction with respect to the dimension n. Let first $n = 1$. Consider the operator $dg(\lambda) = g(\lambda + 1) - g(\lambda)$. This is obviously a $\mathbb{C}$-linear operator from the space of polynomials of order $m + 1$ into the space of polynomials of order m. Its kernel consists only of constants and hence is one-dimensional. Counting dimensions, we conclude that the operator is surjective; in other words, the statement is proved for $n = 1$.

Now suppose that the statement is correct for the dimension n and prove it for $n + 1$. We denote by d_i the operator

$$d_i g(\lambda) = g(\lambda_1, \dots, \lambda_{i-1}, \lambda_i + 1, \lambda_{i+1}, \dots, \lambda_n) - g(\lambda).$$

Then we have to solve the system

$$d_i g = g_i \quad (i = 1, \dots, n),$$

where $d_i g_j = d_j g_i$. The operators d_i obviously commute. Expanding into the powers of $\lambda_2, \dots, \lambda_n$ and applying d_1 to each coefficient (which is a polynomial in λ_1), we get a series of one-dimensional problems that are solvable. Thus, there exists a polynomial $h_1(\lambda)$ such that $d_1 h_1 = g_1$. Let us try to find a polynomial $h_2(\lambda_2, \dots, \lambda_n)$ such that $g = h_1 + h_2$ solves our problem. First of all, $d_1 h_2 = 0$, so we have to satisfy only the system

$$d_i h_2 = g_i - d_i h_1 \text{ for } i = 2, \dots, n.$$

The polynomials in the right hand side do not depend on λ_1, since

$$d_1(g_i - d_i h_1) = d_i g_1 - d_i d_1 h_1 = d_i g_1 - d_i g_1 = 0 \text{ for } i > 1.$$

Now, let us denote by g'_i the polynomial $g_i - d_i h_1$ $(i = 2, \dots, n)$. Then

$$d_j g'_i = d_j g_i - d_j d_i h_1 = d_i g_j - d_i d_j h_1 = d_i g'_j.$$

This means that our problem of finding h_2 is a $(n-1)$-dimensional problem of the same type, as the original one. Due to the induction hypothesis, it is solvable. The lemma is proved.

1.6. Fredholm operators that depend holomorphically on a parameter

We shall briefly present in this section some of basic facts about holomorphic Fredholm operator functions. The proofs mostly will not be provided. The reader can find all necessary references in section 1.9.

Theorem 1.6.1. *Let A be a Banach algebra with unit e, X be a Stein manifold, and $T: X \longrightarrow A$ be a holomorphic mapping such that $T(z)$ is right (left, two-sided) invertible for all $z \in X$. Then there is a right (left, two-sided) inverse $T^1(z)$ that depends holomorphically on $z \in X$.*

Proof. Let $T(z)$ be right invertible. The proof of the theorem 1.4.6 shows that some holomorphic right inverse $T^1(z)$ can be constructed locally. Let $\{U_\alpha\}$ be an acyclic cover of X such that in each of U_α there is a right inverse $T^1_\alpha(z)$ depending holomorphically on $z \in U_\alpha$. Let $z \in U_\alpha \cap U_\beta$, then $T(z)(T^1_\alpha(z) - T^1_\beta(z)) = 0$. In this connection we introduce for each z the subspace N_z of all elements $s \in A$ for which $T(z)s=0$. Then the operator $P_\alpha(z)$ $(z \in U_\alpha)$ of left multiplication by the element $e - T^1_\alpha(z)T(z)$ is obviously a (holomorphic) projection onto N_z. Therefore, the family N_z forms a holomorphic subbundle N of the trivial bundle $X \times A$ over X. The functions $T_{\alpha\beta}(z) = T^1_\alpha(z) - T^1_\beta(z)$ define a one-dimensional cocycle. By theorem 1.5.25, this cocycle is cohomologous to zero, that is, there are local sections $G_\alpha(z)$ $(z \in U_\alpha)$ of N such that

$$G_\alpha(z) - G_\beta(z) = T^1_\alpha(z) - T^1_\beta(z) \quad (z \in U_\alpha \cap U_\beta).$$

Then the function $T^1(z) = T^1_\alpha(z) - G_\alpha(z)$ defines a global right inverse of $T(z)$. Because the situation with left inverses is completely similar, this proves the theorem.

For an arbitrary analytic manifold the statement of the theorem does not hold. Let us consider the operator $T_0(z_1, z_2) : \mathbf{C} \longrightarrow \mathbf{C}^2$ defined by $T_0(z_1, z_2)z = (z_1 z, z_2 z)$

$((z_1, z_2) \in \mathbf{C}^2)$. Let E be a Banach space such that $E \oplus \mathbf{C}$ is isomorphic to E. The operator $1 \oplus T_0(z_1, z_2)$ maps $E \oplus \mathbf{C}$ into $E \oplus \mathbf{C}^2$, and by the assumed isomorphism generates an operator on E that depends holomorphically on $(z_1, z_2) \in \mathbf{C}^2$. It is left invertible for $(z_1, z_2) \in \mathbf{C}^2 \backslash 0$, but not for $z_1 = z_2 = 0$. If it had a global holomorphic left inverse on $\mathbf{C}^2 \backslash 0$, then this inverse could be holomorphically extended to the whole $\mathbf{C}^2$ (because $\mathbf{C}^2$ is the holomorphic hull of $\mathbf{C}^2 \backslash 0$). This means that the original operator would be left invertible at the point $z_1 = z_2 = 0$, which is impossible.

Note that for a (two-sided) invertible holomorphic element $T(z)$ defined on arbitrary complex manifold, the inverse $T^{-1}(z)$ is analytic. In this case we do not need the manifold of parameters to be Stein.

Theorem 1.6.2. *Let $T: X \longrightarrow \Phi(E)$ be a holomorphic map of an analytic manifold X. Then there exists a global holomorphic regularizer $T^1: X \longrightarrow \Phi(E)$.*

Proof. Let π: $L(E) \longrightarrow Cal\,(E)$ be the standard projection. The element $\pi(T(z)) \in Cal\,(E)$ is invertible by theorem 1.2.10. Consider the (holomorphic) inverse element $t(z) = (\pi(T(z)))^{-1}$. Due to the remark after theorem 1.5.22, there exists a holomorphic map T^1: $X \longrightarrow L(E)$ such that $\pi(T^1(z)) = t(z)$. This map obviously provides the regularizer we need. This proves the theorem.

The next theorem results from theorem 1 by the same arguments.

Theorem 1.6.3. *Let $T: X \longrightarrow \Phi_r(E)$ $(\Phi_l(E))$ be a holomorphic map of a Stein manifold X. Then there exists a global holomorphic right (left) regularizer $T^1: X \longrightarrow \Phi_l(E)$ $(\Phi_r(E))$.*

As in the case of theorem 1, the last statement is not valid without the condition that X is a Stein manifold. The next theorem can be proved using Bungart's results [69].

Theorem 1.6.4. *Let T: $X \longrightarrow \Phi(E)$ be a holomorphic map of a Stein manifold X. For the existence of a holomorphic equivalent regularizer T^1: $X \longrightarrow GL(E)$ it is necessary and sufficient that there is a continuous such regularizer.*

Corollary 1.6.5. *Let X be a Stein manifold, K a holomorphically convex compact subset of X, T: $X \longrightarrow \Phi(E)$ a holomorphic map. If $\operatorname{index}(T|_K) = 0$, then on some neighborhood U of K there is a holomorphic equivalent regularizer T^1: $U \longrightarrow GL(E)$.*

The difference in the behavior of two-sided and one-sided inverses and regularizers arises from the different geometries of corresponding sets of operators.

Definition 1.6.6. A domain D in a complex Banach space E is said to be *pseudoconvex* if the function

$$-\log dist(z,\partial D)$$

is plurisubharmonic in this domain.

Theorem 1.6.7. *A domain $D \subset E$ is pseudoconvex if and only if its intersection with any finite-dimensional affine manifold in E is a domain of holomophy in this manifold.*

It will be shown below that the domains $GL(E)$ and $\Phi(E)$ in $L(E)$ are pseudoconvex. Examples show (see, for instance, the counter-example after theorem 1) that $\Phi_l(E)$, $\Phi_r(E)$ and $GL_l(E)$ are not.

Theorem 1.6.8. *Let A be a Banach algebra, J a two-sided ideal of A, and D the set of all elements $a \in A$ such that the corresponding coset $\hat{a} \in A/J$ is invertible. Then D is pseudoconvex in A.*

Proof. Consider the set G of all invertible elements in $B{=}A/J$, and show its pseudoconvexity. Let $\mathfrak{L}^n$ be an affine submanifold of B. Let also $G^n{=}G \cap \mathfrak{L}^n$ be non-empty. We define a holomorphic map $\varphi : G^n \longrightarrow B$ by the formula $\varphi(b){=}b^{-1}$. It is clear that $||\varphi(b)|| \longrightarrow \infty$ as $b \longrightarrow \partial G$ and that φ has no holomorphic extension outside G. This means holomorphic convexity of G^n. Now theorem 7 shows that G is pseudoconvex.

Let now π: $A \longrightarrow B$ be the canonical mapping. Due to the equality $D = \pi^{-1}(G)$, we have the following obvious identities:

$$-\log dist(z,\partial D) = -\log dist(\pi(z),\partial G) =$$
$$= (-\log dist(\cdot,\partial G)\circ\pi)(z)(z \in D).$$

Consequently, $-\log dist(z,\partial D)$, being the superposition of a plurisubharmonic and a linear function, is plurisubharmonic. This proves the theorem.

Corollary 1.6.9. *In any Banach algebra the set of invertible elements is pseudoconvex.*

Corollary 1.6.10. *$\Phi(E)$ is pseudoconvex in $L(E)$.*

As we have already mentioned before, the counter-examples to the existence of one-sided inverses and regularizers show that the sets $\Phi_l(E)$ and $\Phi_r(E)$ are not pseudoconvex.

Thus, the set of invertible (Fredholm) operators has a better complex structure than that of one-sided invertible (semi-Fredholm) operators.

It can be also noted that $\Phi^{(0)}(E)$ is a homogeneous space of an analytic Banach-Lie group. Let us introduce a group structure on the Banach manifold $\Gamma = GL(E) \times C(E)$ as follows:

$$(g_1, k_1) \cdot (g_2, k_2) := (g_1 g_2,\ g_2^{-1} k_1 + k_2)$$

This makes Γ an analytic Banach-Lie group. The action of Γ on $\Phi^{(0)}(E)$ is defined by the formula

$$(g, k) \cdot\ f = g^{-1} f - k,$$

where $(g, k)\ \in \Gamma$, $f \in \Phi^{(0)}(E)$. It is not hard to verify that this action is holomorphic and transitive. The stabilizer of the point I is the image of $GL_C(E)$ under the anti-homomorphic embedding

$$\theta\colon GL_C(E) \longrightarrow \Gamma, \quad \theta(g) = (g^{-1}, g - I).$$

Thus, $\Phi^{(0)}(E)$ is the homogeneous space $\Gamma/\theta(GL_C(E))$.

Dealing with Fredholm operators depending on a parameter, it is very often necessary to understand the behavior of the kernel of the operator. We shall present now a simple result that plays an important role in such considerations.

Theorem 1.6.11. *The set $S_p\Phi(E,F)$ is an analytic subset of $\Phi(E,F)$. In particular, the set*

$$\Phi^{(0)}(E)\backslash GL(E)$$

is a principal analytic subset in $\Phi^{(0)}(E)$. *The same is true for the sets* $CS_p\Phi(E,F)$. (*We remind the reader that the sets* $S_p\phi$ *and* $CS_p\phi$ *were defined in section 1.2, and a subset is principal analytic if it can be locally described as the set of zeros of a single analytic function.*)

Proof. Let $A_0 \in S_p\Phi(E,F)$. Without loss of generality we may assume that *ind* $A_0 = 0$ (this can be achieved by extending E or F). Then there is a finite-dimensional operator $K \in L(E,F)$ such that $K + A_0$ is invertible. Then $A + K$ is invertible for all A in some neighborhood U of A_0. We represent $A \in U$ in the form $A= (A + K)\ (1-Q(A))$, where $Q(A) = (A + K)^{-1}K$. The map Q: $U \longrightarrow L(E)$ is obviously analytic, and $Ker\ A = Ker(I-Q(A))$. The operator K can be represented as

$$Kx = \sum_{i=1}^{n} f_k(x)\ e_k,$$

where $f_k \in E^*, e_k \in F$. Hence, $Q(A)$ has the form

$$Q(A)x=\sum_{k=1}^{n} f_k(x)y_k(A),$$

where $y_K(A)$ are analytic functions of A with values in E. We transform the equation $Ax = 0$ into $x-Q(A)\ x = 0$, or

$$x=\sum_{k=1}^{n} f_k(x)y_k(A).$$

We write $f_k(x) = \xi_k$ and substitute into the equation. As a result we get

$$\xi_k=\sum_{q=1}^{u} \xi_q f_k(y_q(A)).$$

We are interested in those A for which this system has at least p linearly independent solutions. This happens when all the minors of order $n-p+1$ of the matrix

$$\{a_{k,q}\} = \{f_k(y_q(A))\}$$

vanish. Since these minors are analytic functions of $A \in U$, the set $S_p\Phi(E,F) \cap U$ is the set of common zeros of some collection of analytic functions on U. In the case when $A_0 \in \Phi^{(0)}$, $p=1$, we need only one such function - the determinant. This proves the theorem.

A direct generalization of a holomorphic Fredholm operator function is the notion of a Fredholm morphism of analytic Banach bundles.

Definition 1.6.12. Let $\mathcal{E}_j$ $(j=1,2)$ be holomorphic Banach bundles over a complex manifold X, and A: $\mathcal{E}_1 \longrightarrow \mathcal{E}_2$ a morphism of these bundles. The morphism A is called a *Fredholm morphism* if its restriction to each fiber is a Fredholm operator in corresponding Banach spaces.

In the case of trivial bundles this reduces to the notion of a holomorphic Fredholm operator function.

The analog of theorem 1.4.10 can be proved also in holomorphic situation.

Theorem 1.6.13. *Let A: $\mathcal{E}^1 \longrightarrow \mathcal{E}^2$ be a Fredholm morphism of holomorphic bundles over a complex manifold X such that dim Ker $A(z)$ is constant for all $z \in X$. Then*

$$\bigcup_{z \in X} \mathit{Ker}\ A(z)$$

and

$$\bigcup_{z \in X} \mathit{Im}\ A(z)$$

are holomorphic subbundles of $\mathcal{E}^1$ and of $\mathcal{E}^2$ correspondingly.

The statement is local, and locally the proof coincides with the proof of theorem 1.4.10.

Let A: $\mathcal{E}_1 \longrightarrow \mathcal{E}_2$ be a Fredholm morphism of holomorphic Banach bundles. We denote by

$$A^0: \mathcal{O}^{\mathcal{E}_1} \longrightarrow \mathcal{O}^{\mathcal{E}_2}$$

the corresponding morphism of sheaves.

Theorem 1.6.14. *Let A: $\mathcal{E}^1 \longrightarrow \mathcal{E}^2$ be a Fredholm morphism of holomorphic Banach bundles over a complex manifold X, and $\mathcal{K}=Ker\ A^0$, $\mathcal{R}=Im\ A^0$, $\mathcal{P}=O^{\mathcal{E}_2}/Im\ A^0=$ Coker A^0. Then*

i) the sheaves $\mathcal{K}$ and $\mathcal{P}$ are coherent;

ii) the sheaf $\mathcal{R}$ is a BCAF-sheaf.

In particular, if X is a Stein manifold, then

$$H^p(X,\mathcal{K})=0,\ H^p(X,\mathcal{R})=0\ ,\ H^p(X,\mathcal{P})=0 \text{ for } p \geq 1.$$

Proof. First of all, the statement about cohomologies follows from i), ii), theorem 1.5.9 and theorem 1.5.29, so we have to prove only the first two statements.

By the definition of a coherent sheaf and of a BCAF sheaf, i) and ii) are local properties. We can therefore assume that $\mathcal{E}_j = U \times E_j$, where U is a small neighborhood of $z_0 \in \Omega$, E_j (j=1,2) are Banach spaces, and A is a holomorphic Fredholm operator function $A(z)$: $E_1 \to E_2$ ($z \in U$). By lemma 1.2.11, if the neighborhood U is small, there is a closed subspace $E \subset E_1$ of finite co-dimension such that $Ker A(z) \cap E = 0$ for $z \in U$. Due to theorem 13,

$$\bigcup_{z \in U} A(z)E$$

forms a holomorphic subbundle $\mathcal{F}$ of finite co-dimension in $U \times E_2$. Let M be a complement to $A(z_0)E$ in E_2. Then $A(z)E_1 \oplus M = E_2$ in a sufficiently small neighborhood U of z_0. Let Q: $E_1 \to N$ be the projection onto a complement to E parallel to E, and $P(z)$: $E_2 \to M$ the projection onto M parallel to $A(z)E$. Note that $P(z)$ is a holomorphic operator function on U. Let $dim_{\mathbb{C}} N = n$, $dim_{\mathbb{C}} M = m$. Choosing some bases in N and M, we can consider the holomorphic matrix function $A_1(z)$ defined as follows:

$$A_1(z) = P(z)\ A(z)|_N .$$

This matrix function determines a sheaf homomorphism A_1^0: $O^n \to O^m$. Let K_1, R_1 and P_1 be its kernel, image and cokernel, respectively. Then, by theorem 1.5.8 all these sheaves are coherent. It is trivial to verify that Q generates an isomorphism Q: $\mathcal{K} \cong K_1$, that is, $\mathcal{K}$ is also coherent.

Next we consider the sheaf $\mathcal{R}$. Let $y \in \mathcal{R}$ be a germ of some section, $y = A^0 x$. Then $y = A^0(Q^0 x + (1-Q)^0 x) = P^0 A^0 Q^0 x + A^0(1-Q^0)x + (1-P^0)A^0 Q^0 x$. Clearly, $P^0 A^0 Q^0 x \in R_1$ (by the definition of R_1). Moreover,

$$A^0 (1-Q^0)\, x + (1-P^0)A^0 Q^0 x \in O^{\mathcal{F}}.$$

Thus, $\mathcal{R} \subset O^{\mathcal{F}} \oplus R_1$. We claim that the reverse inclusion also holds. Let $y_1 \in O^{\mathcal{F}}$, $y_2 \in R_1$. Then there exist $x_1 \in O^E$ and $x_2 \in O^N$ such that $A^0 x_1 = y_1$ and $PA^0 x_2 = y_2$ (the existence of x_1 follows from the invertibility of $A(z)$: $E \to A(z)E$, and that of x_2 from the definition of R_1). Then, putting $x = x_1 + x_2$, we obtain $y - A^0 x \in O^{\mathcal{F}}$. Because $O^{\mathcal{F}} \subset \mathcal{R}$, we conclude that $y \in \mathcal{R}$. Thus, $\mathcal{R}$ is a direct sum of a coherent sheaf and a sheaf of the type O^E (E being a Banach space). According to the remarks after definition 1.5.28, $\mathcal{R}$ is a BCAF-sheaf.

The last step is to observe that P^0 generates an isomorphism (a local one) between $\mathcal{P}$ and P_1. The proof is complete.

Corollary 1.6.15. *If A has non-zero kernel on at least one fiber, then there exists some not identically zero section $y \in \Gamma(X, \mathcal{E}^1)$ such that $Ay=0$.*

Proof. In our conditions $\mathcal{K} \neq 0$. Due to the statement ii) of theorem 1.5.9, there exists $y \neq 0$, $y \in \Gamma(X,\mathcal{K})$. This proves the statement.

Let A: $\mathcal{E}^1 \longrightarrow \mathcal{E}^2$ be a morphism of Banach bundles over a base X. Then the sets S(A) and CS(A) are defined as the sets of points of the base over which the morphism has nontrivial kernel (cokernel).

Theorem 1.6.16. *Let A: $\mathcal{E}^1 \longrightarrow \mathcal{E}^2$ be a Fredholm morphism of holomorphic Banach bundles over a complex manifold X. Then the sets $S(A)$ and $CS(A)$ are analytic in X. If A has zero index on each fiber, then these sets are principal analytic ones.*

The statement of the last theorem is local, and locally it coincides with the statement of theorem 11.

1.7. Image and cokernel of a Fredholm morphism in spaces of holomorphic sections

Let Ω be a Stein manifold and let $\mathcal{E}_j \xrightarrow{\pi_j} \Omega$ (j=1,2) be holomorphic Banach bundles over Ω. Consider some Fredholm morphism A: $\mathcal{E}_1 \to \mathcal{E}_2$, and denote as before by $S(A)$ and $CS(A)$ its *spectrum* and *cospectrum* respectively:

$$S(A)=\{z \in \Omega \mid Ker\ A(z) \neq \{0\}\}$$
$$CS(A)= \{z \in \Omega \mid Coker\ A(z) \neq \{0\}\}.$$

We denote here by $A(z)$ the restriction of A onto the fiber $\pi_1^{-1}(z)$ over the point z.

The morphism gives rise to an operator on the spaces of holomorphic sections: $x(z) \to A(z)\ x(z)$, which we denote by the same letter:

$$A:\ \Gamma(\Omega, O^{\mathcal{E}_1}) \to \Gamma(\Omega, O^{\mathcal{E}_2}).$$

We denote by A^0: $O^{\mathcal{E}_1} \to O^{\mathcal{E}_2}$ the corresponding morphism of sheaves, and by

$$M= Coker\ A^0 = O^{\mathcal{E}_2}/Im\ A^0.$$

its cokernel. Due to theorem 1.6.14, M is a coherent analytic sheaf. We denote by

$$C:\ \Gamma(U, O^{\mathcal{E}_2}) \to \Gamma(U, M)$$

the natural projection, where U is any open subset of Ω. Let $\{N_k\} = Assoc\ M$ (see definition 1.5.15, in particular, $\cup N_k = CS(A)$), and ν_k be a Noether operator associated with N_k (see definition 1.5.16).

Our immediate task is to describe the image $Im\ A \subset \Gamma(U, O^{\mathcal{E}_2})$ of the preceding operator and its cokernel

$$Coker\ A = (\Gamma(U, O^{\mathcal{E}_2})\ /Im\ A)^* = (Im\ A)^{\perp} \subset \Gamma(U, O^{\mathcal{E}_2})^*.$$

A solution of this problem results from the following theorem.

Theorem 1.7.1. *Let A: $\mathcal{E}_1 \to \mathcal{E}_2$ be a Fredholm morphism, and $U \subset \Omega$. Let a proper closed analytic subset $S_k \subset N_k$ be defined for any k (for instance, S_k can be empty). Then any element ϕ that belongs to Coker $A \subset (\Gamma(U, O^{\mathcal{E}_2}))^*$ can be represented as a finite sum*

$$\phi(y) = \sum_k \int_{N_k} \langle \nu_k Cy, \mu_k \rangle \, dV_k \tag{1.7.1}$$

where $\mu_k(z)$ are C^∞-functions with values in $\mathbf{C}^{r(N_k)}$ on the set $reg(N_k)$ such that supp μ_k is a compact contained in $(N_k \setminus S_k) \cap U$. Here dV_k is the corresponding Lebesgue measure on the set $reg(N_k)$ of regular points of N_k. Any functional of this form belongs to Coker A.

Proof. Consider the exact sequence

$$0 \to Im\ A \to \Gamma(U, O^{\mathcal{E}_2}) \overset{C}{\to} \Gamma(U,M) \overset{\Pi\nu_k}{\to} \prod_N \Gamma(U \cap (N_k \setminus S_k))^{r(N_k)} .$$

The exactness in the first two terms is obvious from definitions. The exactness in the third one follows from the properties of Noether operators (definition 1.5.16). Due to theorem 1.5.18, the mapping $\prod \nu_k$ is open onto its image, if we consider the space

$$\prod_k [\Gamma(U \bigcap (N_k \setminus S_k)]^{r(N_k)}$$

endowed with the topology of the distribution space

$$\mathfrak{D} := \prod_k \mathfrak{D}'(U \bigcap (N_k \setminus S_k))^{r(N_k)}.$$

Then every element $\phi \in (Im\ A)^{\perp}$ can be represented as

$$\phi = C^* (\prod_k \nu_k)^* \psi$$

for some $\psi \in \mathfrak{D}^*$. Any functional ψ has the form

$$\psi(g) = \sum_k \int_{N_k} \langle g, \mu_k \rangle \, dV_k.$$

This provides the proof of the direct statement. The converse statement is trivial. The proof is complete.

Corollary 1.7.2. *A section $y \in \Gamma(U, O^{\mathcal{E}_2})$ belongs to the image of A if and only if*

$$\langle \phi, y \rangle = 0$$

for all functionals $\phi \in (\Gamma(U, O^{\mathcal{E}_2}))^*$ of the form

$$\langle \phi, y \rangle = \frac{\partial^k}{\partial z^k} \langle t, y \rangle \mid_{z = z_0}$$

that annihilate the image of A, where t is a holomorphic section of the adjoint bundle $\mathcal{E}^{2\,*}$.

1.8. Image and cokernel of a Fredholm morphism in spaces of holomorphic sections with bounds

We shall present now some analogs of the previous results for the case of sections with some special bounds. We consider here only the case of a one-dimensional parameter.

Consider the cylinder $\Lambda = \mathbf{R} \times (i\mathbf{R}/2\pi i\mathbf{Z}) = \mathbf{C}/2\pi i\mathbf{Z}$. The points of Λ will be denoted by $z = x+iy$, where $x \in \mathbf{R}$, $y \in \mathbf{R}/2\pi\mathbf{Z}$.

Let H be a Hilbert space, and $P(z)$ $(z \in \Lambda)$ a holomorphic with respect to z bounded projection in H in such that

$$\|P(z)\| \le C_a \exp a|x|^p, \quad \forall a>0. \tag{1.8.1}$$

We denote by $\mathcal{E}^1$ the subbundle of the trivial bundle $\Lambda \times H$ which is the image of the projector P.

We denote, as before, by $\widetilde{A}_p(\mathcal{E}^1)$ the space of holomorphic sections y: $\Lambda \to \mathcal{E}^1$ with the estimate

$$\|y(z)\|_H \leq C_a(y)\ exp\ a|x|^p, \quad \forall a > 0.$$

This space is equipped with the topology of a Fréchet space by means of semi-norms

$$\gamma_a(y) = \sup_z \|y(z)\|_H\ exp\ (-a|x|^p)\ .$$

Let also $\mathcal{E}^2$ be another analytic subbundle of $\Lambda \times H$ and $\widetilde{A}_p(\mathcal{E}^2)$ is defined in analogous way. (We do not assume here the existence of a good projection onto E^2.)

Let us have now a holomorphic Fredholm morphism A: $\mathcal{E}^1 \longrightarrow \mathcal{E}^2$ such that its restriction to some (and therefore to almost every) fiber is invertible. Assume that the inverse morphism A^{-1} can be represented as

$$A^{-1} = B/\Delta,$$

where $B : \mathcal{E}^1 \to \mathcal{E}^2$ is a holomorphic morphism and Δ is a scalar analytic function (hence A^{-1} is not defined on the fibers over zeros of Δ). We impose the following conditions on A, B, and Δ:

i) A generates continuous mapping

$$A\colon \widetilde{A}_p(\mathcal{E}^1) \longrightarrow \widetilde{A}_p(\mathcal{E}^2); \tag{1.8.2}$$

ii) B generates continuous mapping

$$B\colon \widetilde{A}_p(\mathcal{E}^2) \longrightarrow \widetilde{A}_p(\mathcal{E}^1); \tag{1.8.3}$$

iii)

$$|\Delta(z)| \leq C_a\ exp\ a|x|^p, \ \forall a > 0\ . \tag{1.8.4}$$

The conditions *i)* and *ii)* mean, roughly speaking, estimates of the type (4) for $A(z)$ and

$B(z)$, and they will be verified in applications by such estimates.

The function $\Delta(z)$ is in fact a $2\pi i$-periodic entire function of order p and of minimal type on $\mathbf{C}$.

We are interested now in descriptions of the image and cokernel of the operator

$$A\colon \widetilde{A}_p(\mathcal{E}^1) \longrightarrow \widetilde{A}_p(\mathcal{E}^2).$$

Let $\{\lambda_n, p_n\}$ be the complete set of zeros (with their multiplicities) of the function $\Delta(z)$ in $\mathbf{C}^1$. The numbers $\beta_{i,k,n,m}$ are introduced as in section 1.5 before theorem 1.5.3. There is some natural linear mapping

$$\delta_p\colon \widetilde{A}_p(\mathcal{E}^2) \longrightarrow \widetilde{\Gamma}_p(\{\lambda_n, p_n\}, H) := \Gamma_p(\Delta, H).$$

This map is defined as

$$(\delta_p y)_{m,k} = \frac{\partial^k}{\partial z^k}\big(B(z)y(z)\big)\,|_{z=\lambda_m}.$$

Theorem 1.8.1. *The sequence of linear mappings*

$$0 \longrightarrow \widetilde{A}_p(\mathcal{E}^1) \xrightarrow{A} \widetilde{A}_p(\mathcal{E}^2) \xrightarrow{\delta_p} \Gamma_p(\Delta, H) \tag{1.8.5}$$

is exact. The map δ_p is a topological homomorphism (i.e. open mapping onto its image).

Proof. The exactness on the left is trivial. Indeed, A cannot annihilate non-zero analytic sections, due to the condition that the kernel of A is zero on almost every fiber.

Consider the exactness in the middle. Let $y \in \widetilde{A}_p(\mathcal{E}^2)$ and $y \in Im\ A$. This means that $y = Ax$ for some $x \in \widetilde{A}_p(\mathcal{E}^1)$. Then

$$x = A^{-1}y = By/\Delta.$$

In other words, the H-valued analytic function By is analytically divisible by Δ.

Hence, By has zeros of order at least equal to p_n at every point λ_n. We conclude now that

$$\delta_p y = \{a_{m,k}\} = \left\{ \frac{\partial^k}{\partial z^k} (By) \,|_{z=\lambda_m} \right\} = 0 \,.$$

And back, if $\delta_p y=0$, then By has the corresponding zeros and therefore $x=By/\Delta$ is analytic. Since $By \in \widetilde{A}_p(\mathcal{E}^1)$ and $\Delta(z)$ is an entire function of order p and of minimal type, theorem 1.5.6 implies that $x \in \widetilde{A}_p(\mathcal{E}^1)$. Since $y = A\,A^{-1}\,y = A(By)/\Delta = Ax$, we conclude that $y \in Im\ A$.

The last step is to check that δ_p is an open map (onto its image). To do this means the following. Given a semi-norm γ on $\widetilde{A}_p(\mathcal{E}^2)$, we have to find a semi-norm ρ on $\Gamma_p(\Delta, H)$ such that for every $a \in Im\ \delta_p$ we can solve the equation

$$\delta_p y = a$$

with the estimate $\gamma(y) \le \rho(a)$.

Let us recall that $\Gamma_p(\Delta, H) = \widetilde{\Gamma}_p(\{\lambda_n, p_n\}, H)$. Due to the exactness of the sequence (1.5.15) (corollary 1.5.4) and theorem 1.1.5, there exists a continuous (non-linear) mapping $S: \Gamma_p(\Delta, H) \longrightarrow \widetilde{A}_p(\mathbb{C}, H)$ such that $TSa=a$ for all $a \in \Gamma_p(\Delta, H)$. (Here T is the operator from (1.5.15).) Consider $y_0 = Sa$, and denote by y_1 its projection $y_1 = Py_0$ onto $\mathcal{E}^1$. Due to the estimate (1), $y_1 \in \widetilde{A}_p(\mathbb{C}, H)$, so $y_1 \in \widetilde{A}_p(\mathcal{E}^1)$. We know also that $a \in Im\ \delta_p$, i.e. $a = TBy$ for some $y \in \widetilde{A}_p(\mathcal{E}^2)$. Consider the difference $y_0 - By$. Because $T(y_0 - By) = a - a = 0$, and due to the corollary 1.5.4, we conclude that $y_0 - By = \Delta h$ for some $h \in \widetilde{A}_p(\mathbb{C}, H)$. Therefore,

$$y_1 = Py_0 = PBy + \Delta Ph = By + \Delta Ph.$$

The last equality holds, since $P\,|_{\mathcal{E}^1} = id$. Consider

$$\widetilde{y} := (Ay_1)/\Delta = (A/\Delta)By + (\Delta APh)/\Delta = y + APh.$$

We see that $\widetilde{y} \in \widetilde{A}_p(\mathcal{E}^2)$. Moreover,

$$\delta_p(\widetilde{y}\,) = TB\widetilde{y} \;= TBy + TBAPh = a + T(\,\Delta\, Ph) = a.$$

Because the mapping $y_0 \longrightarrow \widetilde{y} \;= (APy_0)/\,\Delta$ is continuous by the construction, we get a solution of $\delta_p y = a$ with the required estimates. This finishes the proof of the theorem.

The next step is to describe all elements of the dual space $\widetilde{A}_p(\mathcal{S}^2)^*$ that annihilate the image of A.

Theorem 1.8.2. *Under the above conditions there exists a sequence of finite-dimensional subspaces $H_n \subset H^*$ such that every element $\varphi \in \widetilde{A}_p(\mathcal{S}^2)^*$ that is orthogonal to Im A can be represented in the following form:*

$$\langle\varphi, y\rangle = \sum_{i,n} \langle\varphi_{i,n}, \sum_{k,m} \beta_{i,k,n,m} \big(\delta_p(y)\big)_{m,k}\rangle, \tag{1.8.6}$$

where $\varphi_{i,n} \in H_n{}^*$, and

$$\sum_{i,n} ||\varphi_{i,n}||_{H^*}\; exp\;(a_\varphi|\lambda_n|^p) < \infty. \tag{1.8.7}$$

And conversely, every collection $\varphi_{i,n} \in H^*$ satisfying (7) *gives rise by* (6) *to an element $\varphi \in \widetilde{A}_p(\mathcal{S}^2)^*$ orthogonal to Im A.*

The representation (6) *can be written in the following more detailed form:*

$$\langle\varphi, y\rangle = \sum_{i,n} \langle\varphi_{i,n}, \sum_{k,m} \beta_{i,k,n,m} \frac{\partial^k By}{\partial z^k}\Big|_{z=\lambda_m}\rangle. \tag{1.8.8}$$

Proof. Due to the last theorem, all elements φ of $\widetilde{A}_p(\mathcal{S}^2)^*$ that annihilate Im A (and only these elements) have the form $\langle\varphi, y\rangle \;=\; \langle\psi, \delta_p y\rangle$, where $\psi \in \Gamma_p(\,\Delta\,, H)^*$. Therefore, it is sufficient to describe all continuous linear functional ψ on $\Gamma_p(\,\Delta\,, H)$. Consider the space Γ consisting of collections $\beta = \{\beta_i^{(n)}\}$, where $n = 1, 2, \ldots,$ i runs over the set of values described during the definition of $\beta_{i,k,n,m}$, $\beta_i^{(n)} \in H$ and

$$\rho_b(\beta) = \sup_{i,n}\; ||\beta_i^{(n)}||\; exp\;(-b|\lambda_n|^p) < \infty \;\text{ for all } b > 0.$$

The semi-norms ρ_b define the structure of Frechet space on Γ. The mapping

$$a=\{a_{m,k}\}\longrightarrow\beta=\{\beta_i^{(n)}\},$$

(1.8.9) $$\beta_i^{(n)} = \sum_{k,m}\beta_{i,k,n,m}a_{m,k}$$

generates a linear embedding

$$\Gamma_p(\Delta,H)\longrightarrow\Gamma.$$

In fact, $\beta_i^{(n)}$ are the coefficients of the Newton's interpolation polynomial Q_n in U_n (see section 1.5) corresponding to the data $a_{m,k}$. If $\beta=0$, then all $a_{m,k}$ are equal to zero. Due to the definition of topology on $\Gamma_p(\Delta,H)$, we see that our mapping $a\longrightarrow\beta$ is a topological homomorphism. Since that, every continuous linear functional ψ on $\Gamma_p(\Delta,H)$ can be obtained as the restriction of some continuous linear functional on Γ. Now let $\psi\in\Gamma^*$. Consider the restriction of ψ onto the subspace of Γ consisting of collections $\{\beta_i^{(n)}\}$ such that only $\beta_{i_0}^{(n_0)}$ can be non-zero, and $\beta_i^{(n)}=0$ for $(i,n)\neq(i_0,n_0)$. This is a Hilbert space. Hence, the restriction of ψ onto it has the form

$$\langle\psi,\beta\rangle=\langle\psi_{i_0,n_0},\beta_{i_0}^{(n_0)}\rangle$$

for some $\psi_{i_0,n_0}\in H^*$. It is worth mentioning, that we are dealing with a proper subspace

$$\delta_p(\widetilde{A}_p(\mathbf{s}^2))$$

in Γ, and for all elements a from this subspace we have that $\beta_i^{(n)}\in M_n$, where $M_n\subset H$, $dim\ M_n<\infty$. To show this, let us return to the definition of δ_p:

$$a_{m,k}=\frac{\partial^k By}{\partial z^k}|_{z=\lambda_m}\ (m=1,\ ...,\ k=0,\ 1,\ ...,\ p_m-1).$$

If we prove that $a_{m,k}\subset N_m$ $(dimN_m<\infty)$ for all m,k, then we can choose M_n as the linear hull of $\bigcup N_m$, where the union is taken over all m such that $\lambda_m\in U_n$. So, let us restrict consideration to $a_{m,k}$.

Due to lemma 1.2.11, we can choose in a neighborhood of λ_m a holomorphic subbundle $\mathcal{M} \subset \mathcal{E}^1$ of finite codimension such that A has zero kernel on $\mathcal{M}$. Then the image of $A|_{\mathcal{M}}$ over this neighborhood is a subbundle of $\mathcal{E}^2$ of finite codimension (see theorem 1.6.13). Let $S(z)$ be a holomorphic projector in $\mathcal{E}^2$ onto the image of $A|_{\mathcal{M}}$. Given any holomorphic section y of $\mathcal{E}^2$ over the neighborhood, the section

$$x(z) = A^{-1}(z)S(z)y(z)$$

is analytic, since $y \in Im\ (A|_{\mathcal{M}})$, and $A: \mathcal{M} \longrightarrow Im\ (A|_{\mathcal{M}})$ is a bundle isomorphism. This means that $B(z)S(z)y(z)$ is analytically divisible by $\Delta(z)$:

$$\Big(B(z)S(z)y(z)\Big) \Big/ \Delta(z) = A^{-1}(z)S(z)y(z) = x(z).$$

Hence,

$$\frac{\partial^k B\ Sy}{\partial z^k}\Big|_{z=\lambda_m} = 0 \ (0 \le k \le p_m - 1)$$

for any section y. This means that

$$\{\delta_p(y)\}_{m,k} = \frac{\partial^k (B\ (1-S)y)}{\partial z^k}\Big|_{z=\lambda_m}.$$

The projection $(1 - S(z))$, however, maps the whole bundle $\mathcal{E}^2$ into a finite-dimensional subbundle. This obviously means that the space of possible values $a_{m,k}$ is finite-dimensional.

Let us return now to the functional ψ. We created the collection of elements $\{\psi_{i,n}\}$, $\psi_{i,n} \in H^*$. We may assume also that $\psi_{i,n} \in H_n \subset H^*$, where $dim\ H_n < \infty$. The definition of the topology on Γ shows that if $\gamma_{i,n} \in \Gamma$ is defined as

$$\{\ldots, 0 \ldots, 0, \beta_{i,n}, 0, \ldots\},$$

then the series $\sum_{i,n} \gamma_{i,n}$ converges to $\beta = \{\beta_{i,n}\}$. This implies the equality

(1.8.10) $$\langle\psi,\beta\rangle=\sum_{i,n}\langle\psi_{i,n}\beta_{i,n}\rangle.$$

Due to the continuity of ψ, there exist constants $b,c>0$ such that

(1.8.11) $$|\langle\psi,\beta\rangle|\le C\sup_{i,n}||\beta_{i,n}||exp(-b|\lambda_n|^p).$$

Let us transform the series $\sum\langle\psi_{i,n},\beta_{i,n}\rangle$ to the form

$$\sum\langle\psi_{i,n}exp(b|\lambda_n|^p),\ \beta_{i,n}exp(-b|\lambda_n|^p\rangle.$$

Then we get from (11) the estimate

(1.8.12) $$\sum_{i,n}||\psi_{i,n}||exp(b|\lambda_n|^p)<\infty.$$

The converse statement is obviously also correct: the convergence of (12) implies (11). If we plug now (9) into (10), we get the conclusion of the theorem.

Corollary 1.8.3. *Elements*

$$\varphi\in(ImA)^{\perp}\subset\left(\widetilde{A}_p(\mathbb{S}^2)\right)^*$$

(and only such elements) can be represented in the form

(1.8.13) $$\langle\varphi,y\rangle=\sum_{i,k,l,n,m}\varphi_{i,n,l}\beta_{i,k,n,m}\langle h_{n,l},\frac{\partial^k By}{\partial z^k}|_{\lambda_m}\rangle,$$

where the elements $h_{n,l}\in H^*$ are fixed, and $\varphi_{i,n,l}$ are complex numbers satisfying the condition

(1.8.14) $$\sum_{i,n}\Big(\sum_l|\varphi_{i,n,l}|^2\Big)^{1/2}exp(a_\varphi|\lambda_n|^p)<\infty$$

Proof. We choose in $H_n\subset H^*$ an orthonormal basis $h_{n,l}$ and expand $\varphi_{i,n}$ in the last theorem as

$$\varphi_{i,n} = \sum_l \varphi_{i,n,l} h_{n,l}.$$

The proof is complete.

Similar results can be obtained using theorem 1.5.5 instead of 1.5.4. Let us have, as before, two analytic subbundles $\mathcal{E}^j (j=1,2)$ in $\Lambda \times H$, and a projector $P: \Lambda \times H \longrightarrow \mathcal{E}^1$ with an estimate

$$\|P(z)\| \leq C exp\, c|x|^{p_0}.$$

Let us have also a Fredholm morphism $A: \mathcal{E}^1 \longrightarrow \mathcal{E}^2$ such that $A^{-1} = B \Delta^{-1}$. We assume that Δ is an entire function of order p_0, and the mappings

$$A : B_{p,p_0}(\mathcal{E}^1) \longrightarrow B_{p,p_0}(\mathcal{E}^2)$$

$$B : B_{p,p_0}(\mathcal{E}^2) \longrightarrow B_{p,p_0}(\mathcal{E}^1)$$

are continuous (where $B_{p,p_0}(\mathcal{E}) = \widetilde{B}_{p,p_0}(\mathbf{C}, H) \cap \Gamma(\Lambda, \mathcal{O}^{\mathcal{E}})$).

Theorem 1.8.4. *The sequence*

$$(1.8.15) \qquad 0 \longrightarrow B_{p,p_0}(\mathcal{E}^1) \xrightarrow{A} B_{p,p_0}(\mathcal{E}^2) \xrightarrow{\delta_p} \widetilde{\Pi}_{p,p_0}(\{\lambda_n, p_n\}, H)$$

is exact. The mapping δ_p is a topological homomorphism.

Proof. The exactness can be proved in the same way as in the theorem 1. Let us prove now that δ_p is a homomorphism. Due to theorem 1.5.5, the operator

$$T: \widetilde{B}_{p,p_0}(\mathbf{C}, H) \longrightarrow \widetilde{\Pi}_{p,p_0}(\{\lambda_n, p_n\} H)$$

is surjective. The spaces $\widetilde{\Pi}_{p,p_0}$ and $\widetilde{B}_{p,p_0}$ are countable inductive limits of Frechet spaces. Due to theorem 1.1.7, the mapping T is open. Let us show now that $\delta_p: B_{p,p_0}(\mathcal{E}^2) \longrightarrow Im\, \delta_p$ is also open. Let U be a neighborhood of zero in $B_{p,p_0}(\mathcal{E}^2)$. We apply now the morphism B. Let us show that BU is a neighborhood of zero in $B(B_{p,p_0}(\mathcal{E}^2))$. In fact, the inverse to B (on $Im\ B$) is $\Delta^{-1} A$. The operator

$A: B_{p,p_0}(\mathcal{S}^1) \longrightarrow B_{p,p_0}(\mathcal{S}^2)$ is continuous and maps $Im\ B$ onto some subspace $\Phi \subset B_{p,p_0}(\mathcal{S}^2)$. This subspace consists only of sections that are analytically divisible by Δ. Due to the estimates of Δ from below (theorem 1.5.6), the division by Δ is a continuous mapping from Φ into $B_{p,p_0}(\mathcal{S}^2)$. We proved the continuity of B^{-1} on $Im\ B$, so B is a topological homomorphism. Let us show now that

$$T: Im\ B \longrightarrow \widetilde{\Pi}_{p,p_0}(\{\lambda_n, p_n\}H)$$

is also a topological homomorphism (this will finish the proof). Let U_1 be a neighborhood of zero in $Im\ B$ and $U_1 = Im\ B \cap V_1$, where V_1 is a neighborhood of zero in $\widetilde{B}_{p,p_0}(\mathbf{C}, H)$. Consider on the space

$$\mathcal{M} = \Delta \widetilde{B}_{p,p_0}(\mathbf{C}, H) + Im\ B$$

the operator P of multiplication by $P(z)$. Then P maps $\mathcal{M}$ into $Im\ B$. In fact,

$$P(By + \Delta h) = By + \Delta Ph = By + BAPh = B(y + APh) \in Im\ B.$$

Due to the estimates on P, this operator is a continuous projector onto $Im\ B$. Therefore, there is a neighborhood of zero V_1^1 in $\mathcal{M}$ such that $PV_1^1 \subset U_1$. Let $V_1^1 = V_2 \cap \mathcal{M}$, where V_2 is a neighborhood of zero in $\widetilde{B}_{p,p_0}(\mathbf{C}, H)$. Due to theorem 1.5.5, TV_2 is a neighborhood of zero in $\widetilde{\Pi}_{p,p_0}$. This means that $T(V_2 \cap \mathcal{M})$ is a neighborhood of zero in $T(\mathcal{M})$. On the other hand, $T(\mathcal{M}) = T(Im\ B)$. In fact, if $x = By + \Delta h \in \mathcal{M}$, then $Tx - TBy = T(\Delta h) = 0$. Moreover, $Tx = TPx$, because $Px = By + \Delta Ph = x - \Delta(h - Ph)$. Therefore, $T(V_1^1) = T(PV_1^1) \subset T(U_1)$. This shows that $T(U_1)$ is a neighborhood of zero in $T(Im\ B)$. This finishes the proof of the theorem.

The following result can be proved now in the same way as the corollary 3.

Corollary 1.8.5. *Under the above assumptions, elements*

$$\varphi \in (ImA)^{\perp} \subset (B_{p,p_0}(\mathcal{S}^2))^*$$

(and only such elements) can be represented in the form

(1.8.16)
$$\langle \varphi, y\rangle = \sum_{i,\alpha} \langle \varphi_{i,\alpha}, \sum_{k,m} \gamma_{i,k,\alpha,m} (\delta_p y)_{m,k}\rangle =$$
$$= \sum_{i,\alpha} \langle \varphi_{i,\alpha}, \sum_{k,m} \gamma_{i,k,\alpha,m} \frac{\partial^k By}{\partial z^k}|_{\lambda_m}\rangle,$$

where $\varphi_{i,\alpha} \in H_\alpha \subset H^*$ $(\dim H_\alpha < \infty)$ *and*

(1.8.17)
$$\sum_{i,\alpha} \|\varphi_{i,\alpha}\|_{H^*} d_\alpha^{-j} \exp(|z_\alpha|^q) < \infty$$

for any $q > p$ *and* $z_\alpha \in U_\alpha$.

The definitions of U_α, $\gamma_{i,k,\alpha,m}$, and d_α can be found before theorem 1.5.5.

1.9. Comments and references

Section 1.1. The proof of the Bartle-Graves theorem 1.1.1 [36] follows the scheme of E. Michael [293, 294]. These papers also contain more general results on sections of multi-valued maps. Important discussion of lifting problems can be found also in [242]. Theorem 1.1.6 is due to S. Banach [30]. Theorem 1.1.7 was proved by A. Grothendiek [152]; cf. also [339, 340] about open mapping theorems.

Section 1.2. See [93, 140, 141, 199, 210, 211, 331, 341, 363] for the fundamental classes of linear operators that we use and their algebraic and topological properties. Important lemma 1.2.11 can be found in [19]. Theorem 1.2.15 and lemma 1.2.16 are due to H. Widom [402]. Lemmas 1.2.18 and 1.2.19 are standard results of Sobolev spaces theory (see, for instance, [394]). The statements of theorem 1.2.20 can be found in [93, 141, 363]. The statement about the infinite dimensional analyticity of the determinant was not formulated explicitly in these books. It was proved, however, that Δ is continuous on $\mathcal{I}_p(H)$ and analytic on any finite dimensional affine subspace. Standard results of infinite dimensional holomorphy (see, for instance, theorem 14.9 in [73]) imply now the infinite dimensional analyticity.

Section 1.3. We follow [247] in our account of the fundamental notions of the theory of Banach vector bundles; cf. also [411]. The reader should be aware that there is a different type of bundles, also called Banach bundles, in which the transition functions may be families of arbitrary (not necessarily linear) homeomorphisms of the fiber. If E

is homeomorphic to the countable power of itself (this happens, for instance, when E is separable), then all such bundles are trivial in their category [74]. The reader can find a comprehensive account of main properties of (finite dimensional) vector bundles in [19, 176].

Theorem 1.3.3 is due to N. Kuiper [240]. Many results on contractibility of the general linear group of a Banach space can be found in survey of B. Mityagin [301]. Theorem 1.3.4 was proved in [100, 217, 319, 411]. Theorems 1.3.5 - 1.3.8 and 1.3.10 are from [217]; cf also [411]. Theorem 1.3.9, which is a simple generalization of lemma 1.2.11, can be found in [411]. It implies, for instance, that every finite Fredholm complex (see the definition in [351]) is isomorphic to the direct sum of an exact and a finite dimensional ones.

Section 1.4. The index theory for compact families of Fredholm operators on Hilbert spaces (theorems 1.4.2 - 1.4.4) was developed by M. Atiyah and K. Jänich [19, 181] and in Banach case by G. Neubauer [311]. Theorems 1.4.2 - 1.4.5 can be found with proofs in survey [411]. Theorems 1.4.3 and 1.4.5 are due to G. Neubauer [311]. It should be mentioned that the index theory can be also extracted from works of R. Palais [320] and A. Shvartz [362], where the topological properties of $\Phi^{(n)}(E)$ and $GL_C(E)$ were studied. The results of theorems 1.4.6 - 1.4.10 (and many others) about continuous families of Fredholm and semi-Fredholm operators are essentially contained in [95, 96, 100, 101, 146, 311, 411]. See also [238] for some further relations with the K-theory.

Section 1.5.A. Theorems 1.5.1 and 1.5.2 are due to the author [222, 227], and were proved specially for the Floquet theory. Theorem 1.5.3 was essentially proved in [57, 58]. These papers deal with scalar functions. The case of H-valued functions does not create any additional troubles. The continuity of the map T was not mentioned explicitly in [57, 58], but it was proved there. Theorem 1.5.5 is just a reformulation of the main result of [46]. Consideration of H-valued functions instead of scalar ones does not require any additional tricks.

The basics of the theory of entire functions in several variables can be found in [255, 345].

It is well known that interpolation of entire functions is very important for many problems of harmonic analysis. It plays, for instance, crucial role in problems of completeness of exponential solutions for constant coefficient PDEs and convolution equations [39, 42, 45-48, 324]. We shall use theorems 1.5.3 - 1.5.5 in our study of evolution equations in chapter 5. This will lead to Floquet expansion theorems in some classes of solutions that grow faster than exponentially. In the case of periodicity in several variables, however, we have not been able yet to prove analogous expansions (see section 3.3, where only completeness theorems are obtained). The main obstacle is that we could not find n-dimensional versions of theorems 1.5.3-1.5.5 that would suit our needs. There are nice results about interpolation from algebraic sets [324, 347]. In our case, however, we deal with general sets of zeros of entire functions of some prescribed growth. In this general situation C. Berenstein and B.A. Taylor obtained some strong results in [47]. Unfortunately, we have not been able yet to apply these results to Floquet theory due to some conditions that we could not verify in our situation. Theorem 1.5.6 can be found in [55, 256].

Section 1.5.B. The reader can find the required information from complex analysis in several variables in [154, 167, 169, 307]. The necessary information from commutative algebra can be found in [413]. Theorem 1.5.14 was proved in [323]. Theorems 1.5.17 and 1.5.18 are also due to V. Palamodov [326]. Their local versions were obtained earlier in [322-324]. The local versions were the basis of previous author's results on Floquet theory [229]. V.Palamodov has obtained recently the global version, which allowed him to improve the main result of Floquet theory [326].

Section 1.5.C. Theorems 1.5.20 and 1.5.21, which play an important role in any consideration of vector valued analytic functions, were proved by E. Bishop [52] in the more general case of Fréchet spaces. Theorem 1.5.22 is due to A. Gleason [137]. Sometimes the problem arises of lifting holomorphic functions with some prescribed boundary behavior (continuity up to the boundary, prescribed bounds at infinity, etc.). Lifting in such classes is not always possible. For instance, lifting in the class of functions analytic in the unit disk and continuous up to the boundary is impossible in general [237]. It is possible, however, for functions with values in the Calkin algebra [218]. Under some restrictions either on the spaces involved, or on the boundary smoothness some positive results can be also obtained [186, 187, 190, 218].

Theorems 1.5.23 and 1.5.25 were proved by L. Bungart [68, 69]. They generalize the famous finite dimensional results of H. Grauert [151]. Theorem 1.5.26 is due to

M. Shubin [357]. The notion of a BCAF-sheaf was introduced and theorem 1.5.29 was obtained by J. Leiterer [252]. This technic has proved to be very useful in PDEs (see, for instance, this book and works of F. Mantlik [275]). I believe that the statement of lemma 1.5.30 should be known, but I could not find the corresponding reference.

Section 1.6. Theory of holomorphic Fredholm operator functions (in one and several variables) has been developed by many prominent mathematicians (I. Gohberg, B. Gramsch, T. Kato, M. Krein, S. Krein, H. Langer, A. Markus, V. Matsaev and many others). We present here only a small part of this theory that we need for our purposes. We do not touch some large areas at all (meromorphic behavior of the resolvent, kernel functions, factorization, completeness of the kernel vectors, etc.). The reader can look at [31-35, 64, 138, 139, 142, 144-150, 182-185, 188, 189, 199, 214, 253, 254, 275, 280, 281, 409, 411] for additional information and references. Our references here are definitely neither complete, nor up to date.

Theorem 1.6.1 was proved by G. Allan [8, 9]. Theorem 1.6.2 in the case of Stein manifolds was proved by S. Krein and V. Trofimov [214], and in general case by B. Gramsch [144]. Theorem 1.6.3 is also due to B. Gramsch. Theorem 1.6.4 and corollary 1.6.5 were established by M. Zaidenberg [409]. Theorem 1.6.7 was obtained in [59], theorems 1.6.8 - 1.6.10 - in [410]. Theorem 1.6.11 is from [214], and 1.6.16 is its immediate corollary.

Results of **sections 1.7** and **1.8** were obtained by the author [229, 230]. The main result of section 1.7 is based upon V. Palamodov's results about coherent sheaves [322, 323, 326]. The results of section 1.8 employ the interpolation theorems of A. Bratishchev and Ju. Korobeinik [57, 58] and C. Berenstein and B.A. Taylor [46]. As we have already mentioned before, n-dimensional versions of theorems of section 1.8 would enable us to get some new results about Floquet expansions.70

CHAPTER 2

SPACES, OPERATORS AND TRANSFORMATIONS

In this chapter we introduce some functional spaces, classes of differential operators and a transform that will play the role of the Fourier transform. We investigate the properties of the spaces and operators with respect to this transform.

2.1. Basic spaces and operators

We consider here only the spaces that we need for the case of elliptic differential equations. In the cases of hypoelliptic equations, elliptic systems or pseudodifferential equations some simple changes are necessary that will be mentioned in appropriate places.

All spaces introduced in this section consist of functions that locally belong to a Sobolev space H^s and satisfy appropriate decay conditions at infinity (exponential decay, subexponential decay or summability). These spaces will play the role of spaces of test functions.

Let $s \geq 0$, $a > 0$, $p \geq 1$ and $n \in \mathbf{N}$. We define the space $\Phi^{s,a}_{n,p}$ as follows:

$$\Phi^{s,a}_{n,p} = \{ f \in H^s_{loc}(\mathbf{R}^n) : \gamma_b(f) =$$

$$= \sup_{k \in \mathbf{Z}^n} \| f \|_{H^s(\mathfrak{K} - k)} \; exp(b \mid k \mid^p) < \infty \text{ for all } 0 \leq b < a \},$$

where $\mathfrak{K} = \{ x = (x_1, \dots , x_n) : 0 \leq x_j \leq 1 \}$ is the unit cube in $\mathbf{R}^n$ and $\mathfrak{K} - k$ is its translation by the vector k. This is obviously a Frechet space under the semi-norms γ_b. When $a = \infty$, we obtain a Frechet space, which we denote by $\Phi^s_{n,p}$.

We also need the Banach space

$$\Psi^s_n = \{ f \in H^s_{loc}(\mathbf{R}^n) : \sum_{k \in \mathbf{Z}^n} \| f \|_{H^s(\mathfrak{K} - k)} < \infty \}.$$

Now let us introduce the class of operators that we are going to consider here (however, the main results may be generalized to wide classes of differential and pseudodifferential operators, see section 3.4). We denote by $L(x,D)$ a scalar linear elliptic operator of order $2m$ on $\mathbf{R}^n$ that has infinitely smooth, periodic with period 1 in all the variables x_j $(j=1, \dots, n)$ coefficients (this means that the lattice of periods is $\mathbf{Z}^n \subset \mathbf{R}^n$, i.e. $L(x,D) = L(x+k,D)$ for all $k \in \mathbf{Z}^n$). Ellipticity means that the principal symbol $L_0(x,\xi)$ has no real zeros ξ, except $\xi = 0$. The condition of smoothness can be reduced (see section 3.4). The case of arbitrary lattice of periods can be easily reduced to the case of $\mathbf{Z}^n$ (in most of the results) by a change of variables.

Let $L = L(x,D)$ be as above. We are interested in the mapping of our spaces that is generated by L: $u(x) \to L(x,D)u(x)$. According to the assumptions on L, it generates a continuous mapping from the space $H^{2m+\alpha}(\overline{\mathfrak{K}})$ into $H^{\alpha}(\overline{\mathfrak{K}})$, in particular from $H^{2m}(\overline{\mathfrak{K}})$ into $L_2(\mathfrak{K})$. Due to periodicity of the operator $L = L(x,D)$, we have (uniformly with respect to $k \in \mathbf{Z}^n$) the estimate

$$\text{(2.1.1)} \qquad \| L(x,D)\, u(x) \|_{H^{\alpha}(\overline{\mathfrak{K}-k})} \leq C \, \| u(x) \|_{H^{2m+\alpha}(\overline{\mathfrak{K}-k})}.$$

This estimate implies that L is continuous from $\Phi^{2m,a}_{n,p}$ into $\Phi^{0,a}_{n,p}$, from $\Phi^{2m}_{n,p}$ into $\Phi^{0}_{n,p}$ and from Ψ^{2m}_{n} into Ψ^{0}_{n}.

We shall often need the following auxiliary result.

Lemma 2.1.1. *Let $L(x,D)$ be an elliptic differential operator in $\mathbf{R}^n$ with smooth $\mathbf{Z}^n$-periodic coefficients. Then there exists a proper pseudodifferential parametrix B that commutes with all $\mathbf{Z}^n$-translations.*

Remark. We recall that a pseudodifferential operator B is a *parametrix* to L, if the operators $(BL-I)$ and $(LB-I)$ are infinitely smoothing [174, 359]. Operator B is *proper*, if on the support of its Schwartz kernel both projections $\pi_x: \mathbf{R}^n_x \times \mathbf{R}^n_\xi \to \mathbf{R}^n_x$ and $\pi_\xi: \mathbf{R}^n_x \times \mathbf{R}^n_\xi \to \mathbf{R}^n_\xi$ are proper mappings [174, 359].

Proof of the lemma. In the first place, there always exists a proper parametrix B_1 for L (see [174, 359]); the problem is to find a periodic one. We choose a bounded open set $U \subset \mathbf{R}^n$ such that its translations $U_k = U + k$ $(k \in \mathbf{Z}^n)$ cover the whole space $\mathbf{R}^n$. The standard construction of partition of unity provides a partition $\phi_k(x)$ ($supp\ \phi_k \subset U_k$)

such that $\phi_k(x) = \phi_0(x-k)$. Let also $\psi_k(x) = \psi_0(x-k)$, where $\psi_0 \in C_0^\infty(\mathbf{R}^n)$ and $\psi_0 \equiv 1$ on $supp\ \phi_0$. We denote by T_k the translation operator $(T_k f)(x) = f(x+k)$. Now we can construct the required parametrix:

$$Bu = \sum_{k \in \mathbf{Z}^n} \psi_k (T_k^{-1} \circ B \circ T_k)\, \phi_k u.$$

It is not difficult to check all the required properties of this operator. The lemma is proved.

2.2. Fourier transform on the group of periods

Now we need some transforms that play the role of the Fourier transform for the periodic case. In fact, they are just some versions of the Fourier transform on the group $\mathbf{Z}^n$ of periods (it is well known that harmonic analysis on this group is provided by the theory of Fourier series in several variables). Transformations of this kind were used by I. Gelfand in [126] (see also [257]) in his study of periodic differential equations. In similar situations they are frequently used in papers on the quantum theory of solids [18]. Interesting discussions on these transforms can be found in [23, 24, 295].

We will denote by Ω_n the region $(\mathbf{C}\backslash 0)^n \subset \mathbf{C}^n$ and by $\Omega_n(a)$ the region $\{z = (z_1,\dots,z_n) :\ exp(-a) < |z_j| < exp\ a\ ,\ j = 1,\ \dots\ ,\ n\}$ in $\mathbf{C}^n$ (in particular, $\Omega_n = \Omega_n(\infty)$). Let us consider a function $f(x)$ (to begin with, compactly supported) on $\mathbf{R}^n$ and define transforms $\boldsymbol{F}$, $\hat{\boldsymbol{F}}$ and $\mathfrak{F}$ in the following way :

(2.2.1) $$\boldsymbol{F}f(z, x) = \sum_{k \in \mathbf{Z}^n} f(x-k)\, z^k \quad (z \in \Omega_n\ ,\ x \in \mathfrak{X}\ )$$

(2.2.2) $$\hat{\boldsymbol{F}}f(\lambda, x) = \boldsymbol{F}f(e^\lambda\ , x)\ \ (\ \lambda \in \mathbf{C}^n,\ x \in \mathfrak{X},\ e^\lambda = (e^{\lambda_1}, \dots , e^{\lambda_n}) \in \Omega_n\ )$$

(2.2.3) $$\mathfrak{F}f(\lambda, x) = e^{-\lambda x}\, \hat{\boldsymbol{F}}f(\lambda, x) \quad (\lambda \in \mathbf{C}^n,\ x \in \mathfrak{X}\ ).$$

Here $z^k = z_1^{k_1} \cdot z_2^{k_2} \cdot \dots \cdot z_n^{k_n}$ and $\lambda x = \lambda_1 x_1 + \dots + \lambda_n x_n$.

The basic of what follows is the study of the action of these transforms on the spaces introduced in the preceding section.

For each $z \in \Omega_n$ we define the closed subspace $E_z^s \subset H^s(\mathfrak{X}) :\ f(x) \in H^s(\mathfrak{X})$

belongs to E^s_z, if for all $j=1, \dots, n$ and $k=0, \dots, s-1$ it satisfies the cyclic boundary conditions

$$(2.2.4) \qquad \frac{\partial^k f}{\partial x_j^k}\Big|_{x_j=1} = z_j \frac{\partial^k f}{\partial x_j^k}\Big|_{x_j=0}.$$

The closedness of this subspace easily follows from the embedding theorems for Sobolev spaces (see, for example [394]).

Let us consider the trivial Hilbert bundles $\Omega_n \times H^s(\mathfrak{X})$ over Ω_n and $\mathbf{C}^n \times H^s(\mathfrak{X})$ over $\mathbf{C}^n$ and introduce the subspaces $\mathcal{E}^s = \bigcup_z E^s_z$ and respectively $\hat{\mathcal{E}}^s = \bigcup_\lambda E^s_{e^\lambda}$ in the spaces of these bundles.

Theorem 2.2.1. *The space $\mathcal{E}^s$ (respectively, $\hat{\mathcal{E}}^s$), equipped with the natural projection, is a trivial holomorphic Hilbert bundle over Ω_n (respectively, $\mathbf{C}^n$).*

Proof. Let us prove first that $\hat{\mathcal{E}}^s$ is a trivial bundle. We consider the following holomorphic automorphism of the bundle $\mathbf{C}^n \times H^s(\mathfrak{X})$: it multiplies all functions in the fiber above the point λ by $exp(-\lambda x)$. It is easy to see that after this transform the condition (4) (for $z = exp\ \lambda$) simply becomes the periodicity condition. That is, $\hat{\mathcal{E}}^s$ becomes $\mathbf{C}^n \times H^s(\mathbf{T}^n)$, where $\mathbf{T}^n$ is the n-dimensional torus $\mathbf{R}^n/\mathbf{Z}^n$. (Here we do not check the elementary assertion that the condition (4) with $z_j = 1$ ($j = 1, 2, \dots n$) is sufficient for gluing $f(x)$ into a function of class H^s on the torus.)

Thus, we proved that $\hat{\mathcal{E}}^s$ is a trivial bundle. This implies that $\mathcal{E}^s$ is locally trivial, because it is the lifting of $\hat{\mathcal{E}}^s$ from Ω_n to $\mathbf{C}^n$ via the exponential map. Next, Ω_n is a product of planar domains and is, therefore, a Stein manifold. Then, by theorem 1.5.23, the bundle $\mathcal{E}^s$ is also holomorphically trivial. This proves the theorem.

We are able now to describe the images of the spaces, introduced in the section 2.1, under the transformations. The next theorem plays the role of the Paley-Wiener's theorem. Let us denote by $\mathcal{T}^n$ the skeleton of the unit polydisk in $\mathbf{C}^n$, i.e.

$$\mathcal{T}^n = \{\ z \in \mathbf{C}^n : \ |z_j| = 1, j = 1, 2, \dots, n\}.$$

Theorem 2.2.2.

i) *The transform* F *can be extended to an isomorphism of the space* $\Phi^{s,a}_{n,1}$ *onto the space* $\Gamma(\Omega_n(a), \mathcal{E}^s)$.

ii) *The transform* $\widehat{F}$ *can be extended to an isomorphism of the space* $\Phi^{s}_{n,p}$ $(p > 1)$ *onto the space* $A_{p/(p-1)}(\widehat{\mathcal{E}}^s)$.

iii) *The transform* F *can be extended to an isomorphism of the space* Ψ^s_n *onto the space* $W(\mathcal{T}^n, \mathcal{E}^s)$.

(The definitions of the spaces Γ, A, and W can be found in section 1.5.)

Proof. In each case we can define the transformations F and $\widehat{F}$ (originally defined only for compactly supported functions) by the same formulas (1) and (2) on the whole spaces under consideration. We have to prove that these transformations are correctly defined and produce the required isomorphisms.

i) We regard $Ff(z,x)$ as an abstract Laurent series $\sum f_k(x)z^k$ (where $f_k(x) = f(x-k)$) with coefficients $f_k(x)$ that belong to the space $H^s(\mathfrak{X})$. If f belongs to the space $\Phi^{s,a}_{n,p}$ (where $p \geq 1$), then a trivial estimate shows that the series is absolutely convergent in $\Omega_n(a)$ and defines in this region an analytic $H^s(\mathfrak{X})$ -valued function. Thus, our operator F is correctly defined and maps continuously the space $\Phi^{s,a}_{n,p}$ into $\mathcal{H}(\Omega_n(a), H^s(\mathfrak{X}))$. The absence of its kernel is obvious. Let us discuss the image. Direct calculation shows, that for each $z \in \Omega_n$ the function $Ff(z, \cdot)$ satisfies the boundary conditions (4) corresponding to the value z. Therefore, $Ff(z,\cdot)$ is actually a section of the subbundle $\mathcal{E}^s$, and F acts between the spaces, described in i). Let us prove surjectivity of this operator. Consider an element $\phi(z)$ of $\Gamma(\Omega_n(a),\mathcal{E}^s) \subset \mathcal{H}(\Omega_n(a),H^s(\mathfrak{X}))$. Let $\sum \phi_k z^k$ be its Laurent series with coefficients ϕ_k from $H^s(\mathfrak{X})$. The coefficients can be found by means of the standard formula:

$$\phi_k = (2\pi i)^{-n} \int_{\Lambda} \zeta^{-(k+1)} \phi(\zeta)\, d\zeta,$$

where Λ is the product of the circles γ_j $(j = 1,\dots n)$ with the centers at the origin and with radii $exp(-a) < r_j < exp\ a$, and $k+1$ means $(k_1+1, \dots, k_n+1)$. It follows from this representation that

$$\sup_{k \in \mathbb{Z}^n} \|\phi_k\| \; exp(b\,|k|) < \infty$$

for all $b < a$. Let us construct a function $f(x)$ on $\mathbf{R}^n$ in the following way:

$$f(y) = \phi_k(y+k) \text{ for } y \in \mathfrak{K} - k.$$

It is obvious that f belongs to H^s_{loc} in the interior of each cube $\mathfrak{K} - k$ (because that is the property of the coefficients ϕ_k) and satisfies the decay condition from the definition of the space $\Phi^{s,a}_{n,p}$. We only need to prove that f belongs to $H^s_{loc}(\mathbf{R}^n)$, so we need some "gluing" conditions for f on the joins of cubes. Since $\phi(z)$ belongs to E^s_z, it satisfies conditions (4). Substituting into (4) the Laurent expansion of ϕ and using the definition of $f(y)$, we see that the normal derivatives of f up to the order $s-1$ on each face of $\mathfrak{K} - k$, taken from both sides of the face, coincide. Now it is a simple exercise to show that f belongs to H^s_{loc} in the entire space $\mathbf{R}^n$.

ii) We now turn to the proof of the statement ii) of the theorem. Since $\Phi^s_{n,p} \subset \Phi^s_{n,1}$ for $p > 1$, since the space $A_{p/(p-1)}(\hat{\mathcal{E}}^s)$ after the change of variables $z = exp\,\lambda$ becomes a subset of $\Gamma(\Omega_n, \mathcal{E}^s)$ and since the statement i) has been proved already, it is sufficient to show that the estimates

$$\sup_{k \in \mathbb{Z}^n} \|f_k\| \; exp\,(a\,|k|^p) < \infty \text{ for all } a > 0 \tag{2.2.5}$$

go over into

$$\|\widehat{F}f(\lambda)\|_{H^s(\mathfrak{K})} \le C_a \, exp\,(a\,|\lambda|^{p/(p-1)}) \text{ for all } a > 0. \tag{2.2.6}$$

The relevant arguments are quite standard. Let

$$y(\lambda) = \sum y_k \, e^{\lambda k} \in \widetilde{A}_{p/(p-1)}(\mathbf{C}^n, H)$$

for some Hilbert space H. Elementary estimates show that for any k_0

$$y_{k_0} = \lim_{T \to \infty} T^{-n} \int_{t_0}^{T} \cdots \int_{t_0}^{T} y(\nu + i\mu)\, e^{-k_0(\nu + i\mu)} \, d\mu_1 \ldots d\mu_n ,$$

where ν, $\mu \in \mathbf{R}^n$ and the integral and limit are understood in the strong sense. Indeed,

$$y(\lambda) = \sum_{|k| \leq N, k \neq k_0} y_k e^{\lambda k} + \sum_{|k| > N} y_k e^{\lambda k} + y_{k_0} e^{\lambda k_0}.$$

We choose $a > |\nu|$ and consider the integral

$$\frac{1}{T^n}\int_{t_0}^{T} \cdots \int_{t_0}^{T} y(\nu + i\mu)\, e^{-k_0(\nu + i\mu)}\, d\mu = \frac{1}{T^n}\int_{t_0}^{T} \cdots \int_{t_0}^{T} \sum_{|k| \leq N, k \neq k_0} y_k e^{\lambda(k - k_0)}\, d\mu \ +$$

$$\frac{1}{T^n}\int_{t_0}^{T} \cdots \int_{t_0}^{T} \sum_{|k| > N} y_k\, e^{\lambda(k - k_0)}\, d\mu + \frac{1}{T^n}\int_{t_0}^{T} \cdots \int_{t_0}^{T} y_{k_0} d\mu.$$

(Here $\lambda = \nu + i\mu$.) The norm of the second integral can be estimated from above by

$$C \sum_{|k| > N} e^{|k - k_0|\,|\nu| - a|k|}.$$

For $a > |\nu|$ and sufficiently large N this expression can be made arbitrarily small (uniformly with respect to T). The last term obviously tends to y_{k_0}, when T tends to infinity. All integrals in the first term tend to zero, when T tends to infinity (for a fixed N). This follows from the direct computation of the integrals.

This integral representation for coefficients y_k, together with the estimates on $y(\lambda)$, leads to the estimate:

$$\| y_k \| \leq C_a \, exp\, (a|\nu|^{p/(p-1)} - k\nu)$$

for any $a > 0$, $\nu \in \mathbf{R}^n$. Choosing $\nu = (k/|k|)\,(|k|/ap)^{p-1}$, we get

$$\| y_k \| \leq C_a \, exp(-a|k|^p)$$

for any $a > 0$. Conversely, suppose that $\| y_k \| \leq C_a \, exp(-a|k|^p)$ for all $a > 0$. Then

$$\| y(\lambda) \| = \| \sum y_k e^{\lambda k} \| \leq C_a \sum e^{-\frac{a}{2}|k|^p} e^{Re\, \lambda k - \frac{a}{2}|k|^p}.$$

It is not hard to show that

$$exp\,(Re\,\lambda k - \tfrac{a}{2}\,|k|^{p}) \le C\,exp\,(C\,a^{-1}\,|\lambda|^{p/(p-1)}).$$

Hence, $\|y(\lambda)\| \le C_a\,exp\,(a|\lambda|^{p/(p-1)})$ for all $a > 0$. This proves the part ii) of the theorem.

iii) If f belongs to the space Ψ_n^s, then we regard $\boldsymbol{F}f$ as an abstract Fourier series on the torus $\mathcal{T}^n$ with coefficients in $H^s(\mathfrak{X})$. It follows immediately from the definition of the space $W(\mathcal{T}^n, E)$ that $\boldsymbol{F}$ maps Ψ_n^s into the space $W(\mathcal{T}^n, H^s(\mathfrak{X}))$. The arguments of the part i) of the proof show that in fact $\boldsymbol{F}f$ is a section of the bundle $\mathcal{E}^s$, and hence it belongs to the space $W(\mathcal{T}^n,\mathcal{E}^s)$. The injectivity of the operator is obvious, and the surjectivity is proved by the arguments of the part i) .

The proof is now finished.

The meaning of the last theorem is that the transformations introduced above map our spaces isomorphically onto some spaces of (in most cases holomorphic) sections of some holomorphic Hilbert bundles. We are concerned with the action of periodic elliptic operators under these transformations. We shall prove that these operators become holomorphic Fredholm morphisms of the bundles, which will enable us to exploit the general results of chapter 1. We adopt the notations and conditions concerning an operator $L(x,D)$ from section 2.1. We already know that it produces some continuous linear operators between $\Phi_{n,p}^{2m,a}$ and $\Phi_{n,p}^{0,a}$ and also between Ψ_n^{2m} and Ψ_n^0. The next theorem describes the images of these operators under our transformations.

Theorem 2.2.3.

i) *Under the transform* $\boldsymbol{F}$ *the operator*

$$L: \Phi_{n,1}^{2m} \to \Phi_{n,1}^{0}$$

becomes the operator of multiplication by some holomorphic morphism $\mathcal{L}:\mathcal{E}^{2m} \to \mathcal{E}^{0}$ *acting on the spaces of sections*

$$\Gamma(\Omega_n, \mathcal{E}^{2m}) \xrightarrow{L} \Gamma(\Omega_n, \mathcal{E}^0).$$

On each fiber the morphism coincides with the restriction to the fiber of the operator L, acting between $H^{2m}(\mathfrak{X})$ and $L_2(\mathfrak{X})$.

ii) *Under the transform $\mathfrak{F}$ the operator L becomes the operator of multiplication by the operator polynomial $L(x,D-i\lambda)$, considered on the corresponding spaces of functions on the torus $\mathbf{T}^n$.*

Proof. Periodicity of L implies that $F(Lf)(z,x) = L(x,D)\, Ff(z,x)$ for $f \in \Phi^{2m}_{n,p}$. This means that the restriction of the operator $F \circ L \circ F^{-1}$ to the fiber over some point z is the operator from $H^{2m}(\mathfrak{X})$ to $L_2(\mathfrak{X})$, generated by L and by the boundary conditions (4). It follows therefore that $F \circ L \circ F^{-1}$ is the multiplication by the restriction to $\mathcal{E}^{2m}$ of the constant morphism $L: \Omega_n \times H^{2m}(\mathfrak{X}) \to \Omega_n \times L_2(\mathfrak{X})$. This proves the statement i). Now, by the definition, the transformation $\mathfrak{F}$ is the composition of $\hat{F}$ with the trivialization of $\hat{\mathcal{E}}^s$ (i.e., with the multiplication by $exp(-\lambda x)$ on the fiber above λ), so $L(x,D)$ becomes the operator $exp(-\lambda x) L(x,D)\, exp(\lambda x)$ acting on $H^{2m}(\mathbf{T}^n)$. Now,

$$exp(-\lambda x)\ L(x,D)\ exp(\lambda x) = L(x,D-i\lambda).$$

This proves the theorem.

Corollary 2.2.4. *If, under the conditions of the theorem, the operator $L(x,D)$ is elliptic (in the meaning of section 2.1), then the corresponding morphism is a Fredholm one.*

Proof. We expand $L(x,D-i\lambda)$ into powers of λ:

$$L(x,D-i\lambda) = L(x,D) + \sum_{\alpha \neq 0} \frac{1}{\alpha!} (\partial_\xi{}^\alpha L\ (x,D))(-i\lambda)^\alpha,$$

where

$$(\partial_\xi{}^\alpha L)(x,\xi) = \frac{\partial^\alpha L(x,\xi)}{\partial \xi^\alpha}.$$

The operator $L(x,D)$ is elliptic and therefore Fredholm, as operator from $H^{2m}(\mathbf{T}^n)$ to

$L_2(\mathbf{T}^n)$ $(=H^0(\mathbf{T}^n))$ (see [262]). The operators $\partial_\xi{}^\alpha L$ are of order less than $2m$ and so, by the embedding theorem, they are compact as operators from $H^{2m}(\mathbf{T}^n)$ into $L_2(\mathbf{T}^n)$. It follows that the operator $L(x,D-i\lambda): H^{2m}(\mathbf{T}^n) \to L_2(\mathbf{T}^n)$, and hence the morphism $\mathcal{L}$, is Fredholm. The proof is complete.

We shall denote by K the following fundamental domain of the lattice $\Gamma = 2\pi\mathbf{Z}^n$ (the dual lattice to $\mathbf{Z}^n$):

$$K = \{x \in \mathbf{R}^n \mid 0 \le x_i \le 2\pi,\ i = 1, \dots, n\}.$$

We denote, as before, by $\mathfrak{K}$ the unit cube in $\mathbf{R}^n$, which is the fundamental domain for the integer lattice. Let us also recall that $\mathfrak{T}^n$ is the skeleton of the n-dimensional polydisk

$$\mathfrak{T}^n = \{z \in \mathbf{C}^n \mid |z_j| = 1,\ j = 1, \dots, n\}.$$

Theorem 2.2.5.

i) The transform $\mathcal{F}$: $f(x) \to \mathcal{F}f(z,x)$ *is an isometry of* $L_2(\mathbf{R}^n)$ *onto*

$$L_2(\mathfrak{T}^n, L_2(\mathfrak{K})) = \int_{\mathfrak{T}^n}^{\oplus} L_2(\mathfrak{K})$$

and of $H^{2m}(\mathbf{R}^n)$ *onto the subspace* $L_2(\mathfrak{T}^n, \mathcal{E}^{2m})$ *in* $L_2(\mathfrak{T}^n, H^{2m}(\mathfrak{K}))$.

ii) The transform $\mathfrak{F}_1$: $f(x) \to \mathfrak{F}f(ip,x)$ *is an isometry of* $L_2(\mathbf{R}^n)$ *onto*

$$L_2(K, L_2(\mathbf{T}^n)) = \int_K^{\oplus} L_2(\mathbf{T}^n)$$

and of $H^{2m}(\mathbf{R}^n)$ *onto*

$$L_2(K, H^{2m}(\mathbf{T}^n)) = \int_K^{\oplus} H^{2m}(\mathbf{T}^n).$$

iii) $\mathcal{F}L(x,D)\mathcal{F}^{-1} = \int_{\mathfrak{T}^n}^{\oplus} L(x, D)\colon L_2(\mathfrak{T}^n, \mathcal{E}^{2m}) \to \int_{\mathfrak{T}^n}^{\oplus} L_2(\mathfrak{K}).$

$$iv)\ \mathfrak{F}_1 L(x,D)\mathfrak{F}_1^{-1} = \int_{p\in K}^{\oplus} L(x,\ D+p):\ \int_K^{\oplus} H^{2m}(\mathbf{T}^n) \to \int_K^{\oplus} L_2(\mathbf{T}^n).$$

Proof.

i) Let $f(x) \in L_2(\mathbf{R}^n)$, then

$$\|f\|^2_{L_2(\mathbf{R}^n)} = \sum_{k\in \mathbf{Z}^n} \|f(x-k)\|^2_{L_2(\mathfrak{X})}. \tag{2.2.7}$$

On the other hand,

$$Ff(z,x) = \sum_{k\in \mathbf{Z}^n} f(x-k)\ z^k \quad (x\in \mathfrak{X}). \tag{2.2.8}$$

Treating (8) as an n-dimensional Fourier series with coefficients in $L_2(\mathfrak{X})$, we get by the Parseval's equality

$$\|Ff\|^2_{L_2(\mathfrak{T}^n,\ L_2(\mathfrak{X}))} = \sum_{k\in \mathbf{Z}^n} \|f(x-k)\|^2_{L_2(\mathfrak{X})}.$$

Now (7) and (8) show that the operator F is an isometry between the spaces under consideration. We have only to prove its surjectivity. Let $h(z,x) \in L_2(\mathfrak{T}^n,\ L_2(\mathfrak{X}))$ and

$$h(z,x) = \sum h_k(x)\ z^k \quad (h_k \in L_2(\mathfrak{X}))$$

be its Fourier series on $\mathfrak{T}^n$. We define $f(x)$ on $\mathbf{R}^n$ as

$$f(x) = h_k(x) \ \text{ for } x\in \mathfrak{X} - k,\ k\in \mathbf{Z}^n.$$

Then

$$\int_{\mathbf{R}^n} |f(x)|^2\, dx = \sum_{k\in \mathbf{Z}^n} \int_{\mathfrak{X}} |h_k(x)|^2\, dx = \int_{\mathfrak{T}^n} \|h(z,\ \cdot)\|^2_{L_2(\mathfrak{X})}\, dz$$

and $Ff = h$.

Analogous considerations show that F generates an isometry of $H^{2m}(\mathbf{R}^n)$ into $L_2(\mathfrak{T}^n, H^{2m}(\mathfrak{X}))$. In this case, however, the map is not surjective. The proof of theorem

2.2.2 leads to the conclusion that the image of $\boldsymbol{F}$ on the space $H^{2m}(\mathbf{R}^n)$ contains exactly those functions $h(z,x) \in L_2(\mathcal{T}^n, H^{2m}(\mathcal{K}))$ that satisfy the condition

$$h(z,\cdot) \in E_z^{2m}$$

for all $z \in \mathcal{T}^n$.

ii) We observe that transform $\mathfrak{F}_1$ is the composition of the transform $\boldsymbol{F}$, of the substitution $z = e^{ip}$ and of the multiplication by $exp(-ipx)$. The composition of the last two transforms obviously is an isometry of $L_2(\mathcal{T}^n, L_2(\mathcal{K}))$ onto $L_2(K, L_2(\mathcal{K})) = L_2(K, L_2(\mathbf{T}^n))$. It is also an isometry of $L_2(\mathcal{T}^n, \mathcal{E}^{2m})$ onto $L_2(K, H^{2m}(\mathbf{T}^n))$. Combining with i), we get the statement ii). Theorem 3 provides iii) and iv). This finishes the proof.

2.3. Comments and references

As we have already mentioned, the transforms $\boldsymbol{F}$, $\widehat{\boldsymbol{F}}$ and $\mathfrak{F}$ are different versions of a transform that plays the role of the Fourier transform in Floquet theory. In fact, $\boldsymbol{F}$ can be considered as the Fourier transform of the abstract function $f_k(\cdot) = f(\cdot - k)$ on the group $\mathbf{Z}^n$. As the reader will see throughout the book, it is important to allow non-unitary representations, i.e. complex values of quasimomentums, or the values of z that do not belong to $\mathcal{T}^n$. The first known to the author application of these transforms to PDEs in mathematical literature was in I. Gelfand's paper [126]. Physicists have been using these transforms for similar purposes in quantum theory of solids [18].

It is interesting to mention that these transforms (sometimes called *Zak transform*) naturally arise in harmonic analysis on nilpotent homogeneous manifolds, in theory of radars and in wavelets' theory [23, 24, 295].

CHAPTER 3

FLOQUET THEORY FOR HYPOELLIPTIC EQUATIONS AND SYSTEMS IN THE WHOLE SPACE

This chapter is the place where the analytic tools developed in chapters 1 and 2 start to work for periodic partial differential equations. The main notions of Floquet theory (Floquet-Bloch solutions, quasimomentums, etc.) are introduced in section 3.1. In the same section we study analytic properties of the sets of quasimomentums and Floquet exponents. The main theorem that provides a Floquet type expansion for solutions of scalar periodic elliptic equations with smooth coefficients is proved in section 3.2. This theorem deals only with solutions of exponential growth. Solutions of some faster growth are considered in section 3.3. We describe in section 3.4 how to reduce the restrictions on equations. The last section contains some comments and references.

3.1. Floquet-Bloch solutions. Quasimomentums and Floquet exponents

Now our task is to restate the various concepts of Floquet theory in terms of bundle morphisms.

The main notion in the theory is the one of Floquet-Bloch solution. (In the mathematical literature one can mainly find the name "Floquet solution", but physicists use "Bloch solution" instead. We combine here both names and emphasize a more general character of Floquet solutions.)

We denote by $L(x,D)$, as in the preceding chapter, an elliptic operator of order $2m$ with smooth periodic (with period 1 in each variable) coefficients in $\mathbf{R}^n$.

Definition 3.1.1. A *Floquet solution* of the equation $L(x,D)u=0$ on $\mathbb{R}^n$ is a solution of the form

$$(3.1.1) \qquad u(x) = exp\ (i\lambda x)\left(\sum_{k \in \mathbb{Z}^n_+} x^k g_k(x)\right),$$

where the sum is finite and the functions $g_k(x)$ are 1-periodic in each x_j. Here $x^k = x_1^{k_1} \cdot \ldots \cdot x_n^{k_n}$ and $\lambda x = \lambda_1 x_1 + \ldots + \lambda_n x_n$. The complex vector $\lambda \in \mathbb{C}^n$ is the *quasimomentum* and $z = exp\ i\,\lambda \in \Omega_n$ is the *Floquet exponent* (or *Floquet multiplier*) of the solution. (Here $exp(i\lambda) = (e^{i\lambda_1}, \ldots, e^{i\lambda_n})$).

If the sum has only the term with k=0, then $u(x)$ is called a *Bloch solution.*

Bloch and Floquet solutions are related in the same way as eigen and adjoint vectors of a linear operator, and in the case of equations of evolutionary type they really correspond to such vectors for the monodromy operator (see chapter 5).

On the other hand, Bloch and Floquet solutions are analogs of exponential and exponential-polynomial solutions of equations with constant coefficients (see [99, 324]).

We shall interpret solutions of our equation in some weak sense as functionals from the cokernel of the transpose operator L_1, that is, the operator for which $\int (Lu)v dx = \int uL_1 v dx$ for all $u, v \in C_0^\infty(\mathbb{R}^n)$. Clearly, L_1 is also 1-periodic in each x_j and has smooth coefficients. Let $\mathcal{L}_1$ be the bundle morphism corresponding to L_1 by theorem 2.2.3. Let also $u(x)$ be a solution of the equation

$$(3.1.2) \qquad L(x, D)u(x) = 0$$

satisfying the estimate

$$(3.1.3) \qquad |u(x)| \le C\ exp\ a|x|^p$$

for some $p \ge 1$, $a > 0$. We regard $u(x)$ (which is obviously smooth) as a generalized function (that is, as a continuous linear functional on $\mathfrak{D} = C_0^\infty(\mathbb{R}^n)$). As in section 2.2, the operator L_1 maps continuously $\Phi^{2m}_{n,p}$ into $\Phi^0_{n,p}$.

Lemma 3.1.2. *The solutions of* (2) *that satisfy* (3) *are precisely those elements of* $\mathfrak{D}'(\mathbb{R}^n)$ *that extend by continuity from* $C_0^\infty(\mathbb{R}^n)$ *to functionals*

(3.1.4) $$u \in (Im\ L_1)^{\perp} \subset (\Phi^0_{n,p})^*.$$

Proof. First of all, if $u(x)$ satisfies (3), it can obviously be continuously extended to a functional from $(\Phi^0_{n,p})^*$. If $\varphi \in C_0^{\infty}(\mathbb{R}^n)$, then $\int (Lu(x))\varphi(x)dx = \int u(x)L_1\varphi(x)dx$. According to the estimate (3) and conditions on L, both sides of the equality are continuous linear functionals with respect to $\varphi \in \Phi^{2m}_{n,p}$. Hence, the equality holds for all $\varphi \in \Phi^{2m}_{n,p}$. If $u(x)$ satisfies (2), then

(3.1.4) $$u \in (Im\ L_1)^{\perp} \subset (\Phi^0_{n,p})^*.$$

Now let us assume that (4) holds. Since $C_0^{\infty}(\mathbb{R}^n)$ is continuously embedded into $\Phi^0_{n,p}$, $u(x)$ belongs to $\mathfrak{D}'(\mathbb{R}^n)$. Thus, if $u \in (Im\ L_1)^{\perp}$, then u is a weak solution of the equation $Lu = 0$. Due to ellipticity of L, $u(x)$ belongs to $C^{\infty}(\mathbb{R}^n)$ and is a classical solution to (2). We have to prove only the estimate (3). We need some kind of Schauder estimate for this.

According to the condition (4), there exist constants $a > 0$, $C > 0$ such that

$$\left| \int u(x)\ \varphi(x)\ dx \right| \le C\gamma_a(\varphi)$$

for all $\varphi \in \Phi^0_{n,p}$. Let $\mathcal{K}_1$ be a cube containing the unit cube $\mathcal{K}$ in its interior, and $\mathcal{K}_1 + k$ be its translation on $k \in \mathbb{R}^n$. Let $\chi(x)$ be the characteristic function of the cube $\mathcal{K}_1$:

$$\chi(x) = \begin{cases} 1 & \text{for } x \in \mathcal{K}_1 \\ 0 & \text{otherwise.} \end{cases}$$

We define

$$\psi_k(x) = \chi(x-k)\ \overline{u}(x) / \| u \|_{L_2(\mathcal{K}_1 + k)},$$

where u is not identically zero in $\mathcal{K}_1 + k$, and $\psi_k \equiv 0$ otherwise. One can see that $\psi_k \in \Phi^0_{n,p}$ and $\gamma_a(\psi_k) \le C\ exp\ a'\ |k|^p$, so

$$(3.1.5)\qquad ||u||_{L_2(\mathcal{K}_1+k)} = |\int u(x)\,\psi_k\,(x)\,dx| \le C\gamma_a(\psi_k) \le C\,exp\,a'\,|k|^p.$$

In other words, we have an estimate of the type (3), however only in L_2 sense. In the usual way, ellipticity enables us to obtain (3) in the strong sense. Indeed, let B be a proper periodic pseudodifferential parametrix for L (see lemma 2.1.1). This means that $BL = I+T$, where T is a proper infinitely smoothing operator. In particular, T maps continuously $L_{2,loc}(\mathbb{R}^n)$ into $H^\nu_{loc}(\mathbb{R}^n)$ for every ν. Hence, there exists a cube $\mathcal{K}_1 \subset \mathbb{R}^n$, such that

$$||Tf||_{H^\nu(\mathcal{K})} \le C\,||f||_{L_2(\mathcal{K}_1)}.$$

Due to periodicity,

$$||\,Tf||_{H^\nu(\mathcal{K}+k)} \le C\,||\,f\,||_{L_2(\mathcal{K}_1+k)}$$

for all $k \in \mathbb{Z}^n$. Now let us return to the solution $u(x)$. We have $BLu=B0=0$, so $u = -Tu$ and

$$||u||_{H^\nu(\mathcal{K}+k)} = ||\,Tu||_{H^\nu(\mathcal{K}+k)} \le C||u||\;_{L_2(\mathcal{K}_1+k)}.$$

According to (5), for every ν we obtain the estimate

$$||u||_{H^\nu(\mathcal{K})} \le C_\nu\,exp\,a|k|^p.$$

If we choose a sufficiently large ν, then the embedding theorem yields the estimate

$$|u(x)| \le C\;exp\;a|k|^p \quad (x \in \mathcal{K} + k),$$

which is in fact equivalent to (3) (with another value of $a>0$). The lemma is proved.

Thus, in order to describe all the solutions of (2) subject to (3), we have to describe the cokernel of $L_1\colon \Phi^{2m}_{n,p} \to \Phi^0_{n,p}$.

Our next step is to distinguish functionals that correspond to Floquet-Bloch functions among all elements of $(\Phi^0_{n,p})^*$. It is convenient to use first our "Fourier transform" $\boldsymbol{F}$. We shall see that these "Fourier images" of Floquet-Bloch functions behave like the usual Fourier images of exponential-polynomial functions (the latter are just derivatives of δ-functions).

Theorem 3.1.3. *The transform $\boldsymbol{F}$ gives rise to a one-to-one correspondence between the set of Floquet functions* (1) (*considered as elements of* $(\Phi^0_{n,1})^*$) *with Floquet exponent $z_0 \in \Omega_n$ and the set of functionals $\varphi \in \Gamma(\Omega_n, \mathcal{E}^0)^*$ of the form*

$$\langle \varphi, y \rangle = \sum_{l=1}^{r} P_l\left(\frac{\partial}{\partial z}\right) \langle t_l(z), y(z)\rangle \Big|_{z = z_0^{-1}}, \tag{3.1.6}$$

where $P_l\left(\frac{\partial}{\partial z}\right)$ are linear differential operators and $t_l(z)$ are holomorphic sections of $(\mathcal{E}^0)^*$ in a neighborhood of z_0^{-1} (where $z^{-1} = (z_1^{-1}, \ldots, z_n^{-1})$). In the case of Bloch solutions the orders of $P_l\left(\frac{\partial}{\partial z}\right)$ are equal to zero. If the coefficients of P_l depend holomorphically on z_0, then the corresponding Floquet functions (obtained by varying z_0^{-1}) depend holomorphically on the Floquet exponent.

Proof. A function of Floquet-Bloch form is a linear combination of functions of the form $z_0^x x^k g(x)$, where $g(x)$ is periodic (k=0 in the Bloch case). On the other hand, a functional of the form (6) is a linear combination of functionals

$$\frac{\partial^k}{\partial z^k} \langle t, y(z)\rangle \Big|_{z = z_0^{-1}}, \tag{3.1.7}$$

where $t \in L_2(\mathcal{K})$. Putting $t(x) = g(x) z^{-x}$, where $g(x)$ is extended by periodicity to $\mathbb{R}^n$, and applying (7) to $y = \boldsymbol{F}f$, we rewrite this expression in the form

$$\left(\frac{\partial}{\partial z}\right)^k \left(\int_{\mathcal{K}} t(x) \sum_k f(x-k) z^k dx \right) \Bigg|_{z = z_0^{-1}} =$$

$$\left(\frac{\partial}{\partial z}\right)^k \left(\int_{\mathcal{K}} g(x) \sum_k f(x-k) z^{k-x} dx \right) \Bigg|_{z = z_0^{-1}} =$$

$$\left(\frac{\partial}{\partial z}\right)^k \int_{\mathbb{R}^n} g(x) z^{-x} f(x) dx \Bigg|_{z = z_0^{-1}} =$$

$$\int_{\mathbf{R}^n} g(x) z_0^x A_k(x) f(x) dx,$$

where $A_k(x)$ is a polynomial. This means that our functional corresponds to a function that can be expanded in terms of functions $z_0^x x^k g(x)$.

For the converse argument, we observe that the set of functions $A_k(x) = z^x \left(\frac{\partial}{\partial z}\right)^k (z^{-x}) = (-1)^k \; x(x+1) \ldots (x+k-1)$ provides a basis in the space of polynomials. Indeed, the leading term of $A_k(x)$ is $(-1)^k \; x^k$. This means that a function of the form $z_0^x x^k g(x)$ can be expanded in terms of functions $z_0^x A_k(x) g(x)$.

The last two assertions of the theorem - on Bloch functions and holomorphic property - can be deduced easily from the preceding arguments. This proves the theorem.

An immediate consequence of lemma 2 and theorem 3 is the following statement.

Corollary 3.1.4. *The set of all Floquet-Bloch solutions of the equation $Lu=0$ corresponds via the transform F to the set of all functionals $\varphi \in (Im\ \mathcal{L}_1)^{\perp} \subset (\Gamma(\Omega_n, \mathcal{E}^0))^*$ of the form* (6).

When $M \subset \Omega_n$, we denote by M^{-1} the set

$$M^{-1} = \{z \in \Omega_n : z^{-1} = (z_1^{-1}, \ldots, z_n^{-1}) \in M\}.$$

Theorem 3.1.5. *Let N (correspondingly, $\widetilde{N}$) be the set of Floquet exponents of all non-zero Floquet (correspondingly, Bloch) solutions of* (2). *Then*

$$N = \widetilde{N} = S(\mathcal{L}) = (CS(\mathcal{L}_1))^{-1}.$$

Here $S(\mathcal{L})$ (correspondingly, $CS(\mathcal{L}_1)$) is the spectrum (cospectrum) of the bundle morphism $\mathcal{L}$ ($\mathcal{L}_1$) that arises from the operator L (L_1) under the transform F (see theorem 2.2.3).

Proof. The inclusion $N \supset \widetilde{N}$ is obvious, because the set of all Floquet solutions includes all Bloch solutions. We shall prove that $\widetilde{N} \supset S(\mathcal{L}) \supset (CS(\mathcal{L}_1))^{-1} \supset N$. Let $z \in S(\mathcal{L})$, so the morphism $\mathcal{L}$ has non-zero kernel on the fiber over the point z. By the definition of the morphism and the bundle, this means that $L(x,D)$ has non-zero kernel on the space of functions of the type $z^x g(x)$ (with periodic $g(x)$). Hence, $z \in \widetilde{N}$. Now let $z_0 \in (CS(\mathcal{L}_1))^{-1}$. Then the image of the morphism $\mathcal{L}_1$ on the fiber over z_0^{-1} is not dense. Let t be a functional on the fiber, orthogonal to the image. In this case the functional

$$\langle t, y(z)\rangle \Big|_{z=z_0^{-1}}$$

on $\Gamma(\Omega_n, \mathcal{E}^0)$ annihilates the image of $\mathcal{L}_1$. According to corollary 4, there exists a non-zero Bloch solution with Floquet exponent z_0. The existence of non-zero kernel of $\mathcal{L}$ on the fiber over z_0 obviously means the same, so $z_0 \in S(\mathcal{L})$.

Consider now $z_0 \in N$. By corollary 4, there is a functional φ of the form (6) that annihilates $Im\ \mathcal{L}_1$. Hence, the morphism $\mathcal{L}_1$ is not surjective on the fiber above z_0^{-1}. For if $\mathcal{L}_1$ is surjective here, then it is also surjective in a polydisk neighborhood of z_0^{-1} (see theorem 1.6.16). Let U be such a neighborhood. Then the sheaf homomorphism $\mathcal{L}_1^0\colon \mathcal{O}^{\mathcal{E}^{2m}}|_U \to \mathcal{O}^{\mathcal{E}^0}|_U$ is surjective (this easily follows from the theorem 1.6.1). Note that any germ of a section of $\mathcal{O}^{\mathcal{E}^{2m}}|_U$ can be approximated by elements of $\Gamma(\Omega_n, \mathcal{E}^{2m})$ (see theorem 1.5.29). Therefore, any germ in $\mathcal{O}^{\mathcal{E}^0}|_U$ can be approximated by elements of $Im\ \mathcal{L}_1 \subset \Gamma(\Omega_n, \mathcal{E}^0)$. It follows that our functional φ annihilates any germ of section at z_0^{-1}. This trivially implies that φ=0, so the theorem is proved.

Corollary 3.1.6. *The set N of Floquet exponents of* (2) *is a principal analytic subset of Ω_n.*

Proof. By the Atiyah-Singer theorem [22], the Fredholm index of L on $\mathbb{T}^n$ is zero. Therefore, $\mathcal{L}$ is a morphism of zero index (on each fiber). Then, by theorem 1.6.16, $S(\mathcal{L})$ is a principal analytic subset of Ω_n. This proves the corollary.

In fact, a stronger assertion holds, and it will play a very important role later. It is more convenient to pass on to quasimomentums instead of Floquet exponents.

Theorem 3.1.7. *The set $Q \subset \mathbb{C}^n$ of quasimomentums of all non-zero Floquet solutions of* (2) *is the set of all zeros of an entire function Δ of order n (generally speaking, of infinite type). (We do not exclude the case when the function is identically zero, so $Q = \mathbb{C}^n$, see section 4.1.) Increasing the order to $n+1$, we can assume that Δ is $2\pi\mathbb{Z}^n$-periodic.*

Proof. When $Q=\mathbf{C}^n$, the assertion is trivial, so we assume that $Q\neq\mathbf{C}^n$. Due to theorems 5 and 2.2.3,

$$Q=\{\lambda\in\mathbf{C}^n:\ Ker\ L(x,D+\lambda)\neq 0\},$$

where the kernel of $L(x,D+\lambda)$ is considered in the space $H^{2m}(\mathbf{T}^n)$. As it was shown in the proof of corollary 2.2.4, the operator polynomial $L(\lambda)=L(x,D+\lambda)$ has the form

$$L(\lambda)=L_0+\sum_{|j|>0}\lambda^j L_j,$$

where $L_0=L(x,D)$ and the order of L_j is $2m-|j|$. Suppose that the $L(0)=L_0$ is invertible (this can be achieved by a shift with respect to λ, because $Q\neq\mathbf{C}^n$). Then $L(\lambda)=L_0(I+\sum\lambda^j L_0^{-1}L_j)$. The order of $L_0^{-1}L_j$ is $-|j|$. Let S be an invertible pseudodifferential operator of order -1 on $\mathbf{T}^n$. Then $L(\lambda)=L_0(I+\sum\lambda^j B_j S^{|j|})$, where $B_j=L_0^{-1}L_j S^{-|j|}$ are operators of order 0, and therefore bounded on all spaces $H^s(\mathbf{T}^n)$. Due to lemma 1.2.18, S belongs to the class $\mathcal{S}_p$ (for any $p>n$) in $L_2(\mathbf{T}^n)$. It follows that the spectrum of this operator polynomial is the set of all zeros of the regularized determinant $\Delta(\lambda)$ (see lemma 1.2.21), which is an entire function of order p, as required. Now we discuss the problem of periodicity of Δ. The operators $L(\lambda)$ and $L(\lambda+2\pi k)$ are similar, provided $k\in\mathbf{Z}^n$. Indeed, the equivalence is provided by the multiplication by $exp\ 2\pi ik$. Thus, due to the construction of the function Δ, we conclude that $\Delta(\lambda)$ and $\Delta(\lambda+2\pi k)$ have the same zeros (including multiplicities). Hence, the function $s_k(\lambda)=\Delta(\lambda+2\pi k)/\Delta(\lambda)$ is an entire function of order n without zeros. Due to Hadamard's theorem [256, 390], $s_k(\lambda)$ can be represented as $exp\ G_k(\lambda)$, where $G_k(\lambda)$ is a polynomial of degree n. We denote by $g_i(\lambda)$ the polynomial $G_{k_i}(\lambda)$, where k_i $(i=1,\ \dots,\ n)$ is the standard basis in $\mathbf{R}^n$. Lemma 1.5.30 shows that there exists a polynomial $g(\lambda)$ of order $n+1$ such that $g(\lambda+2\pi k_i)-g(\lambda)=g_i(\lambda)$ $(i=1,\ \dots\ ,\ n)$. We define now $\Delta_1(\lambda)=\Delta(\lambda)\ exp\ (-g(\lambda))$. Then obviously $\Delta_1(\lambda)$ has the required periodicity and the same set of zeros as $\Delta(\lambda)$. This proves the theorem.

3.2. Floquet expansions of solutions of exponential growth

We present here an analog of the Floquet theorem for periodic elliptic PDEs. We obtain an expansion into Floquet solutions for solutions of exponential growth that are defined on the whole space $\mathbf{R}^n$. The reader can find another result in this direction in the next section.

Theorem 3.2.1. *Let $L(x,D)$ be a scalar periodic elliptic operator in $\mathbb{R}^n$, and N be the set of Floquet exponents of* (3.1.2). *Then there exist:*

a) a locally finite covering of N by analytic subsets $N_k \subset N$;

b) for each k a finite set of functions $f_{j,k}(z,x)$ $(x \in \mathbb{R}^n, z \in N_k)$ such that:

1) $f_{j,k}(z,x)$ is a holomorphic function on N_k with values in $C^{\infty}_{loc}(\mathbb{R}^n)$;

2) $f_{j,k}(z,x)$ for fixed $z \in N_k$ is a Floquet solution of (3.1.2) *with Floquet exponent z;*

3) any solution of (3.1.2) *on $\mathbb{R}^n$ that satisfies $|u(x)| \leq C \; exp \; a \; |x|$ can be represented as a finite sum*

$$u(x) = \sum_{j,k} \int_{N_k} f_{j,k}(z,x) d\mu_{j,k}(z), \tag{3.2.1}$$

where $\mu_{j,k}$ are finite measures with compact supports in N_k; if $S_k \subset N_k$ are any given closed proper analytic subsets (for instance, empty), then for any $b > a$ the measure $\mu_{j,k}$ can be chosen as $m_{j,k}(z) \; dV$, where $m_{j,k}$ is a smooth compactly supported function on

$$(reg \; N_k \setminus S_k) \cap \Omega_b$$

and dV is the standard Lebesgue measure on $reg \; N_k$;

4) any finite collection of such measures determines (*by* (1)) *a solution of* (3.1.2) *of exponential growth.*

Proof. Let us apply the transform F. By corollary 3.1.4, solutions of (3.1.2) of exponential growth correspond to functionals $\varphi \in (Im \; \mathcal{L}_1)^{\perp} \subset \Gamma(\Omega_n, \mathcal{E}^0)^*$. In fact, when $|u(x)| \leq C \; exp \; a|x|$, we can consider here $\Omega_n(b)$ instead of Ω_n for any $b > a$. By theorem 2.2.3, we are in the conditions of theorem 1.7.1. Due to theorem 3.1.5, the support of the sheaf $Coker \; \mathcal{L}_1$ is $CS(\mathcal{L}_1) = N^{-1}$. Here N is the set of Floquet exponents for our equation. According to theorem 1.7.1, φ can be represented as

$$\varphi(y) = \sum_k \int_{N_k^{-1}} < \nu_k\, Cy,\, \mu_k > dV_k,$$

where $\{N_k^{-1}\}$ is a locally finite cover of N^{-1} by analytic subsets and μ_k are $\mathbb{C}^{r_k}$-valued compactly supported C^∞-functions on $N_k^{-1} \setminus (S_k^{-1} \cup sing\ N_k^{-1})$. Let us denote by $\delta_{z^{-1}}$ the δ-function at the point z^{-1} defined on $C_0^\infty(reg\ N_k^{-1})$. Let also $\nu_k = (\nu_{k,1}, \dots, \nu_{k,r_k})$ be the coordinate representation of ν_k. We define now the following $\mathbb{C}$-linear functional on $\Gamma(\Omega_n(b), \mathcal{E}^0)$:

$$<\psi_{j,k}(z),\, y> = <\delta_{z^{-1}},\, \nu_{k,j}\, Cy> .$$

Due to corollary 1.7.2, $\psi_{j,k}$ corresponds to a solution $f_{j,k}(z,x)$ of equation 3.1.2. Let us investigate the structure of this solution. Consider the subspace of $\Gamma(\Omega_n(b), \mathcal{O}^{\mathcal{E}^0}) = \mathcal{H}(\Omega_n(b), L_2(\mathbb{T}^n))$ consisting of analytic functions that have zero of order $d+1$ at z^{-1}, where d is the order of differential operator ν_k at this point. Then $\psi_{j,k}$ is zero on this subspace. The corresponding quotient space is obviously isomorphic to the space of $L_2(\mathbb{T}^n)$-valued Taylor polynomials of order d at z^{-1}. This means that $\psi_{j,k}$ can be represented as

$$<\psi_{j,k},\, y> = \sum_{|\alpha| \le d} <\frac{\partial^\alpha y}{\partial z^\alpha},\, t_\alpha>|_{z^{-1}},$$

where $t_\alpha \in L_2(\mathbb{T}^n)$. Now theorem 3.1.3 shows that $f_{j,k}$ has the required structure. This finishes the proof.

3.3. Completeness of Floquet solutions in a class of solutions of faster growth

We now turn to the problem of completeness of Floquet solutions in a class of solutions of faster growth. Nothing about expansions is known here (compare with the evolutionary cases in chapter 5, where some expansions are obtained).

Theorem 3.3.1. *Suppose that the set of Floquet exponents N of* (3.1.2) *does not coincide with Ω_n (or, equivalently, that the set of quasimomentums Q does not coincide with $\mathbb{C}^n$). Then, for any $p < 1 + (n-1)^{-1}$ the linear hull of Floquet solutions of* (3.1.2) *is dense in the topology of* $C^{\infty}_{loc}(\mathbb{R}^n)$ *in the set of all solutions with the bound*

$$|u(x)| \leq C \; exp \; a|x|^{\,p} \tag{3.3.1}$$

(here C and a may depend on u).

Proof. It is sufficient to prove that Floquet solutions are complete in the $L_{2,loc}(\mathbb{R}^n)$ topology. Indeed, let B be a proper pseudodifferential parametrix for L (see lemma 2.1.1). If $u(x)$ is a solution of the equation $Lu=0$, then $BLu=0$, that is $u=(I-BL)u$. Since the operator $(I-BL)$ is proper, it maps $L_{2,loc}(\mathbb{R}^n)$ continuously into $H^s_{loc}(\mathbb{R}^n)$ for any $s \geq 0$. By the embedding theorem, completeness in $C^{\infty}_{loc}(\mathbb{R}^n)$ follows.

Now we prove completeness in $L_{2,loc}$ sense. Suppose the contrary. Then there exist $\varphi(x) \in L_{2,comp}(\mathbb{R}^n)$ and a solution $u(x)$ of (3.1.2) satisfying (1) (for some $p < 1+(n-1)^{-1}$), such that $\int \varphi(x)u(x)dx \neq 0$, while $\int \varphi(x)f(x)dx=0$ for any Floquet solution $f(x)$ of (3.1.2).

Note that obviously $\varphi \in \Phi^0_{n,q}$ for any q. We apply the transformation $\boldsymbol{F}$. Then orthogonality of $\varphi(x)$ to all Floquet solutions, by theorem 3.1.3 and corollary 3.1.4, reads as follows: the section $(\boldsymbol{F}\varphi)(z)$ is annihilated by all functionals of the form (3.1.6) that annihilate the image of $\mathcal{L}_1$. But then, by corollary 1.7.2, the section $\boldsymbol{F}\varphi$ lies in the image of $\mathcal{L}_1$. Going back to $\varphi(x)$, we see that $\varphi(x) = L_1\psi(x)$, where $\psi(x) \in \Phi^0_{n,1}$. Now we claim that we can ensure much faster rate of decay for $\psi(x)$. Indeed, when we perform the transformation $\mathfrak{F}$: $\mathfrak{F}\varphi(\lambda, x) = e^{-\lambda x}\boldsymbol{F}\varphi(e^{\lambda}, x)$, $\mathfrak{F}\psi(\lambda, x) = e^{-\lambda x}\boldsymbol{F}\psi(e^{\lambda}, x)$, we obtain the following equality for the entire vector-valued functions $\mathfrak{F}\psi(\lambda)$ and $\mathfrak{F}\varphi(\lambda)$ (with values in $H^{2m}(\mathbb{T}^n)$ and $L_2(\mathbb{T}^n)$, respectively):

$$L_1(x, D-i\lambda)\mathfrak{F}\psi(\lambda, x) = \mathfrak{F}\varphi(\lambda, x).$$

We denote the operator polynomial $L_1(x, D-i\lambda)$ by $L_1(\lambda)$: $H^{2m}(\mathbb{T}^n) \to L_2(\mathbb{T}^n)$. Notice that $\mathfrak{F}\varphi(\lambda)$ belongs to $A_q(\mathbb{C}^n, L^2(\mathbb{T}^n))$ for any $q>1$, since $\varphi \in \Phi^0_{n,q}$ for any $q>1$, and multiplication by $e^{-\lambda x}$ is an entire operator function of order 1 on $L_2(\mathbb{T}^n)$. We shall prove that $\mathfrak{F}\psi \in A_q(\mathbb{C}^n, H^{2m}(\mathbb{T}^n))$ for any $q > n$. For this purpose we note that the index of $L_1(\lambda)$ on $\mathbb{T}^n$ is zero (as it was mentioned earlier, this follows from the Atiyah-

Singer formula). By the hypothesis of the theorem, $CS(L_1) \neq \Omega_n$. Thus $L_1(\lambda)$ is invertible at some point $\lambda_0 \in \Omega_n \backslash CS(L_1)$. Then, due to lemma 1.2.21, $L_1^{-1}(\lambda) = B(\lambda)/\Delta(\lambda)$, where $\Delta(\lambda)$ is a scalar and $B(\lambda)$ is an operator entire function of order n (here $B(\lambda)$ acts from $L_2(\mathbf{T}^n)$ into $H^{2m}(\mathbf{T}^n)$). Hence, $\mathfrak{F}\psi(\lambda)=B(\lambda)\mathfrak{F}\varphi(\lambda)/\Delta(\lambda)$. If now $q>n$, we have $y(\lambda) = B(\lambda)\mathfrak{F}\varphi(\lambda) \in A_q(\mathbf{C}^n, H^{2m}(\mathbf{T}^n))$ and the ratio $y(\lambda)/\Delta(\lambda)$ is an entire function. In this case (see theorem 1.5.6), the ratio (which is in fact equal to $\mathfrak{F}\psi(\lambda)$ belongs to $A_q(\mathbf{C}^n, H^{2m}(\mathbf{T}^n))$. By theorem 2.2.2, $\psi \in \Phi^{2m}_{n,q/(q-1)}$ for any $q>n$. In other words, $\psi \in \Phi^{2m}_{n,p}$ for any $p < 1+(n-1)^{-1}$.

Suppose now that $|u(x)| \leq C\, exp\; a|x|^p$ with $p< 1+(n-1)^{-1}$. We choose an $\epsilon>0$ such that $p_1 = p+\epsilon < 1+(n-1)^{-1}$. Then $\psi \in \Phi^{2m}_{n,p_1}$ and $L_1\psi=\varphi$. This implies that $\int (L_1\psi)\, u\, dx = \int \varphi u\, dx \neq 0$. If we can prove that the integration by parts $\int (L_1\psi) u\, dx = \int \psi(Lu) dx$ is permitted, we arrive at a contradiction: $\int (L_1\psi) u dx = \int \psi(Lu) dx = 0$ (because $u(x)$ is a solution of (3.1.2)).

We show that this integration by parts is possible. Let $\chi(x)$ be a "cap": $\chi \in C_0^\infty(\mathbf{R}^n)$ and $\chi=1$ in a neighborhood of zero. We denote by $\psi_\epsilon(x)$ the function $\chi(\epsilon x)\ \psi(x)$. It is easy to show that $\int (L_1\psi_\epsilon)\ u dx = \int \psi_\epsilon(Lu) dx$. This follows for $\psi \in C^\infty(\mathbf{R}^n)$ from the definition of L_1 and by continuity for $\psi \in H^{2m}_{loc}(\mathbf{R}^n)$: both sides are continuous functionals of $\psi_\epsilon \in H^{2m}_{comp}(\mathbf{R}^n)$. Hence, $\int (L_1\psi_\epsilon) u dx = \int \psi_\epsilon(Lu) dx = 0$. If we can show that $\int (L_1\psi) u dx = \lim_{\epsilon \to 0} \int (L_1\psi_\epsilon) u dx$, we obtain the required contradiction.

We estimate

$$
|\int (L_1\psi) u dx - \int (L_1\psi_\epsilon) u dx| = |\int L_1(\psi - \psi_\epsilon) u dx|
$$

$$
\leq \sum_{k \in \mathbf{Z}^n} |\int_{\mathfrak{K}+k} L_1(\psi - \psi_\epsilon) u dx| \leq C \sum_{k \in \mathbf{Z}^n} \|\psi - \psi_\epsilon\|_{H^{2m}(\mathfrak{K}+k)} \|u\|_{L_2(\mathfrak{K}+k)}.
$$

It is not hard to see that $\|\psi_\epsilon\|_{H^{2m}(\mathfrak{K}+k)} \leq C\|\psi\|_{H^{2m}(\mathfrak{K}+k)}$, where C is independent on ε. Moreover, $\psi \in \Phi^{2m}_{n,p_1}$, that is, $\|\psi - \psi_\epsilon\|_{H^{2m}(\mathfrak{K}+k)} \leq C\ exp(-a|k|^{p_1})$. Note also that, as $\varepsilon \to 0$, $(\psi - \psi_\epsilon)$ is equal to zero on an expanding sequence of neighborhoods of zero (that absorb the whole of $\mathbf{R}^n$ as $\varepsilon \to 0$). Thus, the last sum is bounded from above by the remainders of the convergent series

$$\sum Cexp(-a_1|k|^{p_1}+a_2|k|^{p}),$$

that is

$$\lim_{\epsilon\to 0}\int_{\mathbb{R}^n}(L_1\psi_\epsilon)u\,dx=\int_{\mathbb{R}^n}(L_1\psi)u\,dx.$$

The theorem is proved.

Remark. The condition $N\neq\Omega_n$ is analyzed in section 4.1.

3.4. Other classes of equations

We have so far considered only scalar elliptic equations with smooth coefficients. It is not hard, however, to obtain some analogs of these results for elliptic systems, hypoelliptic equations and systems, pseudodifferential equations and also to reduce the smoothness conditions on coefficients of the equation. We shall briefly describe the changes that are necessary in these cases, and the corresponding results.

A. Elliptic systems. Let r be a natural number and $s=(s_1,\dots,s_r)\in\mathbf{Z}^r_+$ be a multi-index. We can introduce in the natural way H^s – spaces of vector valued functions, i.e. the spaces of vector functions $f=(f_1,\dots,f_r)$ such that each f_j belongs to the corresponding H^{s_j} – space. We can also define the spaces $\Phi^s_{n,p}$ of vector functions.

Now we recall the notion of ellipticity in the Douglis-Nirenberg sense.

Let $L(x,D)$ be a $r\times r$ matrix linear differential operator with smooth coefficients. We denote by L_{ij} the entries of L. The operator L is said to be *elliptic* if there are multi-indices s and t such that the degree of each L_{ij} is less or equal to (t_j-s_i), and if we keep in the symbol matrix $L(x,\xi)$ only the terms L_{ij} of the order (t_j-s_i), the determinant $L_0(x,\xi)$ of the resulting matrix is not equal to zero for nonzero values of $\xi\in\mathbb{R}^n$. Such operator obviously acts from $\Phi^t_{n,p}$ to $\Phi^s_{n,p}$.

All constructions of section 3.1 remain the same. We have to point out only one important difference. For a *matrix* linear partial differential operator on the torus $\mathbf{T}^n$ it is not longer true that its Fredholm index is zero (see [22]). Some of the results, however, were based on this fact.

The statement of corollary 3.1.6 holds only under the assumption that the index of L on $\mathbb{T}^n$ is zero, otherwise we have to drop the word "principal".

An analog of theorem 3.1.7 also holds under the assumption that *index* $L=0$, but with a different order of the entire function.

The result of section 3.2 is valid without any change.

The result of theorem 3.3.1 is valid with a different value of $\gamma>0$ in the bound $p<1+\gamma$.

All the proofs remain the same. We only have to correct the class $\mathfrak{S}_p$ to which the corresponding embedding operators belong.

B. Hypoelliptic equations and systems. Let $L(x,D)$ be again a $r\times r$ matrix linear differential operator with smooth periodic coefficients. We call it *hypoelliptic* [170] if there exist $r\times r$ matrix functions $A(x,\xi)$ and $B(x,\xi)$ and constants ρ,δ $(1-\rho\leq\delta<\rho\leq 1)$ such that the symbol $L(x,\xi)$ satisfies the following condition: for any compact $N\subset\mathbb{R}^n$ and multiindices α,β there exist constants $C_N, C_{\alpha,\beta,N}$, such that

$$\begin{cases} |A(x,\xi)\left(D_\xi^\alpha D_x^\beta L(x,\xi)\right)B(x,\xi)|\leq C_{\alpha,\beta,N}(1+|\xi|)^{-\rho|\alpha|+\delta|\beta|},\ x\in N \\ |B(x,\xi)^{-1}L(x,\xi)^{-1}A(x,\xi)^{-1}|\leq C_N,\ x\in N,\ |\xi|>C_N \\ |A(x,\xi)|+|B(x,\xi)|\leq C_N|\xi|^{m^1}, x\in N, |\xi|>C_N \end{cases}$$

for some real constant m^1.

It is known [170] that for such operator there exists a parametrix (regularizer). It is also known that Douglas-Nirenberg elliptic and Petrovsky $2b$-parabolic systems and equations are hypoelliptic.

If we consider hypoelliptic operators in standard Sobolev spaces, they will not be Fredholm in general: their ranges will not be closed (but the kernels are finite dimensional and the *closures* of the ranges have finite co-dimensions). Hence, we are not able to apply our theory of Fredholm morphisms (and apparently there is no counterpart to this theory for the case of morphisms with non-closed ranges). Fortunately, we deal here with a highly special class of operators. In particular, very important feature is the existence of a pseudodifferential parametrix. We can force (using the technique of pseudodifferential operators and results of [170, 172]) the theory of chapter I to work in this case see [230]. The other (and, possibly, better) approach is to use some special anisotropic Sobolev type spaces, in which our concrete operator is Fredholm. Anyway, the results of sections 3.1 and 3.2 hold in the hypoelliptic situation modulo the same remarks about the nonzero index. Concerning theorem 3.3.1, we consider some narrower class of hypoelliptic operators satisfying the conditions:

$$|D_\xi^\alpha D_x^\beta L_1(x,\xi)L_1^{-1}(x,\xi)| \le C_{\alpha,\beta}(1+|\xi|)^{-\rho|\alpha|+\delta|\beta|}\ x\in\mathbb{R}^n,\ |\xi|>0$$

$$|L_1^{-1}(x,\xi)| \le C|\xi|^{m^1}(|\xi|>C, x\in\mathbb{R}^n),$$

where $L_1(x,D)$ is the transposed operator to $L(x,D)$. In this case the usual Sobolev scale works smoothly and we obtain the result about completeness (see theorem 3.3.1) with the inequality $p<1+(n\rho^{-1}-1)^{-1}$ instead of $p<1+(n-1)^{-1}$.

We will not provide the reader with long, but purely technical considerations that are parallel to ones of chapters 1-3 (see [230]).

C. Pseudodifferential equations. Let $L(x,D)$ be a pseudodifferential operator of the class $L^m_{\rho,\delta}$ $(1-\rho\le\delta<\rho\le 1)$ (see [174, 359]). We assume that it is periodic, i.e. commutes with the 1 – translations T_j in x_j – direction for every $j=1, \dots, n$.

It is easy to carry over the main part of the theory under one additional assumption (which is always fulfilled in differential case). Let $K(x,y)$ be the Schwartz kernel of $L(x,D)$ and $\prod\subset\mathbb{R}^n\times\mathbb{R}^n$ be its support. We shall say that $L(x,D)$ is *proper* if both projections $\prod$ on $\mathbb{R}^n_x$ and $\mathbb{R}^n_y$ are proper mappings. Our new condition is that the operator is proper.

If $L(x,D)$ is hypoelliptic in the sense of part B of this section, we can easily repeat (modulo the remarks about the index) the proofs of the results of sections 3.1 and 3.2 except theorem 3.1.7.

We are not sure whether it is possible to obtain any analogs to theorems 3.1.7 and 3.3.1. The problem here is that the operator function $L(x,D-i\lambda)$ is not polynomial anymore. It would be interesting to understand the situation here.

D. Smoothness of coefficients. We have assumed so far that all coefficients of th equation under consideration are C^∞ – smooth. In fact, we do not need such restrictiv assumption (for instance, potential in the Schrödinger equation usually has som singularities.) We can easily find some milder conditions in which our proofs still work.

Let us restrict our attention to the case of a single elliptic operator $L(x,D)$. Al that we needed in the sections 3.1-3.2 (except theorem 3.1.7) was that there existed th transposed operator $L_1(x,D)$, and $L_1(x,D+\lambda)$ was Fredholm as an operator fron $H^{2m}(\mathbf{T}^n)$ into $L_2(\mathbf{T}^n)$ for all $\lambda \in \mathbf{C}^n$. Suppose that

$$L_1(x,D) = \sum_{|\alpha| \le 2m} b_\alpha(x) D^\alpha,$$

and consider the assumptions on b_α that guarantee the Fredholm property of L_1. Let us assume first that $b_\alpha \in C(\mathbf{T}^n)$ for $|\alpha|=2m$. In this case, due to ellipticity, the principa part

$$L_{1,0} = \sum_{|\alpha| = 2m} b_\alpha D^\alpha$$

of L_1 defines a Fredholm operator from $H^{2m}(\mathbf{T}^n)$ into $L_2(\mathbf{T}^n)$ (see, for instance, [174, 262]). Since the principal parts of $L_1(x,D)$ and $L_1(x,D+\lambda)$ are identical, we have only to guarantee compactness of the lower order terms of $L_1(x,D+\lambda)$. Let us consider first the value $\lambda=0$. In this case the lower order terms are $b_\alpha D^\alpha$ with $|\alpha| < 2m$. The operator D^α is continuous from $H^{2m}(\mathbf{T}^n)$ into $H^{2m-|\alpha|}(\mathbf{T}^n)$. Let us try to find a space $L_r(\mathbf{T}^n)$ such that the embedding $H^{2m-|\alpha|}(\mathbf{T}^n) \subset L_r(\mathbf{T}^n)$ is compact, and the multiplication operator by $b_\alpha(x)$ is continuous from $L_r(\mathbf{T}^n)$ in $L_2(\mathbf{T}^n)$. In this case the operator $b_\alpha D^\alpha$: $H^{2m} \to L_2$ will be compact.

The embedding $H^{2m-|\alpha|} \subset L_r$ is compact, if $\frac{2m-|\alpha|}{n} > \frac{1}{2} - \frac{1}{r}$ (see, for instance, [394], 4.10.1). Let us assume now that $b_\alpha \in L_p(\mathbf{T}^n)$. It is known that if $f \in L_r$, $b_\alpha \in L_p$, then $fb_\alpha \in L_{\frac{rp}{r+p}}$. We need $rp/(r+p) \ge 2$, and therefore

$$r \ge \frac{2p}{p-2}, \text{ or } \frac{1}{2} - \frac{1}{r} \ge \frac{1}{p}.$$

These two conditions are compatible, if

$$\frac{2m-|\alpha|}{n} > \frac{1}{p}, \; p \geq 2. \tag{3.4.1}$$

Let us recall now that $L_1(x,D)$ is transposed to $L(x,D)$, and therefore b_α is a linear combination of expressions $D^s a_{\alpha+s}$ (where $|\alpha+s| \leq 2m$). Assume now that $D^s a_\alpha \in L_{p_{s,\alpha}}(\mathbb{T}^n)$ ($s \leq \alpha$, and $s \neq 0$ when $|\alpha|=2m$) and

$$p_{s,\alpha} > \frac{n}{2m-|\alpha-S|}\,.$$

In this case all the lower order terms define compact operators from $H^{2m}(\mathbb{T}^n)$ into $L_2(\mathbb{T}^n)$.

Consider now $\lambda \neq 0$. In this case each term $(D^s a_\alpha)D^{\alpha-s}$ of L_1 gives rise to the polynomial

$$(D^s a_\alpha)\,(D+\lambda)^{\alpha-s} = \sum_{\beta\leq\alpha} C_{\alpha,\beta}(D^s a_\alpha)D^{\beta-s}.$$

In the case $\beta=\alpha$ we obtain the terms we discussed above, and in the case when $\beta<\alpha$ the operator $(D^s a_\alpha)D^{\beta-s}$ is compact under the same conditions that guarantee compactness of $(D^s a_\alpha)D^{\alpha-s}$. Repeating now the considerations of sections 2.3 through 3.2, we obtain the following result.

Theorem 3.4.1. *Let* $a_\alpha \in H^{|\alpha|}_{loc}(\mathbb{R}^n)$ *and* $D^s a_\alpha \in L_{p_{s,\alpha},loc}(\mathbb{R}^n)$ *for* $s \leq \alpha$, *where*

$$p_{s,\alpha} > \frac{n}{2m-|\alpha|+|s|},\quad p_{s,\alpha} \geq 2$$

for $|\alpha|<2m$, *and* $a_\alpha \in C^{2m}(\mathbb{R}^n)$ *for* $|\alpha|=2m$. *In this case the statements of theorems* 3.1.2-3.1.6 *and* 3.2.1 *hold.*

Remark. Let us consider, for example, the stationary Schrödinger operator

$$-\Delta + q(x)$$

with a periodic potential q. The condition of theorem 1 in this case is that $q \in L_{p,loc}(\mathbb{R}^n)$ with $p > \frac{n}{2}$ for $n \geq 4$, and $p=2$ for $n=2,3$.

A little bit more delicate thing is to obtain analogs of theorems 3.1.7 and 3.3.1.

We are going to formulate the next result in terms of the coefficients of the transposed operator

$$L_1(x,D)=\sum_{|\alpha|\leq 2m} b_\alpha(x)D^\alpha .$$

Theorem 3.4.2. *Let us assume that:*

i) $b_\alpha \in C(\mathbf{T}^n)$ *for* $|\alpha| = 2m$;

ii) $b_\alpha \in L_{q_\alpha}(\mathbf{T}^n)$, $q_\alpha \geq 2$, $|\alpha| < 2m$, *where*

$$\frac{1}{q_\alpha} < \frac{2m-|\alpha|}{n} . \tag{3.4.2}$$

In this case the assertions of theorems 3.1.7 *and* 3.3.1 *hold.*

Proof. First of all, the same considerations as in the previous theorem, show that $L_1(x,D)$ is a Fredholm operator from $H^{2m}(\mathbf{T}^n)$ into $L_2(\mathbf{T}^n)$. We have to consider now the operator polynomial

$$L_1(x,D+\lambda) = \sum_{|\alpha|\leq 2m} b_\alpha(x)(D+\lambda)^\alpha = \sum_{\substack{|\alpha|\leq 2m\\ 0\leq\gamma\leq\alpha}} C_{\alpha,\gamma}\, b_\alpha(x)D^{\alpha-\gamma}\,\lambda^\gamma .$$

We are interested here only in the terms with $\gamma \neq 0$. Let S be an invertible pseudodifferential operator of order -1 on $\mathbf{T}^n$. Then

$$L_1(x,D+\lambda)S^{2m} = \sum_{\substack{|\alpha|\leq 2m\\ 0\leq\gamma\leq\alpha}} C_{\alpha,\gamma}\,[b_\alpha(x)D^{\alpha-\gamma}S^{-|\gamma|}S^{2m}]\,S^{|\gamma|}\,\lambda^\gamma . \tag{3.4.3}$$

The operator S^{2m} is continuous from $L_2(\mathbf{T}^n)$ into $H^{2m}(\mathbf{T}^n)$, operator $D^{\alpha-\gamma}S^{-|\gamma|}$ is continuous from $H^{2m}(\mathbf{T}^n)$ into $H^{2m-|\alpha|}(\mathbf{T}^n)$, the last space is continuously embedded into $L_{2q_\alpha/(q_\alpha-2)}$, due to (2), and multiplication by b_α is continuous from $L_{2q_\alpha/(q_\alpha-2)}$ into L_2. As the result, the operator $B_{\alpha,\gamma} = [b_\alpha(x)D^{\alpha-\gamma}S^{-|\gamma|}S^{2m}]$ is continuous in L_2. This means that the operator function (3) can be represented as

$$\sum B_\gamma\, S^{|\gamma|}\,\lambda^\gamma$$

with bounded B_γ and $S \in \mathcal{S}_p(L_2(\mathbb{T}^n))$ for $p > n$ (see lemma 1.2.18). Now the rest of the proofs of 3.1.7 and 3.3.1 can be carried over without any troubles.

3.5. Comments and references

We would like to mention that the author's approach to Floquet expansions is based on ideas that were used in [99, 273, 324] in the case of constant coefficient equations.

Section 3.1. The idea of lemma 3.1.2 is similar to one applied in the case of constant coefficient equations [99, 324]. Theorem 3.1.3 was proved by the author (see, for instance, [229]). The statement of corollary 3.1.6 was obtained independently by C. Wilcox [405] for the Schrödinger equation and L. Zelenko and the author [239] for general scalar elliptic equations. Theorem 3.1.7 was proved by the author [229], and rediscovered later for the Schrödinger equation (without the remark about periodicity) in [202]. Some very interesting analysis pertinent to the analytic structure of the sets of quasimomentums and Floquet exponents (including some open problems) can be found in [27, 314]; cf. also [315]. We would like to mention one problem that seems to be not too hard. For an equation with constant coefficients the set of quasimomentums consists of an algebraic surface (the set of zeros of the symbol) and of some of its translations. There are some cases, even for equations with periodic coefficients, when this set contains an algebraic component (theorem 4.1.2). "Everybody knows", however, that for a generic elliptic periodic operator the set of quasimomentums does not contain algebraic irreducible components. This statement has been never proved (at least, the author has never seen any written proof). It should not be too complicated to prove this statement, using an appropriate transversality theorem (see, for instance, papers [7, 53, 396] on generic spectral properties of elliptic operators).

Section 3.2. There are two standard approaches to Floquet expansions in the case of ordinary differential equations [17, 94, 120, 156, 177, 267, 272]. The first one is to consider the spectral expansion for the monodromy operator $x(0) \to x(1)$ (the period is assumed to be equal to 1). This expansion is equivalent to the expansion into Floquet solutions. It is possible to try to carry over this method to evolutionary PDEs, and sometimes (but not always) it works (see section 5.3). It fails, however, even for evolutionary equations, if the corresponding Cauchy problem is ill-posed. The situation is even worse for elliptic equations that are periodic in several variables (for instance, for the periodic Schrödinger equation), when the monodromy operator cannot be defined. The second common approach is to find a periodic substitution (sometimes called Lyapunov's substitution) that reduces the equation to one with constant coefficients, and after that to use the Euler's expansion into exponential-polynomial solutions [106]. The Euler's theorem has been successfully generalized to constant coefficient PDEs in [99, 324]. The problem is that it is hard to understand what could be the analog of Lyapunov's substitutions for PDEs. There are, unfortunately, clear signs that there is no counterpart to this substitutions for periodic elliptic PDEs. Namely, there are properties of the set of quasimomentums that could not occur if a Lyapunov substitution existed. The first obstacle is that this set can cover the whole space (see the discussion after theorem 4.1.6). This cannot happen for constant coefficient elliptic PDEs. Besides, this set is always algebraic in the constant coefficient case, and this is not true for periodic equations. This is, why we have to choose a different approach, if we are going to get a Floquet expansion. The author's approach is analogous to the one adopted in [99, 273, 274, 324] for constant coefficient equations.

The first version of theorem 3.2.1 for scalar elliptic equations was obtained by L. Zelenko and the author [239] under the additional assumption that the set Q of quasimomentums is not equal to the whole space $\mathbf{C}^n$. This condition is equivalent to the absence of L_2-solutions of the equation (see section 4.1). This restriction was eliminated later by the author [219, 221, 229, 230], who also generalized the theorem for systems, pseudodifferential equations, etc. [229]. These theorems were of more local nature than theorem 3.2.1. Namely, it was proved that N can be covered by polydisks U_k and every intersection $N \cap U_k$ can be covered by analytic sets $N_{j,k} \subset N$ such that every solution of exponential growth can be expanded into Floquet solutions $f_{j,k,l}(z,x)$ with Floquet exponent $z \in U_k \cap N_{j,k}$. The proofs relayed onto a local version of theorem 1.5.18. This local version was proved by V. Palamodov in [322, 323]. Due to this local approach, the proofs involved more handling of coherent and BCAF-sheaves than here and were much more lengthy [229]. V. Palamodov has obtained recently the global version (theorem 1.5.18) and proved theorem 3.2.1 in the form presented here. Moreover, he has generalized the result to the case of non-square elliptic systems. In order to do this, the system must be included into an elliptic complex; this results in some complications in the proof. The reader can find all details in [326].

The results of sections 3.3 and 3.4 were obtained by the author [229]. It is still unclear whether it is possible to prove completeness of Floquet solutions for general elliptic equations in the class of all solutions defined in $\mathbb{R}^n$ (without any growth conditions).

The conditions of theorem 3.3.1 include the assumption that the set of quasimomentums does not cover $\mathbb{C}^n$ (or, equivalently, the equation does not posses non-zero fast decaying solutions, see theorem 4.1.5). This assumption seems to be unnecessary. The author could not, however, get rid of it. There is an analytical obstacle that we could not overcome. Consider the operator polynomial $L(\lambda) = L_0 + \sum \lambda^\alpha L_\alpha S^{|\alpha|}$ $(\lambda \in \mathbb{C}^n)$, where L_0 is Fredholm, L_α are bounded and S belongs to an ideal $\mathcal{I}_p$. Let $y(\lambda)$ be an entire vector function of a finite order such that the equation $L(\lambda)x(\lambda) = y(\lambda)$ has an entire solution $x(\lambda)$. The question is, whether there exists an entire solution $x_1(\lambda)$ of a finite order. If this were true, we would be able to drop the assumption. This question can be answered affirmatively, if the space is finite dimensional (and so we deal with a matrix polynomial instead of an operator one). This is a consequence of L.Hörmander's estimates for the $\bar{\partial}$-problem (see, for instance, theorem 7.6.5 in [169]). We can also give the affirmative answer, if $L(\lambda)$ is invertible at some point (this condition is equivalent to our assumption of non-existence of fast decaying solutions). In this case the answer follows from lemma 1.2.21 (see the proof of theorem 3.3.1 for details). The authors guess was that the answer is always affirmative. However, Dr. F. Mantlik constructed recently counterexample even in the case when $n = 1$ and the operator function is linear: $L_0 + \lambda S$ (private communication). The situation with this assumption remains unclear.

It is possible to reduce the requirement (3.4.2), increasing the order of the entire function in the statement of theorem 3.1.7 and reducing at the same time the allowed rate of growth in the statement of theorem 3.3.1. To do so we just choose an operator S of order less than 1 (see the proof of theorem 3.4.2). The reader can easily add all the necessary details. It should be possible to reduce, using the approach of [243], the smoothness requirements on the coefficients in the principal part of the equations.

CHAPTER 4

PROPERTIES OF SOLUTIONS OF PERIODIC EQUATIONS.

In this chapter we study some relations between the geometry of the set of the quasimomentums (or Floquet multipliers) and properties of the corresponding equation. Among the properties we consider are: existence of fast decaying solutions, solvability of non-homogeneous equations, structure of the spectra, etc.

4.1. Distribution of quasimomentums and decreasing solutions

Let $L(x,D)$ be a scalar elliptic operator with smooth periodic coefficients. (In some cases our results hold also for equations considered in section 3.4. We leave the details to the reader.)

We are interested in the set $N \subset \Omega_n$ of Floquet exponents of the equation

$$Lu=0 \tag{4.1.1}$$

(or, equivalently, in the set $Q \subset \mathbf{C}^n$ of quasimomentums). We shall discuss the question whether it can be empty or equal to the whole region Ω_n (correspondingly, $Q=\emptyset$ or $Q=\mathbf{C}^n$). Both possibilities, $N=\emptyset$, and $N=\Omega_n$ look like pathologies. Indeed, they can occur neither in the case of ordinary differential equations (i.e. $n=1$), nor in the case of elliptic equations with constant coefficient. In the ordinary differential case, as it is well known [17, 94, 156, 177, 272], N is a non-empty finite subset of $\mathbf{C}\backslash 0=\Omega_1$. In the case of an elliptic constant coefficients equation $L(D)u=0$ it is more convenient to deal with the set Q of quasimomentums. One can easily check, using Fourier series expansions, that in this case Q consists of the zero set $\{\xi \in \mathbf{C}^n |\ L(\xi)=0\}$ of the symbol and of all its translations by elements of $2\pi\mathbf{Z}^n$. Hence, neither $Q=\emptyset$, nor $Q=\mathbf{C}^n$ can occur.

The situation in periodic case is different. It is not known yet whether $Q=\emptyset$ can occur in PDE case, but there exists an example of a nice abstract parabolic equation in

Hilbert space with no Floquet solutions at all (see theorem 5.3.4).

We start with some simple results about existence of Floquet solutions (i.e. non-emptiness of Q).

Theorem 4.1.1. *If zero belongs to the spectrum of $L(x,D)$ on $L_2(\mathbf{R}^n)$ (with domain $\mathfrak{D} = H^{2m}(\mathbf{R})$), then $N \cap \mathfrak{T}^n \neq \emptyset$ (in particular, $N \neq \emptyset$).*

This is a direct consequence of theorem 4.5.1, which will be proved later.

Now we shall describe one situation, when the perturbation of an elliptic constant coefficient equation by some periodic lower order terms does not disturb the set of quasimomentums. It is more convenient to include the spectral parameter λ here, i.e. to consider the whole Bloch variety $B_{\mathbf{C}}(L)$ (see details in section 4.4):

$$B_{\mathbf{C}}(L) = \{(\xi,\lambda) \in \mathbf{C}^{n+1}\colon\ Lu=\lambda u \text{ has a nonzero Floquet solution with the quasimomentum } \xi\}.$$

In particular, $Q=B_{\mathbf{C}}(L) \cap \{\lambda=0\}$. Analytic properties of Bloch variety will be considered in section 4.4.

Let K be a convex cone in $\mathbf{R}^n$ with the vertex at the origin. We assume that K does not contain entire lines. We denote by K^0 the interior of K.

Theorem 4.1.2. *Let $L_0(D)$ be an elliptic constant coefficient operator of order $2m$ and $L_1(x,D)$ be an operator with smooth periodic coefficients that has the order less than $2m$. Suppose that the Fourier series of all coefficients of $L_1(x,D)$ contain only harmonics $\exp 2\pi ikx$ with $k \in K^0$. In this case $B_{\mathbf{C}}(L_0) \subset B_{\mathbf{C}}(L_0+L_1)$.*

Proof. Let $B_0=B_{\mathbf{C}}(L_0)$ and $B=B_{\mathbf{C}}(L_0+L_1)$, then obviously

$$B_0=\{(\xi,\lambda) \in \mathbf{C}^{n+1} \mid \exists q \in \mathbf{Z}^n,\ L_0(\xi+ 2\pi q) = \lambda\}.$$

Since B is invariant with respect to translations by elements of $2\pi\mathbf{Z}^n$ in ξ-variable, it is sufficient to prove that $M \subset B$, where

$$M = \{(\xi,\lambda) \in \mathbf{C}^{n+1} \mid L_0(\xi) = \lambda\}.$$

Let the subset $M^1 \subset M$ be given by

$$M^1=\{(\xi,\lambda) \in M|\ (\xi+2\pi q,\ \lambda) \notin M \quad \forall q \in \mathbb{Z}^n \backslash 0\}.$$

Lemma 4.1.3. *M^1 is dense in M.*

Proof of the lemma. Let $q \in \mathbb{Z}^n\backslash 0$ and $M^{(q)}=\{(\xi,\lambda) \in M|\ (\xi+ 2\pi q,\lambda) \in M\}$. It is obvious that

$$M^1 = M\backslash \bigcup_q M^{(q)},$$

so the lemma would be proved if we knew that $M^{(q)}$ is a closed and nowhere dense subset of M. The closedness of $M^{(q)}$ is obvious, because it is the intersection of zero sets of two polynomials: $(L_0(\xi)-\lambda)$ and $(L_0(\xi+2\pi q)-\lambda)$. Now, M is the graph of the polynomial $L_0(\xi)$, and therefore a purely n-dimensional algebraic subvariety of $\mathbb{C}^{n+1}$. Hence, if $\dim M^{(q)} < n$, then $M^{(q)}$ is nowhere dense in M. Let us assume the opposite: $\dim M^{(q)}=n$. In this case the zero sets of $L_0(\xi)-\lambda$ and $L_0(\xi + 2\pi q)-\lambda$ have a common irreducible component. These polynomials of (ξ,λ) are irreducible, so $L_0(\xi)$ identically coincides with $L_0(\xi + 2\pi q)$. This is impossible for $q \neq 0$, so the lemma is proved.

Now we return to the proof of the theorem. The set B is analytic and hence closed (theorem 4.4.2). According to the lemma, it is sufficient to prove that $M^1 \subset B$.

Let $(\xi,\lambda) \in M^1$. We have to prove the existence of a nonzero solution

$$u(x)=g(x)\ exp\ i\xi x$$

of the equation $L(x,D)u=\lambda u$, where the function $g(x)$ is 1-periodic. Here we denote $L=L_0 + L_1$. We shall find this solution in the form

$$u(x)= exp\ i\xi x\ (1+ \sum_{k \in \mathbb{Z}^n \cap K^0} g_k\ exp\ 2\pi ikx) = (exp\ i\xi x)g(x).$$

Let us plug this function into the equation $Lu-\lambda u=0$:

$$(L_0(D) + L_1(x,D) - \lambda)\,(exp\; i\xi x\; g(x)) = (L_0(\xi) - \lambda)\, e^{i\xi x} +$$

$$e^{i\xi x}(L_0(D + \xi) - \lambda) \sum_{k \in \mathbb{Z}^n \cap K^0} g_k\; exp\; 2\pi ikx + e^{i\xi x} L_1(x, D + \xi) g(x) = 0.$$

Since $L_0(\xi)=\lambda$, we obtain the following equation for $g(x) = 1 + \sum g_k\; exp\; 2\pi ikx$:

$$(L_0(D + \xi) - \lambda)(g(x) - 1) + L_1(x, D + \xi)g(x) = 0. \tag{4.1.2}$$

Define a norm on $\mathbb{R}^n$ such that $||k_1 + k_2|| \geq ||k_1|| + 1$ for all $k_1 \in K$, $k_2 \in K^0 \cap \mathbb{Z}^n$. This can always be done. Indeed, let $f(x)$ be a linear functional on $\mathbb{R}^n$ that is strictly positive on $K \backslash 0$. In this case

$$\min_{k \in \mathbb{Z}^n \cap K^0} f(k) = \delta > 0.$$

Let us introduce a norm $||\cdot||$ on $\mathbb{R}^n$ such that $||x|| = \delta^{-1} f(x)$ for $x \in K$ (it is obviously possible, just consider the corresponding unit ball). In this case, for $k_1 \in K$, $k_2 \in \mathbb{Z}^n \cap K^0$ we have $||k_1 + k_2|| = \delta^{-1} f(k_1) + \delta^{-1} f(k_2) \geq \delta^{-1} f(k_1) + 1 = ||k_1|| + 1$. We shall also denote by $|k|$ the standard Euclidian norm on $\mathbb{R}^n$.

Now we need some lattice of function spaces. For $j \in \mathbb{Z}^+$ we denote by $V^{s,j}$ the closed subspace of all functions $\varphi \in H^s(\mathbb{T}^n)$ such that their Fourier series contain only harmonics $exp\; 2\pi ikx$ with $k \in K$ and $||k|| \geq j$. Obviously, $V^{s,j_1} \subset V^{s,j_2}$ for $j_1 > j_2$ and $L_0(D)$ maps $V^{s,j}$ into $V^{s-2m,j}$. Let us consider the operator $L_1(x,D + \xi)$. According to the conditions of the theorem, L_1 maps $V^{s,j}$ into $V^{s-2m+1,j+1}$. Indeed, elements of $V^{s,j}$ have in their Fourier series only harmonics with $k \in K$ such that $||k|| > j$. Differentiations do not shift the Fourier series. Multiplication by coefficients of L_1 shift the wave vector k into $k + k_1$, where $k_1 \in K^0$. Due to the choice of the norm $||\cdot||$ on $\mathbb{R}^n$, this exactly means that L_1 maps $V^{s,j}$ into $V^{s-2m+1,j+1}$. The main point here is that L_1 increases the index j at least by 1.

Now, due to the condition $(\xi,\lambda) \in M^1$, the operator $[L_0(D + \xi) - \lambda]$ is invertible on $V^{s,j}$ for $j>0$, so we can define on $V^{s,0}$ the operator $T=[L_0(D+\xi) - \lambda]^{-1} L_1(x,D+\xi)$.

According to preceding considerations, this operator transforms each $V^{s,j}$ into $V^{s,j+1}$, in particular $V^{0,j}$ into $V^{0,j+1}$. Let us estimate the norms of the operator on these subspaces. We consider the invertible operator $A=\sqrt{(1-\Delta)}$ of order 1 on $\mathbb{T}^n$ and represent T in the form $T=[L_0(D+\xi)-\lambda)^{-1}L_1(x,D+\xi)A]A^{-1}$. The operator $L_1(x,D+\xi)A$: $V^{0,j}\to V^{-2m,j}$ is bounded (uniformly with respect to j) as the restriction of the bounded operator $L_1(x,D+\xi)A$: $L_2(\mathbb{T}^n)\to H^{-2m}(\mathbb{T}^n)$ onto the subspace $V^{0,j}$. The operator $[L_0(D+\xi)-\lambda]^{-1}$: $V^{-2m,1}\to V^{0,1}$ is bounded (and therefore bounded on $V^{-2m,j}$ uniformly with respect to $j\geq 1$). Indeed, the subspace $V^{0,1}$ is orthogonal to the (consisting only of constants) kernel of $L_0(x,D+\xi)-\lambda$ on L_2. Due to ellipticity, the operator is an isomorphism on $V^{0,1}$. On the other hand, its range on $V^{0,1}$ is dense in $V^{-2m,1}$(because it contains all trigonometric polynomials from this space). Therefore, our operator is an isomorphism between $V^{0,1}$ and $V^{-2m,1}$, and so its inverse is bounded.

Now we are able to estimate the norm of T on $V^{0,j}$ in the following way:

$$||T||_{V^{0,j}\to V^{0,j}}\leq$$

$$||L_0(x,D+\xi)^{-1}||_{V^{-2m,j}\to V^{0,j}}\,||L_1(x,D+\xi)A||_{V^{0,j}\to V^{-2m,j}}\,||A^{-1}||_{V^{0,j}\to V^{0,j}}\leq$$

$$C\,||A^{-1}||_{V^{0,j}\to V^{0,j}}.$$

Let us estimate the latter expression: for $g\in V^{0,j}$

$$||A^{-1}g||=||A^{-1}\sum_{||k||\geq j} g_k e^{2\pi ikx}||_{L_2}=$$

$$=||\sum_{||k||\geq j}\frac{1}{(1+|k|^2)^{1/2}}\,g_k exp 2\pi ikx||_{L_2}\leq Cj^{-1}||g||_{L_2}.$$

We use here the equivalence between the norms $||\cdot||$ and $|\cdot|$. Thus, we have the estimate: $||T||_{V^{0,j}\to V^{0,j}}\leq Cj^{-1}$. Because T maps $V^{0,j}$ into $V^{0,j+1}$, we have

$$||T^j||_{V^{0,0}\to V^{0,0}}\leq\frac{C^j}{(j-1)!}.$$

Let us return to the equation

(4.1.3) $$(L_0(D+\xi)-\lambda)\,(g-1) = -L_1(x,D+\xi)g,$$

where $g \in V^{0,0}$. Since $g-1 \in V^{0,1}$ and $L_1(x,D+\xi)g \in V^{0,1}$, we can invert the operato on the left hand side, and write down the equation in the form

(4.1.4) $$g = 1 - Tg.$$

We have to prove the existence of a solution $g \in V^{0,0}$. Let us define

(4.1.5) $$g = \sum_{j=0}^{\infty} (-1)^j T^j 1$$

If we prove the convergence of this series in $V^{0,0}$ (i.e. in $L_2(\mathbb{T}^n)$), it obviously will give the solution. Let us mention that $1 \in V^{0,0}$ and $T^j 1 \in V^{0,j}$, therefore

$$\sum_{k=0}^{\infty} ||T^k 1|| \leq 1 + \sum_{1}^{\infty} \frac{C^j}{(j-1)!} < \infty.$$

This proves the theorem.

We now pass on to the question whether the set N of Floquet exponents can coincide with Ω_n (or, equivalently, whether the set Q of quasimomentums can coincide with $\mathbb{C}^n$). There is a trivial case when it is possible. Let us assume here that either L is a matrix elliptic operator, or a pseudodifferential operator.

Theorem 4.1.4. *If the Fredholm index of $L(x,D)$ on $\mathbb{T}^n$ is positive:*

$$\text{index}_{\mathbb{T}^n} L > 0,$$

then $N = \Omega_n$ (respectively, $Q = \mathbb{C}^n$).

Proof. Under the conditions of the theorem, the Fredholm index of the corresponding morphism $\mathcal{L}$ (see theorem 2.2.3) is positive on each fiber, so $S(\mathcal{L})=\Omega_n$ and by theorem 3.1.5 $N=\Omega_n$. This proves the theorem.

The Atiyah-Singer theorem shows that scalar elliptic operators cannot satisfy the conditions of the theorem. There are, however, elliptic systems with positive index. Let us denote by $\wedge^{p,q}$ the bundle of (p,q)-forms on $\mathbf{T}^2 = \mathbf{C}/\mathbf{Z}^2$. We consider a holomorphic line bundle F over $\mathbf{T}^2$ of degree 1 (i.e. the corresponding divisor has degree 1). According to the classical Riemann-Roch theorem, the Fredholm index of the $\bar\partial$-operator

$$\bar\partial : \Gamma(\wedge^{0,0} \otimes F) \to \Gamma(\wedge^{0,1} \otimes F)$$

is equal to one. Since $\wedge^{0,0}$ and $\wedge^{0,1}$ are trivial, we can consider $\bar\partial$ as an operator in $\Gamma(F)$. Let M be a line (not holomorphic) bundle on $\mathbf{T}^2$ such that $M \oplus F$ is trivial. We can introduce the Laplace operator Δ on sections of M, corresponding to the Riemannian metric on $\mathbf{T}^2$ and some Hermitian metric on M. In this case $L = \Delta \oplus \bar\partial^2$ is an example of an elliptic matrix operator on $\mathbf{T}^2$ with Fredholm index 2. For the corresponding periodic system on $\mathbf{R}^2$ we have $N = \Omega_n$.

The arguments of the last theorem do not work in scalar elliptic case, but even in this case the pathology $N = \Omega_n$ can occur. The following theorem shows an important relation between this question and properties of the equation in $\mathbf{R}^n$.

Theorem 4.1.5. *The following assertions are equivalent:*

a) $N = \Omega_n$*;*

b) the equation $Lu=0$ *has a non-zero solution in* $L_2(\mathbf{R}^n)$*;*

c) the equation has a non-zero solution with the bound

$$|u(x)| \le C_a \, exp\,(-a|x|) \quad (\forall a>0). \tag{4.1.6}$$

Proof. If $N = \Omega_n$, then by theorem 3.1.5 $S(\mathcal{L}) = \Omega_n$ (where $\mathcal{L}$ is the morphism described in section 2.2). Let $z_0 \in \mathcal{T}^n$ be a point at which

$$\min_{z \in \mathcal{T}^n} dim\ Ker\ \mathcal{L}(z)$$

is attained (this minimum is strictly positive, because $Ker\ \mathcal{L}(z) \ne 0\ \forall z \in N = \Omega_n$). Then, by corollary 1.2.14, *dim Ker* $\mathcal{L}(z)$ is constant in a neighborhood of z_0 in $\mathcal{T}_n$, that is, *Ker* $\mathcal{L}(z)$ forms a vector bundle over this neighborhood (theorem 1.6.13). Let $\varphi(z)$ be a compactly supported continuous section of this bundle over the neighborhood. We can regard φ as an element of $L_2(\mathfrak{T}^n, H^{2m}(\mathbf{T}^n))$ (with $\varphi(z) \in E_z^{2m}$). According to theorem 2.2.5, $\varphi = \boldsymbol{F}f$ for some function $f \in H_{loc}^{2m}(\mathbf{R}^n)$ such that

$$\sum_{k \in \mathbf{Z}^n} \|f\|^2_{H^{2m}(\mathfrak{X}+k)} < \infty . \tag{4.1.7}$$

Furthermore, $\mathcal{L}\varphi=0$ implies $Lf=0$, so the statement *b*) is proved.

Let now $0 \neq f \in L_2(\mathbf{R}^n)$ and $Lf=0$. Using the same Schauder type estimates as in the proof of lemma 3.1.2, we easily obtain that f belongs to $H_{loc}^{2m}(\mathbf{R}^n)$ and satisfies (7). Applying the transform $\boldsymbol{F}$, by theorem 2.2.5 we obtain a non-zero L_2 – section $\varphi = \boldsymbol{F}f$ of the kernel of $\mathcal{L}$ over $\mathfrak{T}^n$. This means that $Ker\ \mathcal{L}(z) \neq 0$ for all points z from some subset of positive n-dimensional measure in $\mathfrak{T}^n$. In other words, the n-dimensional measure of $\mathfrak{T}^n \cap N$ is positive. Due to corollary 3.1.6 and theorem 1.5.19, this means that $N = \Omega_n$, which proves a).

If $S(\mathcal{L}) = N = \Omega_n$, then by corollary 1.6.15 there is a holomorphic section $\varphi(z) \neq 0$ of $\mathcal{E}^{2m}$ such that $\mathcal{L}(z)\varphi(z) \equiv 0$. According to theorem 2.2.2, $\varphi(z) = (\boldsymbol{F}u)(z)$ for some function u satisfying the estimate (6). Since $\mathcal{L}\varphi=0$, we have $Lu=0$.

The implication c)⇒b) is trivial. The theorem is proved.

Under some additional conditions we can guarantee the existence of solutions that decay even faster than (6). Let us recall that an operator L has the *weak uniqueness* property, if any solution $u(x)$ of the equation $Lu=0$ that is zero on an open set, is identically zero.

Theorem 4.1.6. *Suppose that L, in addition to the previous conditions, has the weak uniqueness property and $index_{\mathbf{T}^n}\ L=0$. Then we can add to the equivalent assertions of theorem 5 the following one:*

d) equation (3.1.2) has a nonzero solution $u(x)$ with the bound

$$|u(x)| \leq C exp(-a|x|^{1+\gamma}) \quad \textit{for some } \gamma > 0 .$$

Proof. The implication d)=>c) is obvious. We shall prove the implication a)=>d). Let us consider the operator polynomial

$$L(x,D-i\lambda)=L(x,D)+\sum_{1\leq|\alpha|\leq 2m}(\alpha!)^{-1}(\partial_\xi{}^\alpha L)(x,D)(-i\lambda)^\alpha.$$

We consider all these operators as elements of $L(H^{2m}(\mathbf{T}^n),\ L_2(\mathbf{T}^n))$. All values of the polynomial are Fredholm operators of index zero. Consider the linear subspace $M\subset L(H^{2m}(\mathbf{T}^n),\ L_2(\mathbf{T}^n))$ that consists of all operators of multiplication by functions $\varphi(x)$ from $C^\infty(\mathbf{T}^n)$. We would like to prove the existence of $\varphi\in C^\infty(\mathbf{T}^n)$ such that the operator $L(x,D)+\varphi(x)$ is invertible. Suppose that this is not true. Then, due to theorem 1.2.15, there exist elements $g(x)\in Ker\ L(x,D)$ and $f(x)\in Ker\ L^t(x,D)$, such that

$$\int\varphi(x)g(x)f(x)\ dx=0$$

for all $\varphi\in C^\infty(\mathbf{T}^n)$. Due to ellipticity, both g and f are smooth; by the weak uniqueness property, the set of zeros of $g(x)$ has empty interior. This implies the existence of an open subset $\mathcal{U}\subset\mathbf{T}^n$ such that $g(x)\neq 0$ and $f(x)\neq 0$ for $x\in\mathcal{U}$. We can now define $\varphi(x)=\psi(x)(f(x)g(x))^{-1}\in C^\infty(\mathbf{T}^n)$, where $\psi\in C_0{}^\infty(\mathcal{U})$ and $\psi\geq 0$. In this case

$$\int_{\mathbf{T}^n}\varphi gf dx=\int_{\mathbf{T}^n}\psi dx>0\ .$$

The contradiction shows the existence of a function $\varphi\in C^\infty$ such that the operator $L+\varphi$ is invertible. Now we apply the parameter continuation method and consider the operator polynomial $L(\lambda,\mu)=L(x,D-i\lambda)+\mu\varphi(x)$, where $\mu\in\mathbf{C}^1$. This is an analytic Fredholm operator function in $\mathbf{C}^{n+1}$, which is invertible at the point (0,1). Hence, it is invertible everywhere except of a proper analytic subset in $\mathbf{C}^{n+1}$ (theorem 1.6.16). Due to the condition $Q=\mathbf{C}^n$ (which is equivalent to a)), this subset contains the hyper-plane $\mathbf{C}^n\times\{0\}\subset\mathbf{C}^{n+1}$. Consider now the inverse operator $L^{-1}(\lambda,\ \mu)$. According to lemma 1.2.21, $L^{-1}(\lambda,\ \mu)\ =\ B(\lambda,\ \mu)/\Delta\ (\lambda,\ \mu)$, where $B(\lambda,\ \mu):\ L_2(\mathbf{T}^n)\to H^{2m}(\mathbf{T}^n)$ and $\Delta(\lambda,\ \mu)$ are an operator and a complex valued entire functions of a finite order (which is any $p>n$). By theorem 3.1.7, increasing the order of the functions to $n+1$, we can achieve $2\pi\mathbf{Z}^n$-periodicity of Δ in λ. We can now represent $L^{-1}(\lambda,\ \mu)$ in the form $B_1(\lambda,\mu)/\Delta_1(\lambda,\mu)$, where B_1 and Δ_1 are entire functions of finite order and Δ_1 is $2\pi i\mathbf{Z}^n$-periodic with respect to λ .

We know that $\Delta_1(\lambda,0) \equiv 0$, and $\Delta_1(\lambda, \mu)$ is not identically equal to zero (because $\Delta_1(0, 1) \neq 0$). Hence, $\Delta_1(\lambda, \mu) = \mu^k \Delta_2(\lambda,\mu)$, where $k>0$, Δ_2 is an entire function of finite order and $\Delta_2(\lambda,0)$ is not identically equal to zero. Let also $B_1(\lambda,\mu) = \mu^r B_2(\lambda,\mu)$, where $r \geq 0$ and $B_2(\lambda,0)$ is not identically equal to zero. Let us mention that necessarily $r<k$, otherwise the function $L^{-1}(\lambda, \mu)$ would be holomorphic at some points $(\lambda,0)$, which contradicts the existence of nonzero kernel of $L(x,D-i\lambda)$. We can cancel μ^r and assume that $r=0$. So we assume that $B(\lambda_0,0) \neq 0$ for some $\lambda_0 \in \mathbf{C}^n$ and

$$\Delta(\lambda, \mu) = \mu^k \Delta_1(\lambda, \mu).$$

Let $\lambda_0 \in \mathbf{C}^n$ and $y \in L_2(\mathbf{T}^n)$ be such that $B(\lambda_0, 0)y \neq 0$. We consider the following entire vector function (with values in $H^{2m}(\mathbf{T}^n)$):

$$f(\lambda) = B(\lambda, 0)[e^{-(\lambda-\lambda_0)x} y(x)].$$

Then

$$L(\lambda, 0)f(\lambda) = L(\lambda, 0)\ B(\lambda, 0)[e^{-(\lambda-\lambda_0)x} y(x)] =$$

$$\frac{L(\lambda, \mu)\ B(\lambda, \mu)}{\Delta(\lambda, \mu)}\ \Delta(\lambda, \mu)\ e^{-(\lambda-\lambda_0)x} y(x)\,|_{\mu=0} =$$

$$\Delta(\lambda, 0)\ e^{-(\lambda-\lambda_0)x} y(x) \equiv 0.$$

We see that $f(\lambda) \in Ker\ L(x,D-i\lambda)$ for all $\lambda \in \mathbf{C}^n$ and f is not identically equal to zero. Let us investigate now, what is going on under $2\pi i\mathbb{Z}^n$ – translations of $f(\lambda)$:

$$B(\lambda+2\pi iq, \mu) = L^{-1}(\lambda+2\pi iq, \mu)\ \Delta(\lambda, \mu) =$$

$$= [e^{-2\pi iqx} L(\lambda, \mu)\ e^{2\pi iqx}]^{-1}\ \Delta(\lambda, \mu) = e^{-2\pi iqx} B(\lambda, \mu)\ e^{2\pi iqx}.$$

Hence,

$$f(\lambda+2\pi iq) = B(\lambda+2\pi iq, 0)\ e^{-2\pi iqx}\ e^{-(\lambda-\lambda_0)x} y(x) = e^{-2\pi iqx} f(\lambda).$$

We now consider the following $H^{2m}(\mathfrak{X})$ – valued vector function:

$$f_1(\lambda, x) = e^{\lambda x} f(\lambda, x).$$

This is obviously an entire vector function of a finite order. Besides, due to periodicity of $f(\lambda, x)$ with respect to x (remember that $f(\lambda) \in H^{2m}(\mathbb{T}^n)$), this function $f_1(\lambda, x)$ satisfies the boundary conditions (2.2.4) for $z = exp\lambda \in \Omega_n$. Let us check the behavior of $f_1(\lambda)$ under $2\pi i\mathbb{Z}^n$ – translations. If $q \in \mathbb{Z}^n$, we have:

$$f_1(\lambda + 2\pi iq, x) = e^{(\lambda + 2\pi iq)x} f(\lambda + 2\pi iq, x) =$$

$$= e^{(\lambda + 2\pi iq)x}\, e^{-2\pi iqx} f(\lambda, x) = f_1(\lambda, x).$$

Hence,

$$f_1(\lambda) \in A_{p/(p-1)}(\hat{\mathcal{E}}^{2m})$$

for some $p>1$. This means, by theorem 2.2.2, that $f_1 = \hat{\mathbf{F}}u$ *for* some $u \in \Phi^{2m}_{n,p}$. Then

$$\hat{\mathbf{F}}(Lu)(\lambda, x) = L(x, D)\,(\hat{\mathbf{F}}u)\,(\lambda, x) = L(x, D) f_1\,(\lambda, x) =$$

$$L(x, D)\, e^{\lambda x} f(\lambda, x) = e^{\lambda x}\, L(x, D - i\lambda)\, f(\lambda, x) \equiv 0.$$

We conclude that $u \in \Phi^{2m}_{n,p}$ is a solution of the equation $Lu = 0$. The rate of decay of $u(x)$ coincides with the rate we require in d), but in the H^{2m} – norm instead of the C – norm. If $2m > n/2$, we can use embedding theorems immediately, otherwise we apply first the Schauder type estimates (as in lemma 3.1.2) and increase $2m$ to a number $s > n/2$. Then we can apply the embedding theorem. The theorem is proved.

We can show that the "pathology" described in the last two theorems really occurs. It is well known (see [335, 302]) that there exist "good" elliptic equations that possess non-zero compactly supported solutions. For example, there exists an elliptic operator with smooth coefficients on $\mathbb{R}^3$ of the form

(4.1.8) $$P(x,\ D) = \chi(x)\ R(x,D)\ + (1-\chi(x))\ \Delta^2$$

such that

(4.1.9) $$P(x,D)\ u(x) \equiv 0$$

for some $u \in C_o^\infty(\mathbf{R}^3)$, $u \neq 0$. Here Δ is the Laplacian on $\mathbf{R}^3$, $R(x,D)$ is an elliptic operator of fourth order with smooth coefficients and $\chi(x) \in C_o^\infty$. Formula (8) for $P(x,D)$ shows that we can extend its coefficients periodically (for some sufficiently large period), while preserving (9). Namely, we choose some cube that contains the supports of $\chi(x)$ and of $u(x)$. Consider the lattice for which this cube is the fundamental domain. If we cut the portion of the operator P that sits inside the cube, and extend it periodically with respect to the group, we get the desired operator. Let us denote the resulting periodic elliptic operator by $L(x,D)$. By the construction, $Lu = 0$. Because u obviously satisfies any of the conditions formulated in b) – d), we have $N = \Omega_3$ for the operator L.

We can use theorems 5 and 6 in two different ways. First, if we manage to find a complex vector λ which is not a quasimomentum (i.e. $Q \neq \mathbf{C}^n$), then by theorem 5 we obtain the nonexistence of L_2–solutions of the equation. This question is often very important in mathematical physics. We emphasized that λ may be complex, because usually it is much easier to construct such complex vector, than a real one.

There is also another possibility. The last two theorems show that the equality $N = \Omega_n$ can occur, but it is still a "pathology", because it is equivalent to the existence of very fast decaying solutions. There exists an extensive literature devoted to studying of possible rate of decay (see, for instance, [61, 245, 246, 288-290]). These papers contain several results that prohibit this pathology in many "normal" situations. Using these results we can sometimes prove that $N \neq \Omega_n$, and correspondingly the nonexistence of L_2–solutions.

We shall present here some theorems of these two sorts. We start with an important example from the quantum solid state theory, namely the Schrödinger operator with a periodic potential: $H_q u = -\Delta u + qu$. Here the potential $q(x)$ is a 1-periodic function in each variable. We assume first that $q(x)$ is smooth, to avoid technical details. The case of more general potentials will be considered later. Let us emphasize also that the potential may be complex. The domain of H_q in $L_2(\mathbf{R}^n)$ is assumed to be the Sobolev space $H^2(\mathbf{R}^n)$.

Theorem 4.1.7. *The discrete spectrum of H_q in $L_2(\mathbf{R}^n)$ is empty (i.e. the equation $H_q u = Eu$ for $E \in \mathbf{C}$ has no square-integrable solutions in $\mathbf{R}^n$).*

Proof. It is sufficient to consider the problem of the existence of L_2-solutions of the equation $Hu = 0$, since the term Eu that corresponds to a non-zero energy level E can be included into the potential.

We shall produce a vector $\lambda \in \mathbf{C}^n$ such that the kernel of the operator $H_q(x, D - i\lambda)$ on $\mathbf{T}^n$ is zero. Then e^λ $(= (e^{\lambda_1}, \ldots, e^{\lambda_n}))$ does not belong to the set N of Floquet exponents, so by theorem 5 the equation $H_q u = 0$ has only zero solution in $L_2(\mathbf{R}^n)$. The vector will be of the form $\lambda = i(\alpha + i\beta, 0, \ldots, 0)$. If

$$u(x) = \sum_{k \in \mathbf{Z}^n} u_k e^{2\pi i k \cdot x},$$

then

$$H_q(x, D - i\lambda) u(x) = H_q(x, D + (\alpha + i\beta, 0, \ldots, 0))\, u(x) =$$

$$= \sum_k ((2\pi k_1 + \alpha + i\beta)^2 + 4\pi^2 k_2^2 + \ldots + 4\pi^2 k_n^2)\, u_k e^{2\pi i k \cdot x} + q(x)\, u(x).$$

Consider the expression

$$(2\pi k_1 + \alpha + i\beta)^2 + \sum_{j \geq 2} 4\pi^2 k_j^2 = (2\pi k_1 + \alpha)^2 - \beta^2 + \sum_{j \geq 2} 4\pi^2 k_j^2 + 2i\beta(2\pi k_1 + \alpha)$$

Its imaginary part is $2\beta(2\pi k_1 + \alpha)$. We chose now $\alpha = \pi$. Then

$$|\, 2\beta(2\pi k_1 + \alpha)\,| \geq 2\pi\, |\beta|.$$

If $\beta = \| q \|_{L_\infty(\mathbf{R}^n)}$, we obtain (by Parseval's equality)

$$\| H_q(x, D - i\lambda) u - qu \|_{L_2(\mathbf{T}^n)} \geq 2\pi \| q \|_{L_\infty} \| u \|_{L_2} .$$

Hence,

$$\| H_q(x, D - i\lambda)u \|_{L_2} \geq (2\pi - 1) \| q \|_{L_\infty} \| u \|_{L_2}.$$

This shows evidently that the kernel of H_q on $L_2(\mathbf{T}^n)$ is zero and finishes the proof of the theorem.

The condition of smoothness of the potential is too strong, and we really do not need it for validity of theorem 7. First of all, it is sufficient to assume that $q \in L_\infty(\mathbf{R}^n)$, because in this case the operator $H_q(x, D - i\lambda)$ is still Fredholm for all $\lambda \in \mathbf{C}^n$ and we may use theorem 5. Fredholm property is obvious because $H_q(x,D - i\lambda)u = H_0(x,D - i\lambda)u + qu$, and the multiplication operator $u \to qu$ is compact as an operator from $H^2(\mathbf{T}^n)$ into $L_2(\mathbf{T}^n)$. The condition of boundedness of the potential is, however, rather restrictive (it excludes, in particular, Coulomb singularities). Some additional efforts allow us to include a wide class of potentials.

Theorem 4.1.8. *The assertion of theorem 7 holds if the sequence of Fourier coefficients q_k of the potential $q(x)$ belongs to the space l_s, where $s<(n-1)/(n-2)$ for $n \geq 3$ and $s = 2$ for $n = 2$.*

Proof. The idea of the proof remains the same: to obtain the Fredholm property of $H_q(x, D - i\lambda)$, to use theorem 5 and to show that H_0 is "more powerful" than q, so for an appropriate "large" λ they cannot cancel on periodic functions.

Since $H_q u = H_0 u + qu$ and $H_0 : H^2(\mathbf{T}^n) \to L_2(\mathbf{T}^n)$ is Fredholm, in order to get the Fredholm property for H_q, it is sufficient to prove that the multiplication operator $u \to qu$ is compact as an operator between these spaces. In the cases $n = 2, 3$ the space $H^2(\mathbf{T}^n)$ is compactly embedded into $L_\infty(\mathbf{T}^n)$ and for $q \in L_2$ the multiplication operator $u \to qu$ is continuous from L_∞ into L_2. This yields the compactness. Let now $n>3$. We use here standard Hölder and Young type estimates [341] to obtain the compactness we need. The idea is to split the multiplication operator into the composition of the following operators: compact embedding of $H^2(\mathbf{T}^n)$ into $H^\alpha(\mathbf{T}^n)$ for some $0<\alpha<2$, continuous embedding of $H^\alpha(\mathbf{T}^n)$ into the space Λ_γ of functions u on $\mathbf{T}^n$ whose Fourier coefficients u_k belong to l_γ and the continuous convolution operator $u_k \to \{u_k\} * \{q_k\}$ from Λ_γ into $\Lambda_2 = L_2(\mathbf{T}^n)$. The first embedding is compact for any $\alpha<2$. If $u \in H^\alpha(\mathbf{T}^n)$, then

$$\sum(1+|k|^\alpha)^2\,|u_k|^2<\infty.$$

Hence, by Hölder inequality

$$\sum|u_k|^\gamma=\sum(1+|k|^\alpha)^\gamma|u_k|^\gamma(1+|k|^\alpha)^{-\gamma}\le$$
$$(\sum(1+|k|^\alpha)^2\,|u_k|^2)^{\gamma/2}\,(\sum(1+|k|^\alpha)^{-\frac{2\gamma}{2-\gamma}})^{(2-\gamma)/2}.$$

The last series is convergent if and only if

$$\frac{2\gamma\alpha}{2-\gamma}>n.$$

We can achieve this inequality for any $\gamma>\frac{2n}{4+n}$, choosing an appropriate $\alpha<2$. The Young inequality

$$\|\{u_k\}*\{q_k\}\|_2\le\ \|\{u_k\}\|_\gamma\|\{q_k\}\|_s\text{, where }\frac{1}{2}+1=\frac{1}{\gamma}+\frac{1}{s}$$

shows that we can find an appropriate $\gamma>\frac{2n}{4+n}$ if $\frac{1}{s}>\frac{3}{2}-\frac{4+n}{2n}=\frac{n-2}{n}$, or $s<\frac{n}{n-2}$. Let us notice now that our initial condition $s<\frac{n-1}{n-2}$ is even stronger than what we need here, so the Fredholm property is proved.

Now we know that theorem 5 can be applied. The last step is to find a (complex) vector λ such that the operator $H_0(D+i\lambda)+q$ has zero kernel in $H^2(\mathrm{T}^n)$. We are going to find such vector λ that the operator $qH_0(D+i\lambda)^{-1}$ is a contraction in $L_2(\mathrm{T}^n)$. This will obviously yield the triviality of the kernel: $(H_0+q)u=0\ \ =>$ $(1+qH_0{}^{-1})H_0u=0\ =>\ H_0u=0\ =>\ \ u=0$. The estimate on $qH_0{}^{-1}$ follows the same route as in the beginning of the proof: according to Hölder inequality, $H_0{}^{-1}$ maps L_2 into a space Λ_γ and is a contraction, the convolution $\{u_k\}\to\{u_k\}*\{q_k\}$ is continuous as the operator from Λ_γ into $\Lambda_2=L_2(\mathrm{T}^n)$. Let us consider a number $\gamma<2$ such that $\gamma^{-1}+s^{-1}=3/2$. In this case, according to Young inequality, $\|qu\|_{L_2}\le$ $\|u\|_{\Lambda_\gamma}\|q\|_{\Lambda_s}$ and the multiplication by q is a continuous operator from Λ_γ into L_2.

Now we estimate the norm of operator $H_o(D-i\lambda)^{-1}$, as an operator from L_2 into Λ_γ, where $\lambda = i(\pi + i\beta,\ 0, \ldots, 0)$. An easy calculation shows that

$$(H_o(D-i\lambda)^{-1}\, u)_k = ((2\pi k_1+\pi)^2 - \beta^2 + 4\pi^2 k'^2 + i2\beta\,(2\pi k_1+\pi))^{-1} u_k\,.$$

Here $k' = (k_2, k_3, \ldots, k_n) \in \mathbb{Z}^{n-1}$. If we denote $(H_o(D-i\lambda)^{-1}\, u)_k$ by $h(k,\beta)^{-1}\, u_k$, then

$$Im\ h(k,\beta) = 2\pi\beta(2k_1+1),$$

$$Re\ h(k,\beta) = \pi^2(2k_1+1)^2 + 4\pi^2 k'^2 - \beta^2$$

and

$$|\,Im\ h(k,\beta)\,| \geq \pi\,|\beta|\,(1+|k_1|)$$

$$|\,Re\ h(k,\beta)\,| \geq |k|^2, \text{ if } |k| \geq |\beta|.$$

We can estimate now:

$$\|\, H_o(D-i\lambda)^{-1}\, u\, \|^\gamma_{\Lambda_\gamma} = \sum |u_k|^\gamma\, |h\,(k,\beta)|^{-\gamma} \leq$$

$$(\sum |u_k|^2)^{\gamma/2} (\sum |h(k,\beta)|^{\frac{-2\gamma}{2-\gamma}})^{\frac{2-\gamma}{2}} = \|u\|^\gamma_{L_2} (\sum |h(k,\beta)|^{\frac{-2\gamma}{2-\gamma}})^{\frac{2-\gamma}{2}}.$$

Now we show that under our choice of γ the last sum is finite and tends to zero when $\beta \to \infty$:

$$\sum |h(k,\beta)|^{\frac{-2\gamma}{2-\gamma}} = \sum_{|k| \leq \beta} \cdots + \sum_{|k| > |\beta|} \cdots \leq$$

$$\leq \frac{C}{|\beta|^{\frac{2\gamma}{2-\gamma}}} \sum_{|k| \leq \beta} \frac{1}{(1+|k_1|)^{\frac{2\gamma}{2-\gamma}}} + \sum_{|k| > |\beta|} \frac{1}{|k|^{\frac{4\gamma}{2-\gamma}}}\,.$$

Since $s<(n-1)/(n-2)$, we have $\gamma>2(n-1)/(n+1)$, which guarantees the convergence. Splitting the first sum into summations over k_1 and then over k', we easily estimate the expression from above as

$$C \frac{|\beta|^{n-1}}{|\beta|^{\frac{2\gamma}{2-\gamma}}} + \sum_{|k|>|\beta|} \frac{1}{|k|^{\frac{4\gamma}{2-\gamma}}} .$$

The inequality for γ guarantees that both terms tend to zero when $\beta \to \infty$. We conclude that for sufficiently large β the operator $qH_0(D-i\lambda)^{-1}$ is a contraction in L_2. The theorem is proved.

The next example is the well known *Dirac operator* in $\mathbf{R}^n$:

$$\mathfrak{D} = -i\sum \alpha_j \frac{\partial}{\partial x_j} + m\beta + V(x) = \mathfrak{D}_0 + V, \tag{4.1.10}$$

where α_j $(j=1,..,n)$ and β are Hermitian $r\times r$-matrices satisfying the conditions

$$\alpha_j^2=\beta^2=I, \quad \alpha_j\alpha_k = -\alpha_k\alpha_j \; (j\neq k) , \; \alpha_j\beta = -\beta\alpha_j.$$

We assume that $V(x)$ is a 1-periodic in all variables scalar function. In the next theorem we shall assume that $V(x)$ is bounded. In this case we can define $\mathfrak{D}$ as a continuous operator from $H^1(\mathbf{R}^n,\mathbf{C}^r)$ into $L_2(\mathbf{R}^n,\mathbf{C}^r)$, or as a closed operator in $L_2(\mathbf{R}^n,\mathbf{C}^r)$ with the domain $H^1(\mathbf{R}^n,\mathbf{C}^r)$.

Theorem 4.1.9. *Let $V=V_1 + V_2$, where V_1 is periodic and continuous and $\| V_2\|_{L_\infty} <\pi$. Then the operator $\mathfrak{D}$ has no discrete spectrum in $L_2(\mathbf{R}^n,\mathbf{C}^r)$.*

We start with some discussion. The operator $\mathfrak{D}$ is elliptic (though a matrix one) , and the considerations of theorem 5 are valid in this case (see the system case in section 3.4). In order to prove theorem 9, we have to find a vector $\lambda \in \mathbf{C}^n$ such that the operator $\mathfrak{D}(x,D-i\lambda)$ has zero kernel on the torus $\mathbf{T}^n$. Now theorem 9 obviously follows from the next result.

Theorem 4.1.10. *For any number $0<s<\pi$ and any continuous periodic function $V(x)$ there is a vector $\lambda \in \mathbf{C}^n$ such that for any $\phi \in H^1(\mathbf{T}^n, \mathbf{C}^r)$*

(4.1.11) $$\|(\mathfrak{D}_0(D-i\lambda) + V)\,\phi\|_{L_2(\mathbf{T}^n,\,\mathbf{C}^r)} \geq s\,\|\phi\|_{L_2(\mathbf{T}^n,\,\mathbf{C}^r)}\,.$$

Remark. In the case of Schrödinger operator we could achieve any positive constant s in the estimate (11) (see the proof of theorem 7). In the case of Dirac operator we can have (11) only for $s < \pi$.

Proof. Let us notice first that we may assume $V(x)$ to be continuously differentiable (and even a trigonometric polynomial). In fact, given $0 < \epsilon < s$, we can always find a trigonometric polynomial $V_1(x)$ such that $\| V - V_1 \|_{L\infty} < \epsilon$ and then if

$$\|(\mathfrak{D}_0(D-i\lambda) + V_1)\,\varphi\| \geq s\|\varphi\|,$$

we have

$$\|(\mathfrak{D}_0(D-i\lambda) + V)\,\varphi\| \geq \|(\mathfrak{D}_0(D-i\lambda)+ V_1)\,\varphi\| - \|(V-V_1)\,\varphi\| \geq (s-\epsilon)\,\|\varphi\|\,.$$

We shall assume therefore that V is a trigonometric polynomial (in particular, it is continuously differentiable). If

$$\varphi(x) = \sum_{k\in\mathbf{Z}^n} \varphi_k\, exp\,(2\pi ik\cdot x)$$

is the Fourier series of φ, then

$$\mathfrak{D}_0(D-i\lambda)\varphi(x) = \sum_k \Big(\sum_{j=1}^n (2\pi k_j - i\lambda_j)\alpha_j + m\beta\Big)\,\varphi_k\, exp\,(2\pi ik\cdot x)\,.$$

We choose $\lambda = (i\pi + \kappa, 0,0,\ldots,0)$, then

$$[\mathfrak{D}_0(D-i\lambda)\,\varphi]_k = \Big(\pi(2k_1 + 1)\,\alpha_1 + 2\,\pi \sum_{j=2}^n k_j\alpha_j + m\beta - i\kappa\alpha_1\Big)\varphi_k$$

Let us denote

$$d_k(\kappa) = \pi\,(2k_1+1)\,\alpha_1 + 2\,\pi \sum_{j=2}^n k_j\alpha_j + m\beta - i\kappa\alpha_1,$$

i.e. $[\mathfrak{D}_o(D\text{-}i\lambda)\,\varphi]_k = d_k(\kappa)\,\varphi_k$. Simple calculations show:

$$
\begin{aligned}
d_k^2(\kappa) &= (\pi(2k_1+1)-i\kappa)^2 + 4\pi^2\sum_2^n k_j^2 + m^2 = \\
&= \pi^2(2k_1+1)^2 - \kappa^2 + 4\pi^2\sum_2^n k_j^2 + m^2 - 2i\kappa\pi\,(2k_1+1),
\end{aligned}
\tag{4.1.12}
$$

$$
d_k^*(\kappa) = d_k(\kappa) + 2i\kappa\alpha_1. \tag{4.1.13}
$$

From (12) we obtain that

$$
\|\, d_k^2(\kappa)\, c\| \geq 2\kappa\,\pi\,\| c\| \tag{4.1.14}
$$

for all $c \in \mathbf{C}^r$. Let us note that an estimate like (14) for the matrix d_k itself (instead of d_k^2) would give the statement of the theorem with arbitrary $s>0$ (see the proof of theorem 7). Unfortunately, such estimate does not take place, and we have to find another trick. Let us choose a positive number b (that we shall intend to make sufficiently large) and fix a function $\varphi = \sum \varphi_k \exp 2\pi ikx \in H^1(\mathbf{T}^n,\mathbf{C}^r)$. We split the grid $\mathbf{Z}^n$ into the disjoint union of following two sets. The first (the "good" one) is:

$$
A = \{k \in \mathbf{Z}^n \mid \|\, d_k(\kappa)\varphi_k\| \geq b\, \|\varphi_k\|\} \tag{4.1.15}
$$

(we denote here by $\|\cdot\|$ the Euclidian norm on $\mathbf{C}^r$). In other words, if $k \notin A$, then

$$
\|\, d_k(\kappa)\,\varphi_k\| < b\|\varphi_k\|\,. \tag{4.1.16}
$$

Now $B = \mathbf{Z}^n \setminus A$ is the "bad" set. (In the case of Schrödinger operator for any b we can achieve the equality $A = \mathbf{Z}^n$, choosing sufficiently large κ.)

We shall denote by $\varphi = \varphi_{(1)} + \varphi_{(2)}$ the decomposition of $\varphi(x)$ corresponding to the splitting $\mathbf{Z}^n = A \cup B$, i.e.

$$
\varphi = \varphi_{(1)} + \varphi_{(2)} = \sum_{k\in A} \varphi_k \exp(2\pi ik\cdot x) + \sum_{k\in B} \varphi_k \exp(2\pi ik\cdot x).
$$

In the sequel we denote by $\|\cdot\|$ either the L_2 - norm on $\mathbf{T}^n$ or l_2 - norm in $\mathbf{C}^r$. Now we have :

$$b\|\varphi_{(1)}\| \le \|\mathfrak{D}_o(D-i\lambda)\varphi_{(1)}\| =$$

$$\|[(\mathfrak{D}(D\text{-}i\lambda)-V))\varphi]_{(1)}\| \le \|[\mathfrak{D}(D\text{-}i\lambda)\varphi]_{(1)}\| + \|V\|_\infty\|\varphi_{(1)}\| + \|V\|_\infty\|\varphi_{(2)}\|$$

or

$$(4.1.17) \qquad (b-\|V\|_\infty)\|\varphi_{(1)}\| \le \|[\mathfrak{D}(D\text{-}i\lambda)\varphi]_{(1)}\| + \|V\|_\infty\|\varphi_{(2)}\|.$$

Let us pass on to estimates of $\varphi_{(2)}$. This is obviously a more complicated task, because the Fourier image of $\varphi_{(2)}$ is supported on the "bad" set B. By (14), we have for $k\in B$:

$$2\pi\kappa\|\varphi_k\|^2 \le |(\varphi_k, d_k^2(\kappa)\varphi_k)| = |(d_k^*(\kappa)\varphi_k, d_k(\kappa)\varphi_k)| =$$

$$= |((d_k(\kappa)+2i\kappa\alpha_1)\varphi_k, d_k(\kappa)\varphi_k)| = |((d_k(\kappa)+2i\kappa\alpha_1)\varphi_k, [(\mathfrak{D}(D-i\lambda)-V)\varphi]_k)| \le$$

$$\le |((d_k(\kappa)+2i\kappa\alpha_1)\varphi_k, [\mathfrak{D}(D-i\lambda)\varphi]_k)| + |((d_k(\kappa)+2i\kappa\alpha_1)\varphi_k, [V\varphi_{(1)}]_k)| +$$

$$+ |(d_k^*(\kappa)\varphi_k, [V\varphi_{(2)}]_k)$$

Taking into account the definitions of the set B and of $\varphi_{(2)}$, we obtain that the first two terms in the last sum do not exceed

$$(b+2\kappa)\|\varphi_k\|(\|[\mathfrak{D}(D\text{-}i\lambda)\varphi]_k\| + \|[V\varphi_{(1)}]_k\|).$$

Let us transform the third term:

$$|(d_k^*(\kappa)\varphi_k, [V\varphi_{(2)}]_k)| \le$$

$$|(\varphi_k, [V\mathfrak{D}_o(D-i\lambda)\,\varphi_{(2)}]_k)| + |(\varphi_k, [[\mathfrak{D}_o(D-i\lambda), V]\,\varphi_{(2)}]_k)|,$$

where $[\cdot\ ,\cdot\ ]$ denotes the operator commutator. Adding these inequalities for all $k\in B$ together, calculating the commutator and using Cauchy inequality, we get:

$$2\pi\kappa\|\varphi_{(2)}\|^2 \le (b+2\kappa)\|\varphi_{(2)}\|(\|[\mathfrak{D}(D\text{-}i\lambda)\,\varphi]_{(2)}\| + \|[V\varphi_{(1)}]_{(2)}\|) +$$

$$\|\varphi_{(2)}\|\ \|[V\mathfrak{D}_o(D-i\lambda)\,\varphi_{(2)}]_{(2)}\| + \|\varphi_{(2)}\|\ \|[[\mathfrak{D}_o(D-i\lambda), V]\,\varphi_{(2)}]_{(2)}\| \le$$

$$(b+2\kappa)\|\varphi_{(2)}\|(\|[\mathfrak{D}(D\text{-}i\lambda)\,\varphi]_{(2)}\| + \|V\|_\infty\|\varphi_{(1)}\|) +$$

$$\|\varphi_{(2)}\|^2(b\|V\|_\infty + C\|\nabla V\|_\infty),$$

or

$$(\pi - \frac{b\|V\|_\infty + C\|\nabla V\|_\infty}{2\kappa})\ \|\varphi_{(2)}\| \le$$

(4.1.18)

$$(1+\frac{b}{2\kappa})\ (\|[\mathfrak{D}(D-i\lambda)\,\varphi]_{(2)}\| + \|V\|_\infty\|\varphi_{(1)}\|)\,.$$

Let us plug into this inequality the estimate (17) for $\|\varphi_{(1)}\|$ (assuming that $b>\|V\|_\infty$):

$$\|\varphi_{(1)}\| \le \frac{1}{b-\|V\|_\infty}\ \|[\mathfrak{D}(D\text{-}i\lambda)\,\varphi]_{(1)}\| + \frac{\|V\|_\infty}{b-\|V\|_\infty}\ \|\varphi_{(2)}\|\,. \tag{4.1.19}$$

Then we obtain:

$$(\pi - \frac{b\|V\|_\infty + C\|\nabla V\|_\infty}{2\kappa} - (1+\frac{b}{2\kappa})\ \frac{\|V\|_\infty^2}{b-\|V\|_\infty})\|\varphi_{(2)}\| \le$$

$$\le (1+\frac{b}{2\kappa})\ \|[\mathfrak{D}(D-i\lambda)\,\varphi]_{(2)}\| + \frac{(1+\frac{b}{2\kappa})\|V\|_\infty}{b-\|V\|_\infty}\ \|[\mathfrak{D}(D\text{-}i\lambda)\,\varphi]_{(1)}\| \le$$

$$\le (1+\frac{b}{2\kappa})\ (1+\frac{\|V\|_\infty}{b-\|V\|_\infty})\ \|\mathfrak{D}(D-i\lambda)\,\varphi\|\,.$$

It is obvious that for any $\epsilon > 0$ an appropriate choice of large b and κ leads to:

$$(\pi - \epsilon) \| \varphi_{(2)} \| \leq (1 + \epsilon) \| \mathfrak{D}(D - i\lambda) \varphi \| . \tag{4.1.20}$$

If we plug (20) into (19), we obtain even better estimate for $\| \varphi_{(1)} \|$:

$$\| \varphi_{(1)} \| \leq \delta \| \mathfrak{D}(D\text{-}i\lambda) \varphi \|$$

with a small δ. We observe that κ does not depend on the choice of φ. This proves the theorem.

As we have already mentioned, the question about existence of rapidly (sub-exponentially) decreasing solutions of "normal" elliptic equations has been considered in many works. Though such pathologies do occur, for some "nice" equations we are able to prove non-existence of sub-exponential solutions, which leads by theorems 5 and 6 to corresponding results for periodic equations. The next result is an example of this sort.

Theorem 4.1.11. *Let*

$$Lu = \sum_{i,j=1}^{n} \frac{\partial}{\partial x_i} (a_{i,j}(x) \frac{\partial u}{\partial x_j}),$$

where $a_{i,j}(x) \in C^{\infty}(\mathbf{R}^n)$ *are periodic real functions and* $\sum a_{i,j}(x)\, \xi_i \xi_j \geq C |\xi|^2$ *for all* $\xi \in \mathbf{R}^n$ *and for some* $C>0$. *Then,*

1) $N \neq \Omega_n$,

2) the equation

$$Lu=0 \tag{4.1.21}$$

has no non-zero solutions in $L_2(\mathbf{R}^n)$.

Proof. By theorem 5, statements 1) and 2) are equivalent to each other and to the non-existence of non-zero solutions of (21) with the estimate

$$|u(x)| \leq C_a \exp(-a|x|) \quad \forall a>o.$$

Theorem 4.1 from [245] says, however, that such solutions do not exist. This completes the proof.

We shall provide another example of this type for elliptic boundary problems in section 5.3.

4.2. Solvability of non-homogeneous equations

It is well known that in the case of periodic ordinary differential equations the knowledge of Floquet exponents distribution enables one to answer many questions about properties of the equation (see [17, 177]). For instance, the absence of Floquet exponents on the unit circle is equivalent to the (unique) solvability of non-homogeneous equation in square-integrable or bounded functions. The same is true for PDEs.

We regard, as before, the torus $\mathcal{T}^n$ in Ω_n as the skeleton of the unit polydisk: $\mathcal{T}^n = \{ z \in \Omega_n \mid |z_j| = 1,\ j = 1, \ldots, n\}$. The set of multipliers, or Floquet exponents, of the equation $Lu = 0$ is denoted by N. Here L is a scalar elliptic operator of order $2m$ with smooth periodic coefficients.

Theorem 4.2.1. *The following assertions are equivalent:*

i) $N \cap \mathcal{T}^n = \emptyset$.

ii) The equation $Lu = f$ *is uniquely solvable in* $L_2(\mathbf{R}^n)$.

iii) For any function $f \in L_\infty(\mathbf{R}^n)$ *there exists a unique solution* $U \in H^{2m}_{loc}(\mathbf{R}^n)$ *with the bound*

$$(4.2.1) \qquad \sup_{k \in \mathbf{Z}^n} \| u \|_{H^{2m}(\mathcal{K} + k)} < \infty .$$

In particular, for $2m > n/2$ *there exists a unique bounded solution.*

iv) For any function f with $|f(x)| \leq C \exp(-a|x|)$ *there exists a solution of the equation* $Lu=f$ *with the bound* $|u(x)| \leq C_1 \exp(-a_1 |x|)$.

Proof. The equivalence i)<=>ii) follows immediately from theorem 2.2.5. Namely, if $N \cap \mathcal{T}^n = \emptyset$, then $L(z) = FLF^{-1}$ has zero kernel for all $z \in \mathcal{T}^n$. Because $L(z)$ is Fredholm operator of index zero, it is invertible for all $z \in \mathcal{T}^n$. Due to analyticity, $L(z)$ is invertible in a neighborhood of $\mathcal{T}^n$, so $L^{-1}(z)$ is a continuous bounded operator function on $\mathcal{T}^n$ with values in $L(L_2(\mathbf{T}^n))$. Then the operator

$$\int_{\mathfrak{T}^n}^{\oplus} L^{-1}(z)$$

is inverse to L, which gives the unique solvability of the equation. And conversely, assume that the equation $Lu=f$ is uniquely solvable in $L_2(\mathbf{R}^n)$. If now $N\bigcap\mathfrak{T}^n\neq\emptyset$, we have two possibilities: a) N does not contain $\mathfrak{T}^n$, and b) $N=\Omega_n$ (see theorem 1.5.19). Let us consider first the case a). This means that $L(z)$ is invertible for almost all (except an analytic subset), but not for all values of $z\in\mathfrak{T}^n$. We have to show that in this case our operator L is not invertible in $L_2(\mathbf{R}^n)$. Let us choose a point $z_0\in\mathfrak{T}^n$ such that $Ker\ L(z_0)\neq 0$. Choose a non-zero vector $u\in Ker\ L(z_0)$ and consider the function $g(z)=L(z)u$. This function is obviously analytic, $g(z_0)=0$ and $g(z)$ is not identically zero. We can find a scalar function $f(z)\in L_2(\mathbf{T}^n)$ such that $\|g(z)\|^{-1}f(z)$ is not square integrable. Define now $\varphi(z)=\|g(z)\|^{-1}f(z)u$. This function does not belong to L_2. On the other hand, applying to it the operator

$$\int_{\mathfrak{T}^n}^{\oplus} L(z),$$

we get the function $y(z)=L(z)\varphi(z)=\|g(z)\|^{-1}f(z)g(z)$. This function is measurable and $\|y(z)\|=|f(z)|\in L_2(\mathfrak{T}^n)$, so $y\in L_2(\mathfrak{T}^n,L_2(\mathbf{T}^n))$. If our operator was invertible in $L_2(\mathbf{R}^n)$, we would have a function $\psi\in L_2(\mathfrak{T}^n,\ H^{2m}(\mathbf{T}^n))$ such that $L(z)\psi(z)=y(z)$. In this case $\psi(z)$ would be almost everywhere equal to $\varphi(z)$, which is a contradiction because $\varphi\notin L_2$. In the case when $N=\mathbf{C}^n$, by theorem 4.1.5, there exists a non-zero solution $u\in L_2(\mathbf{R}^n)$ of the equation $Lu=0$, which contradicts to ii).

Let us prove now the equivalence i) <=> iii). Let $f\in L_\infty(\mathbf{R}^n)$. Then f determines a continuous linear functional on $\overset{o}{\Psi}_n$ by the formula

$$<\varphi,f> = \int_{\mathbf{R}^n} f(x)\ \varphi(x)\ dx\ .$$

Suppose now that i) is satisfied and define a linear functional u by the formula $<u,L_1\varphi>:=\ <f,\ \varphi>$ for all $\varphi\in\Psi_n^{2m}$ (here L_1 is the transposed operator to L). If we can prove that this functional is well-defined and that u is continuous on the image of $L_1:\Psi_n^{2m}\to\overset{o}{\Psi}_n$, then, extending u by the Hahn-Banach theorem to the whole of $\overset{o}{\Psi}_n$, we obtain a generalized function $u(x)$ that satisfies the equation $Lu=f$ and that can be continuously extended to the whole of Ψ_n^o. The latter property implies an estimate of the type described in iii). Let us implement this scheme.

The first thing to prove is the invertibility of $L_1:\Psi_n^{2m} \to \Psi_n^o$. By theorems 2.2.2 and 2.2.3, under the "Fourier" transform $\boldsymbol{F}$ we obtain the operator

$$\int_{\mathcal{T}^n}^{\oplus} L_1(z)\colon W(\mathcal{T}^n, \mathcal{S}^{2m}) \to W(\mathcal{T}^n, \mathcal{S}^o)$$

between "Wiener-type" spaces. Due to i), the operator $L_1(z)\colon H^{2m}(\mathbf{T}^n) \to L_2(\mathbf{T}^n)$ (which is analytic and Fredholm) is surjective for all $z \in \mathcal{T}^n$ (because $CS(L_1) = S(L) = N$). Having zero index, it is also invertible for all $z \in \mathcal{T}^n$. Let us consider the operator function $L_1^{-1}(z)$. It is analytic in a neighborhood of $\mathcal{T}^n$. This means that

$$\int_{\mathcal{T}^n}^{\oplus} L_1^{-1}(z)$$

is a bounded operator from $W(\mathcal{T}^n, \mathcal{S}^0)$ into $W(\mathcal{T}^n, \mathcal{S}^{2m})$ (see the beginning of section 1.5) and is inverse to L_1. This shows that our definition of u is correct and u is a continuous linear functional on Ψ_n^0. We need only to obtain necessary estimates on u. First of all, due to ellipticity, $u \in H^{2m}_{loc}(\mathbf{R}^n)$. Since u belongs to $\{\Psi_n^o\}^*$, we have the following estimate:

$$|\int u(x)\, \varphi(x)\, dx| \leq C\Big(\sum_{k \in \mathbf{Z}^n} \|\varphi\|_{L_2(\mathcal{K}+k)}\Big). \tag{4.2.2}$$

Choosing $\psi \in L_2(M)$, where M is any compact in $\mathbf{R}^n$, and plugging into (2) translations of ψ instead of φ, we obtain that

$$\sup_{k \in \mathbf{Z}^n} \|u\|_{L_2(M+k)} < \infty. \tag{4.2.3}$$

Let now B be a periodic proper pseudodifferential parametrix for L (see lemma 2.1.1). Then there exists a compact $M \subset \mathbf{R}^n$ such that

$$\|Bv\|_{H^{2m}(\mathcal{K}+k)} \leq C\|v\|_{L_2(M+k)}$$

for all $k \in \mathbb{Z}^n$ and $v \in L_{2,\,loc}(\mathbf{R}^n)$. Together with (3) and the equation $Lu=f$, this gives:

$$u = Bf + (1 - BL)u,$$

$$\|u\|_{H^{2m}(\mathcal{K}+k)} \le C(\|f\|_{L_2(M+k)} + \|u\|_{L_2(M+k)}).$$

This obviously implies (1), so the implication i)⇒iii) is proved. If now i) is not satisfied, i.e. $N \cap \mathcal{T}^n \neq \emptyset$, then there is a bounded Bloch solution of the homogeneous equation. This contradicts iii).

Let us prove now the equivalence i) <=> iv). By theorems 2.2.2 and 2.2.3, the assertion iv) means the existence of a positive number a such that the operator $L(z)$ is invertible for all $z \in \Omega_n(a) = \{z \in \Omega_n \mid e^{-a} < |z_j| < e^{a}\}$. Since the open sets $\Omega_n(a)$ for various $a > 0$ form a basis of neighborhoods of $\mathcal{T}^n$ in Ω_n, the existence of an $a > 0$ such that $N \cap \mathcal{U}_a = \emptyset$ is equivalent to $N \cap \mathcal{T}^n = \emptyset$. This proves the theorem.

4.3. Bloch property

There is a famous theorem of F. Bloch [54] (cf. also [136]): If the stationary (self-adjoint) Schrödinger equation $\Delta u + qu = 0$ with a periodic potential q has a non-zero bounded solution in $\mathbf{R}^n$, then it has a bounded Bloch solution, i.e. a Bloch solution $exp(i\lambda \cdot x)\, v(x)$ with a real quasimomentum λ. In other words, the existence of a non-zero bounded solution implies (by theorem 4.2.1) that the L_2-spectrum of the operator contains zero.

By a result of Shnol' [356, 136], instead of the existence of a bounded solution it suffices to require that for any $a>o$ there exists a non-zero solution $u_a(x)$ satisfying

$$|u_a(x)| \le C_a \exp a|x|.$$

In the form that involves L_2-spectra instead of Bloch solutions, Bloch property holds in a wide class of equations, not necessarily periodic ones (see [360]). We shall see here, how to strengthen (even for the Schrödinger operator) and generalize Bloch's theorem in the periodic case.

Theorem 4.3.1. *Let $L(x,D)$ be a scalar periodic elliptic operator with smooth coefficients in* $\mathbf{R}^n$. If *$Lu=0$ has a non-zero solution satisfying*

$$|u(x)| \leq C \exp a|x| = C \exp a\sum |x_i| \tag{4.3.1}$$

for some $a>o$, then it also has a Bloch solution with a quasimomentum $\lambda \in \mathbf{C}^n$ such that

$$|Im\ \lambda_j| \leq a \quad \text{for all } j=1,\ldots, m$$

(that is, a Bloch solution that also satisfies (1)).

Proof. Suppose that there are no Bloch solutions with quasi-momentums λ for which $|Im\ \lambda_j| \leq a$ for all j. This means that N does not intersect the closure of the poly-annulus $\Omega_n(a)$. Hence, $N \cap \Omega_n(a+\epsilon) = \emptyset$ for some $\epsilon>0$. We can conclude that operator

$$L_1 = \int_{\mathfrak{T}^n}^{\oplus} L_1(z)$$

is an isomorphism of the space $\Phi^{2m,a+\epsilon}_{n,1}$ onto $\Phi^{0,a+\epsilon}_{n,1}$ (here L_1 is the transposed operator L, as usual). On the other hand, our solution $u(x)$ determines some continuous linear functional on $\Phi^{0,a+\epsilon}_{n,1}$ by the formula

$$<u,\varphi> = \int_{\mathbf{R}^n} u(x)\varphi(x)dx.$$

It is easy to observe that $<u, L_1\psi> = <Lu,\psi>$ for any $\psi \in \Phi^{2m,a+\epsilon}_{n,1}$. It follows that $<u,\varphi> = <Lu, L_1^{-1}\varphi> = 0$ for any $\varphi \in \Phi^{2m,a+\epsilon}_{n,1}$. Since

$$\mathfrak{D}(\mathbf{R}^n) \subset \Phi^{2m,a+\epsilon}_{n,1},$$

we deduce that $u \equiv 0$. This contradiction proves the theorem.

4.4. Quasimomentum dispersion relation. Bloch variety

So far we have not introduced the spectral parameter. We shall now discuss some problems that require this. Let us consider the Schrödinger operator $H = -\Delta + q(x)$ with a real periodic potential $q \in L_{2,loc}(\mathbf{R}^n)$. For each $k \in \mathbf{R}^n$ the self-adjoint problem

$$Hu = \lambda u,$$

$$u(x+\gamma) = e^{ik\cdot\gamma}u(x) \quad (\forall \gamma \in \mathbf{Z}^n)$$

has discrete spectrum $E_1(k) \leq E_2(k) \leq \ldots$. The eigenvalue $E_n(k)$, as a function of quasimomentum k, is the *n-th band function*. The set of all band functions is the energy-quasimomentum *dispersion relation*.

It is convenient to work with the graph of the dispersion relation. This leads to the notion of Bloch variety.

Definition 4.4.1. Let $L(x,D)$ be a smooth periodic elliptic operator on $\mathbf{R}^n$. The *Bloch variety* of the operator is the following set:

$$B(L)=\{(k, \lambda) \in \mathbf{C}^n \times \mathbf{C} \mid \text{the equation } Lu = \lambda u \text{ has a non-zero Bloch solution with the quasimomentum } k\}.$$

The Bloch variety $B(L)$ is a subset of $\mathbf{C}^{n+1}$. The next results describes the analytic structure of $B(L)$.

Theorem 4.4.2. *$B(L)$ is the set of all zeros of an entire function of order n (generally speaking, of infinite type) on $\mathbf{C}^{n+1}$. Increasing the order to $n+1$, we can achieve $2\pi\mathbf{Z}^n$-periodicity of this function with respect to k.*

The proof coincides with the proof of theorem 3.1.7. We just have the additional argument in the operator polynomial $L(k,\lambda)$, but this circumstance does not change anything in the proof.

Let us mention that definition 4.4.1 and theorem 4.4.2 are valid for general periodic elliptic operators, and do not require self-adjointness. For the self-adjoint Schrödinger operator, however, much more can be said about the structure of the Bloch variety.

We shall give just a brief account of known results here. The reader can find further references in section 4.7.

Let us start with the one-dimensional case of the operator

$$Ly = -y'' + q(x)y, \quad x \in \mathbf{R}, \tag{4.4.1}$$

where $q \in L^2(\mathbf{R}/\mathbf{Z})$. Consider the solutions $y_j(x,\lambda)$ $(j=1, 2)$ of the equation

$$Ly = \lambda y, \tag{4.4.2}$$

satisfying the initial conditions

$$\begin{aligned} y_1(0, \lambda) &= y_2'(0, \lambda) = 1 \\ y_1'(0,\lambda) &= y_2(0, \lambda) = 0. \end{aligned} \tag{4.4.3}$$

Then, any solution of (2) can be written as $y=y(0)\, y_1 + y'(0)\, y_2$. Consider the matrix

$$\phi(\lambda) = \begin{pmatrix} y_1(1,\lambda) & y_2(1,\lambda) \\ y_1'(1,\lambda) & y_2'(1, \lambda) \end{pmatrix}$$

and set $\Delta(\lambda) = \mathit{trace}\ \phi(\lambda) = y_1(1, \lambda) + y_2'(1, \lambda)$. This is an entire function of order 1/2. It is trivial to check that a Bloch solution of (1) with a quasimomentum k exists if and only if $\mathit{exp}\ ik$ is an eigenvalue of $\phi(\lambda)$, and hence

$$\det\ (\phi(\lambda) - e^{ik} I) = 0. \tag{4.4.4}$$

The characteristic polynomial in (4) can be evaluated as

$$\det\ (\phi(\lambda) - zI) = z^2 - \Delta(\lambda)\, z + \det \phi(\lambda) = z^2 - \Delta(\lambda)\, z + 1,$$

since, due to Liouville's formula, $det\ \phi(\lambda) = 1$. Therefore,

$$B(L) = \{(k,\lambda) \in \mathbb{C}\times\mathbb{C} \mid z = exp\ ik \text{ is a root of } z^2 - \Delta(\lambda)z + 1 = 0\} = \{(k,\lambda) \in \mathbb{C}\times\mathbb{C} \mid 2\cos k - \Delta(\lambda) = 0\}. \tag{4.4.5}$$

So we get the following equation for the Bloch variety $B(L)$:

$$2\cos k - \Delta(\lambda) = 0. \tag{4.4.6}$$

This provides the entire function from the statement of theorem 2. However, the reader can notice that the order of the entire function in the left hand side of (6) is lower than the one guaranteed by the theorem.

We shall summarize now some beautiful results about the analytic structure of the Bloch variety in the one-dimensional case.

We shall need some known properties of the function $\Delta(\lambda)$ (see, for instance, [94]). There exists a sequence of real numbers tending to infinity

$$-\infty < \lambda_0 < \mu_0 \le \mu_1 < \lambda_1 \le \lambda_2 < \mu_2 \le \mu_3 < \dots$$

such that

(*i*) the spectrum $\sigma(L)$ of L in $L_2(\mathbb{R})$ is $\bigcup_{k=0}^{\infty} [\lambda_k, \mu_k]$;

(*ii*) $\Delta(\lambda) = 2$ only at $\lambda = \lambda_j$, $\Delta(\lambda) = -2$ only at $\lambda = \mu_j$;

(*iii*) $\Delta(\lambda)$ is monotonically decreasing on $(-\infty, \mu_0]$ and $[\lambda_j, \mu_j]$, and monotonically increasing on $[\mu_j, \lambda_j]$;

(*iv*) If $\mu_{2k} \ne \mu_{2k+1}$, then $\Delta'_\lambda(\mu_{2k}) \ne 0$, $\Delta'_\lambda(\mu_{2k+1}) \ne 0$. The same for $\lambda_{2k-1} \ne \lambda_{2k}$. If $\mu_{2k} = \mu_{2k+1}$ or $\lambda_{2k-1} = \lambda_{2k}$, then $\Delta'_\lambda = 0$ at these points, but $\Delta''_{\lambda\lambda}$ is not zero.

Theorem 4.4.3. *If $n = 1$, then the Bloch variety $B(L)$ is irreducible up to the* *natural periodicity, i.e. if $V \subset B(L)$ is an irreducible component, then all other components are $2\pi\mathbf{Z}$ – translations (with respect to k) of V.*

It is better to eliminate the symmetry by considering the "Floquet variety"

$$F(L) = \{(z,\lambda) \in \mathbf{C}^* \times \mathbf{C} \mid Lu=\lambda u \text{ has a non-zero Bloch solution with Floquet exponent } z\} = B(L)/2\pi\mathbf{Z}. \tag{4.4.7}$$

The statement of theorem 3 is equivalent to the following one:

Theorem 4.4.3′. *If $n = 1$, then the Floquet variety $F(L)$ is irreducible.*

Proof. According to (5) and (7),

$$F(L)=\{(z,\lambda) \in \mathbf{C}^* \times \mathbf{C} \mid z^2 - \Delta(\lambda)z+1=0\}.$$

In other words, $F(L)$ is the graph of the multivalued analytic function

$$z=\frac{\Delta(\lambda) \pm \sqrt{\Delta(\lambda)^2-4}}{2}. \tag{4.4.8}$$

Let us investigate the structure of the Riemann surface of this function. According to the properties (*i*) and (*ii*) of $\Delta(\lambda)$, the function $z(\lambda)$ has quadratic branch points at the point λ_0, and at all points μ_j, λ_j, provided the corresponding "instability zones" (= gaps in the spectrum) do not collapse into single points (i.e. $\mu_{2k} \neq \mu_{2k+1}$, or $\lambda_{2k-1} \neq \lambda_{2k}$). If the zone does collapse, then due to the property (*iv*) the Riemann surface has simple intersection. In other words, we can obtain this surface as follows: Take two copies of $\mathbf{C}$, cut both of them along the spectrum $\sigma(L)$, and paste them together along $\sigma(L)$. The pasting has to be done in the following way: the northern bank of each cut $[\lambda_j, \mu_j]$ on the upper copy is connected with the southern bank of the lower one, and vice-versa. Due to the type of branching described above, we obviously see that the set of regular (non-branching) points of the resulting surface is connected. This means that the graph of the function (8) (i.e. $F(L)$) is irreducible. The proof is complete.

Theorem 4.4.4. *Let n=1. Then the following statements are equivalent:*

(i) $q(x) = const;$

(ii) there are no gaps in the spectrum $\sigma(L)$ *(i.e.* $\mu_{2j}=\mu_{2j+1}, \lambda_{2j+1}=\lambda_{2j+2}\ \forall j$*);*

(iii) the Bloch variety $B(L)$ *contains the graph of an entire function* $\lambda(k)$ *(i.e.* $B(L) \supset \{(k,\lambda) \mid \lambda = \lambda(k)\}$*).*

Proof. The equivalence of *(i)* and *(ii)* constitutes the famous Borg's theorem (see [56]), so we shall not repeat its proof here. Let us prove the implication *(ii)*⇒*(iii)*. Consider the projection $B(L) \to \mathbf{C}$, $(k,\lambda) \to k$. We are going to show that for every branch of $B(L)$ this projection has local analytic section at every point. Then analytic continuation would give the entire section $\lambda(k)$ we are looking for. Considering the equation $2\cos k = \Delta(\lambda)$ for $B(L)$, we see that the troubles with this projection could happen only at the points where either $\Delta'(\lambda) = 0$, or $\Delta^2(\lambda)=4$. Under the condition *(ii)* these two sets almost coincide: $\Delta'(\lambda)=0$ only at the points λ_j $(j \geq 1)$ and $\mu_j (j \geq 0)$, and $\Delta^2(\lambda) = 4$ exactly at the same points, and also at the point λ_0 (where $\Delta'(\lambda_0) \neq 0$). At every point $(k,\lambda) \in B(L)$ where neither $\Delta'(\lambda)$ nor $\Delta^2(\lambda) - 4$ have zeros, the projection onto $\mathbf{C}$ has rank one, because the derivative $\partial k/\partial\lambda$ that we find from (6) is not zero. Let us consider first the points where $\Delta'(\lambda)=0$ and $\Delta^2(\lambda)-4=0$. As we have already mentioned before, $\Delta''_{\lambda\lambda} \neq 0$ at these points, so we get from (6) that

$$k'_\lambda = -\frac{\Delta'(\lambda)}{\sqrt{4-\Delta^2(\lambda)}} \neq 0 \, .$$

We have only one of "suspicious" points left: $\Delta(\lambda_0) = 2$, $\Delta'(\lambda_0) \neq 0$. In some local coordinates we may assume that $\Delta(\lambda) = 2 - \lambda$, and $\lambda_0 = 0$, so we get the equation

$$2\cos k = 2 - \lambda \, ,$$

that obviously gives analytic dependence λ on k.

Let us prove now the implication *(iii)*⇒*(ii)*. Suppose we have an entire function $\lambda(k)$ such that its graph Λ is a part of $B(L)$. In this case Λ is an irreducible component of $B(L)$. Due to theorem 3, its $2\pi\mathbf{Z}$ – translations (with respect to k-variable) cover the whole Bloch variety. This means that the spectrum $\sigma(L)$ consists of all values of the function $\lambda(k)$ on the real axes. This set of values obviously cannot have any gaps. The proof is complete.

A new important analytic tool has been introduced recently in order to investigate properties of Bloch varieties. This is the so called *directional compactification* of $B(L)$ (see [202, 39]). We will provide just a brief description of this construction in the case $n=1$. Let us start with the case of zero potential, i.e. $L_0=-d^2/dx^2$. In this case the reader can easily find that $B(L_0) = \bigcup_{m\in 2\pi\mathbf{Z}} B_m(L_0)$, where

$$B_m(L_0) = \{(k,\lambda) \in \mathbf{C}\times\mathbf{C} \mid (k+m)^2 = \lambda\}.$$

We consider the following embedding of $\mathbf{C}\times\mathbf{C}$ into the projective space $\mathbf{P}^2$:

$$(k,\lambda) \to (k, \lambda - k^2, 1).$$

We use here the homogeneous coordinates (α,β,γ) on $\mathbf{P}^2$. The closures of the components B_m coincide with the lines

$$2\alpha m + \gamma m^2 - \beta = 0 \; (m\in 2\pi\mathbf{Z}).$$

These lines intersect with the line "at infinity" $L_\infty = \{(\alpha, \beta, 0)\} \subset \mathbf{P}^2$ at the points $(1,2m, 0)$, $m\in 2\pi\mathbf{Z}$. For a general potential we have the equation

$$2 \cos k = \Delta(\lambda)$$

instead of $\cos k = \cos\sqrt{\lambda}$, and it behaves not too well along L_∞. It was suggested in [202] to consider a "directional compactification" of $B(L)$. Namely, let $0<\theta<\pi$, and

$$\Sigma(\theta) = \{(k,\lambda) \in \mathbf{C}\times\mathbf{C} \mid \arg k^2 \notin (-\theta,\theta)\}.$$

It can be proved that the closure of $B(L)\cap\Sigma(\theta)$ in $\mathbf{P}^2$ (after the embedding described above) intersects $L_\infty\backslash\{0,1,0\}$ exactly at the points $(1,2m,0)$, $m\in 2\pi\mathbf{Z}$, and in a neighborhood of each of them this closure looks like the cone

$$\{z\in\mathbf{C} \mid |z|<1, \arg z^2 \in (-\theta,\theta)\}.$$

This type of directional compactification was also constructed in [202, 39] for $n=2$, 3. We refer the reader to these works for the details of the construction. A detailed study of behavior of the Bloch variety at infinity led the authors of these papers to some impressive generalizations of the theorems stated above. We shall state these results without proofs here.

Theorem 4.4.5. *Let $n=2$, $q \in L^2(\mathbf{R}^2/\mathbf{Z}^2)$ be real valued and*

$$L = -\Delta + q.$$

Then for any two irreducible components C_1, C_2 of $B(L)$ there is $m \in 2\pi\mathbf{Z}^2$ such that $m \cdot C_1 = C_2$ (where $m\cdot$ means the translation with respect to the k-variable onto the vector m). In other words, the Floquet variety $F(L)=B(L)/2\pi\mathbf{Z}^2$ is irreducible.

It is well known that the direct analog of Borg's theorem (i.e. the equivalence of (*i*) and (*ii*) in theorem 4) does not hold for $n>1$. In fact, in this case the spectrum $\sigma(L)$ has no gaps for any periodic potential with sufficiently small L_∞ – norm (see sections 4.5, 4.7). However, the equivalence of (*i*) and (*iii*) still holds, and represents a multi-dimensional version of Borg's theorem.

Theorem 4.4.6. *Let $q \in L^2(\mathbf{R}^2/\mathbf{Z}^2)$ be a real valued potential. Then the following assertions are equivalent:*

(i) q is constant;

(ii) there exists an entire function $\Lambda : \mathbf{C}^2 \to \mathbf{C}$ such that the Bloch variety of $(-\Delta + q)$ is the union of the graph of Λ and of its translations by $2\pi\mathbf{Z}^2$, that is

$$B(L) = \bigcup_{m \in 2\pi\mathbf{Z}^2} \{(k,\lambda) \mid \lambda = \Lambda(k+m)\}.$$

Some analogs of these results have been also obtained for $n=3$ (see [39]).

4.5. Some problems of spectral theory

In this section we are going to study the structure of spectra of periodic elliptic operators. In fact, it is impossible to cover the whole subject in one section. It would take a large book to discuss this area, and several books have already been completely or partially devoted to it [94, v.4 of 341, 372, 389]. Thus, we give here and in section 4.6 only a brief sketch of this topic. The proofs of most serious results will be either skipped, or just sketched. The reader can find the references in section 4.7. Some familiarity with direct integrals of Hilbert spaces is assumed (see [341]).

Let us consider an elliptic operator

$$L(x,D) = \sum_{|\alpha| \leq 2m} a_\alpha(x)\, D^\alpha \tag{4.5.1}$$

with smooth $\mathbb{Z}^n$-periodic coefficients in $\mathbb{R}^n$. We shall not go into mostly (but not always) simple technical details related to the smoothness reduction or general lattices of periods. The expression (1) defines the closed operator L in $L_2(\mathbb{R}^n)$ with the domain $H^{2m}(\mathbb{R}^n)$. Its closedness follows easily from Schauder estimates. Let us denote by $\sigma(L)$ the spectrum of L in $L_2(\mathbb{R}^n)$. We also denote by L_p the Fredholm operator

$$L_p = L(x, D+p)\colon H^{2m}(\mathbb{T}^n) \to L_2(\mathbb{T}^n), \tag{4.5.2}$$

and by $\sigma_p(L)$ the spectrum of L_p in $L_2(\mathbb{T}^n)$.

The basic result of the theory is the following

Theorem 4.5.1. *The following equality holds:*

$$\sigma(L) = \bigcup_{p \in \mathfrak{K}} \sigma_p(L). \tag{4.5.3}$$

Proof. Our consideration will be based on theorem 2.2.5. Let us assume that $\lambda \notin \cup \sigma_p(L)$ and show that $\lambda \notin \sigma(L)$. After the transform $\mathfrak{F}$ we deal with the operator

$$\int_K^{\oplus} (L_p - \lambda)$$

instead of $L - \lambda$. The assumption

$$\lambda \notin \bigcup_{p \in K} \sigma_p(L) = \bigcup_{p \in K} \sigma(L_p),$$

implies that operators $L_p - \lambda : H^{2m}(\mathbb{T}^n) \to L^2(\mathbb{T}^n)$ are invertible for all $p \in K$. Due to ellipticity and embedding theorem, each of the operators is Fredholm, and the whole family is holomorphic with respect to $p \in \mathbb{C}^n$. This means that $L_p - \lambda$ is invertible for p in some neighborhood U of K in $\mathbb{C}^n$, and $M_p = (L_p - \lambda)^{-1}$: $L_2(\mathbb{T}^n) \to H^{2m}(\mathbb{T}^n)$ is a holomorphic family in U of bounded operators in $L_2(\mathbb{R}^n)$. Then

$$\int_K^{\oplus} M_p$$

is obviously the bounded inverse to $\int_K^{\oplus} (L_p - \lambda)$, so $L - \lambda$ is invertible, and $\lambda \notin \sigma(L)$.

The second part of the statement (if $\lambda \in \cup \sigma_p(L)$, then $\lambda \in \sigma(L)$) is slightly more delicate. Suppose that $\lambda \in \cup \sigma_p(L)$. We can assume for the sake of simplicity that $\lambda = 0$ (otherwise we include $-\lambda$ into L). Consider the following cases.

I) *dim Ker* $L_p \neq 0 \ \forall p \in K$. Choose a point $p_0 \in K$ at which

$$\dim \operatorname{Ker} L_{p_0} = \min_{p \in K} \dim \operatorname{Ker} L_p > 0.$$

Then for some neighborhood U of p_0 in K the kernels of L_p form a trivial vector bundle $G = \cup_{p \in U} \operatorname{Ker} L_p$ over U (see theorem 1.4.10). Choose a continuous non-trivial section $\varphi(p)$ of G with $supp\ \varphi \subset U$. Extending φ by zero outside U, we get a continuous function $\varphi \in L_2(K, H^{2m}(\mathbb{T}^n))$ such that $L_p\varphi(p) = 0 \ \forall p \in K$. In this case the function $\varphi_1 = \mathcal{F}^{-1}\varphi \in H^{2m}(\mathbb{R}^n)$ provides a non-trivial solution of the equation $Lu = 0$ in $L_2(\mathbb{R}^n)$, so $\lambda(=0) \in \sigma_d(L) \subset \sigma(L)$.

II) *Codim Im* $L_p \neq 0 \ \forall p \in K$. In this case we can construct the cokernel bundle

$$G^1 = \cup_{p \in U} \operatorname{Coker} L_p,$$

and obtain in an analogous way a non-trivial function $\varphi_2 \in L_2(\mathbb{R}^n)$ that is orthogonal to *Im* L, so again $\lambda(=0) \in \sigma(L)$.

III) Assume now that neither of the cases I) and II) is applicable. This means that L_p is invertible for almost all (except an analytic subset), but not for all values of $p \in K$. Now the construction used in the proof of equivalence i)$\Leftrightarrow$ii) in theorem 4.2.1 finishes the proof.

Remark. If *index* $L = 0$ (which is always true for scalar operators), the cases I) and II) occur simultaneously.

Let us emphasize that we do not require self-adjointness of the operator for validity of the theorem. However, in the self-adjoint case the structure of $\sigma(L)$ can be described in more details using the so called band functions.

We restrict ourself to the case of Schrödinger operator

$$L = -\Delta + V(x) \tag{4.5.4}$$

with real $\mathbb{Z}^n$-periodic potential $V \in L_\infty(\mathbb{R}^n)$ (in all following results the restriction on the potential can be reduced in order to include, for instance, potentials with Coulomb-type singularities). Some more general principal parts can be also considered.

The operators L_p in this case have the form

$$L_p = \sum_{j=1}^{n} \left(i\frac{\partial}{\partial x_j} - p_j\right)^2 + V. \tag{4.5.5}$$

These operators are selfadjoint for all $p \in \mathbb{R}^n$, bounded from below and with compact resolvent in $L_2(\mathbb{T}^n)$. The last assertion follows from the embedding theorem, because the resolvent continuously maps $L_2(\mathbb{T}^n)$ into $H^2(\mathbb{T}^n)$. So the spectrum $\sigma(L_p) = \sigma_p(L)$ consists of a sequence $\Lambda_m(p)$ of real numbers such that $\Lambda_m(p) \underset{m\to\infty}{\to} \infty$. The standard perturbation theory shows that each of the "band functions" $\Lambda_m(p)$ is continuous with respect to p and $2\pi\mathbb{Z}^n$-periodic. Periodicity follows from the fact that L_p and $L_{p+2\pi k}$, where $k \in 2\pi\mathbb{Z}^n$, are unitarily equivalent with the equivalence provided by the multiplication by $\exp 2\pi i k \cdot x$. Let us denote by S_m the set of values of the m-th band function Λ_m. Due to the properties of band functions, each S_m is a closed bounded interval $[a_m, b_m]$ in $\mathbb{R}^m$, and $b_m \underset{m\to\infty}{\to} \infty$. Theorem 1 says that

$$\sigma(L) = \bigcup_{m=1}^{\infty} [a_m, b_m].$$

The intervals $S_m = [a_m, b_m]$ are said to be the ***stability zones*** of the operator L. The spectrum $\sigma(L)$ is now a half line $[a, \infty)$ without a (finite or infinite) sequence (α_m, β_m) of open intervals such that $\alpha_{m+1} > \beta_m$ and $\lim_{m\to\infty} \alpha_m = \infty$ (if the sequence is infinite).

These intervals are so called ***gaps***, ***lacunas***, or ***instability zones*** of the spectrum. In the case n=1 the stability zones satisfy the following additional property:

$$a_{m+1} \geq b_m, \tag{4.5.6}$$

i.e. they do not overlap. This means that the nonexistence of one of the gaps $a_{m+1} = b_m$ is a rare occasion. In particular, the presence of only a finite number of gaps implies analyticity of the potential, and all such potentials can be explicitly described, which is one of the main considerations in the inverse scattering method of integration of non-linear PDEs ([216, 313, 314, 286, 287]). The famous Borg's theorem ([56], see also theorem 4.4.4) says that in one-dimensional case, if there are no gaps, then the potential is constant. When $n>1$, the inequality (6) is no longer valid. Moreover, it has been guessed long time ago by Bethe and Sommerfeld that in the case $n>1$ the number of gaps is *always* finite. This conjecture was proved by M. Skriganov [336, 368-372] (see also [81, 194-198, 399, 400]). In fact, if the potential is small enough, there are no gaps in the spectrum at all [312].

Theorem 4.5.2. *Let the Fourier coefficients V_m of the potential V on $\mathbf{R}^3$ satisfy the condition*

$$\sum_{\substack{m \in \mathbf{Z}^3 \\ m \neq 0}} |V_m|^2 \, |m|^{-1+\beta} < \infty \tag{4.5.7}$$

for some positive β. Then there is only a finite number of lacunas in the spectrum of the operator (4) in $L_2(\mathbf{R}^3)$.

Remarks. 1) Condition (7) allows singularities of the type $|x - x_\sigma|^{-\alpha}$ for $\alpha < 2$, for instance, Coulomb ones. 2) The analogous theorem holds for any dimension $n > 1$ with $(-4+n+\beta)$ instead of $(-1+\beta)$ in (9).

There are different approaches to the proof. The original one, provided by M. Skriganov (see the survey [372]), employs arithmetic and geometric properties of lattices. Another one, by O. Veliev and Yu. Karpeshina [194-198, 399, 400], develops some perturbation theory with respect to parameter α for the operator

$$L^{(\alpha)} = -\Delta + \alpha V.$$

O. Veliev constructed first terms of asymptotics for the spectrum of $L_p^{(\alpha)}$ (apart from a "thin" bad set). Yu. Karpeshina managed to obtain convergent perturbation series and explicit formulas for all their coefficients. Any reasonable complete account of these works would go beyond our goals, so we provide only a proof (due to M. Skriganov) for the two-dimensional case in the assumption of boundedness of the potential.

The basic idea of all these proofs is that the stability zones S_m not only intersect, but even strongly overlap for $n > 1$. The reason is that the length of the overlapping portions for the zero potential tends to infinity when we move to the right along the spectrum. This means that any perturbation which is bounded from above cannot improve the situation and cannot open any gaps that are sufficiently far to the right. Hence, the main part of the consideration should deal with the "free" operator $L^{(0)} = -\Delta$. We shall study some properties of the band functions and stability zones S_m for this operator.

Let us assume that $\lambda \in \sigma(L_p^{(0)})$, i.e. there exists a non-trivial $\mathbf{Z}^n$-periodic solution u of the equation

$$-\Delta e^{ip\cdot x}u(x) = \lambda e^{ip\cdot x}u(x)$$

or

$$-\Delta_p u = \lambda u. \tag{4.5.8}$$

Expanding $u(x)$ into Fourier series

$$u(x) = \sum_{\gamma \in 2\pi\mathbf{Z}^n} c_\gamma e^{i\gamma\cdot x},$$

we get

$$\sum_\gamma \Big(\sum_{j=1}^n (\gamma_j + p_j)^2 - \lambda\Big) c_\gamma e^{i\gamma\cdot x} = 0,$$

or

$$\sum_{j=1}^{n} (\gamma_j + p_j)^2 = \lambda \tag{4.5.9}$$

for such $\gamma \in 2\pi\mathbf{Z}^n$ that $c_\gamma \neq 0$. This leads us to the conclusion that $\lambda \in \sigma(L_p^{(0)})$ if and only if there exists a vector $\gamma \in \Gamma := 2\pi\mathbf{Z}^n$ that satisfies (9). The maximal number m such that $\lambda \geq \Lambda_m(p)$ can be obtained as

$$N(p, \lambda) = \# \{\gamma \in \Gamma \mid \sum_{j=1}^{n} (\gamma_j + p_j)^2 \leq \lambda\}. \tag{4.5.10}$$

It is convenient to introduce some auxiliary functions

$$N_+(\lambda) = \max_{p \in \mathbf{R}^n} N(p,\lambda) \tag{4.5.11}$$

$$N_-(\lambda) = \min_{p \in \mathbf{R}^n} N(p,\lambda). \tag{4.5.12}$$

The function $N_+(\lambda)$ gives the maximal number m of stability zone S_m containing some points below λ. The function $N_-(\lambda)$ gives the minimal number m of stability zone S_m containing some points above λ. For instance, the inequality

$$N_+(\lambda - t) > N_-(\lambda + t) \tag{4.5.13}$$

would mean that the point λ is contained in two overlapping stability zones, both of which contain the interval $[\lambda - t, \lambda + t]$. Due to standard perturbation theory, this situation can not be destroyed by perturbations that have the operator L_2-norm less than t, so λ will be in the spectrum of the perturbed operator. Let us imagine now that we managed to prove that for each t there exists C such that (13) holds for *all* $\lambda > C$. This would mean that for *any* bounded potential only a finite number of lacunas can be present. (Some slight modification would allow some unbounded potentials.) Hence, we should prove the following statement:

Theorem 4.5.3. *For every $t>0$ there exists a constant $C>0$ such that the inequality (13) holds for all $\lambda > C$.*

Proof. We shall actually prove that

$$\lim_{\lambda\to\infty} [N_+(\lambda-t)-N_-(\lambda+t)] = \infty, \tag{4.5.14}$$

which obviously implies the statement of the theorem.

We shall prove the statement only for $n=2$, which is the simplest case. The reader can find the general case in [372, 194-198, 399, 400].

Let us denote by $\chi_\lambda(x)$ the characteristic function of the disk $B_{\sqrt{\lambda}}(0)$ of the radius $\sqrt{\lambda}$ and with the center at the origin in $\mathbb{R}^2$. Obviously

$$N(p,\lambda) = \sum_{\gamma\in\Gamma} \chi_\lambda(p+\gamma)\,. \tag{4.5.15}$$

This function is Γ-periodic with respect to p, so it can be expanded into the Fourier series

$$N(p,\lambda) = \sum_{\gamma^*\in\mathbb{Z}^2} c_{\gamma^*}\, e^{-i\gamma^*\cdot p}, \tag{4.5.16}$$

where

$$c_{\gamma^*} = (2\pi)^{-2}\int_{B_{\sqrt{\lambda}}(0)} e^{i\gamma^*\cdot x}\,dx, \tag{4.5.17}$$

in particular

$$c_0 = (4\pi)^{-1}\lambda\,. \tag{4.5.18}$$

Let us consider formula (17) in more details:

$$c_{\gamma^*} = (2\pi)^{-2}\int_{B_{\sqrt{\lambda}}(0)} e^{i\gamma^*\cdot x}\,dx = \tag{4.5.19}$$

$$= \lambda(2\pi)^{-2} \int_{B_1(0)} e^{i\sqrt{\lambda}\,\gamma^*\cdot x}\, dx =$$

$$= \lambda(2\pi)^{-2} \int_0^1 r dr \int_0^{2\pi} e^{i\sqrt{\lambda}\,|\gamma^*|\, r\cos\theta}\, d\theta =$$

$$= \lambda(2\pi)^{-1} \int_0^1 J_0(\sqrt{\lambda}\,|\gamma^*|r)\, rdr =$$

$$= \frac{\lambda(2\pi)^{-1}}{(\sqrt{\lambda}\,|\gamma^*|)^2} \int_0^{\sqrt{\lambda}|\gamma^*|} J_0(r)\, rdr =$$

$$= \frac{\sqrt{\lambda}\,|\gamma^*|}{2\pi|\gamma^*|^2} J_1(\sqrt{\lambda}|\gamma^*|) = \frac{\sqrt{\lambda}}{2\pi|\gamma^*|} J_1\,(\sqrt{\lambda}|\gamma^*|)\,.$$

Taking into account (16),(18) and (19), we get

$$N\,(p,\lambda) = \frac{\lambda}{4\pi} + \frac{\sqrt{\lambda}}{2\pi} \sum_{\gamma^*\in \mathbf{Z}^2\setminus 0} \frac{J_1(\sqrt{\lambda}|\gamma^*|)}{|\gamma^*|} = (4\pi)^{-1}\lambda + R(p,\lambda). \tag{4.5.20}$$

Let us introduce the notations

$$R_+(p,\lambda) = max\,(R(p,\,\lambda),\,0),\;\; R_-(p,\lambda) = -min(R(p,\lambda),0),$$

$$R_\pm(\lambda) = \max_p\; R_\pm(p,\lambda),\; R(\lambda) = \max_{p\in\mathbf{R}^2} |R(p,\lambda)|.$$

The following statement follows from known estimates of the number of integer points in plane domains [244].

Lemma 4.5.4. *The following estimate holds:*

$$R(\lambda) = O(\lambda^{1/3}). \tag{4.5.21}$$

Let us notice that

$$\int_{F_\Gamma} R_+(p,\lambda)\, dp - \int_{F_\Gamma} R_-(p,\lambda) dp = \int_{F_\Gamma} R(p,\lambda)\, dp = 0. \tag{4.5.22}$$

Here F_Γ is the fundamental cube

$$\{p : \ |p_j| \le 2\pi,\ j = 1,2\}$$

for the lattice Γ. Taking into account (22), we get

$$\|R\|^2_{L_2(F_\Gamma)} = \int_{F_\Gamma} R^2(p,\lambda)\, dp = \int_{F_\Gamma} (R_+(p,\lambda)^2 + R_-(p,\lambda)^2) dp \le \tag{4.5.23}$$

$$\le \max_p |R(p,\lambda)| \left(\int_{F_\Gamma} (R_+(p,\lambda) + R_-(p,\lambda)) dp\right) =$$

$$2\, R(\lambda) \int_{F_\Gamma} R_\pm(p,\lambda) dp\ .$$

This means that

$$R_\pm(\lambda) \ge \frac{1}{4\pi^2} \int_{F_\Gamma} R_\pm(p,\lambda) dp \ge C\, \|R\|^2_{L_2(F_\Gamma)} (2R(\lambda))^{-1}. \tag{4.5.24}$$

Now we apply the Parseval's formula to R and get

$$\|R\|^2_{L_2(F_\Gamma)} = C\lambda \sum_{\gamma^* \in \mathbb{Z}^2\backslash 0} \frac{|J_1(\sqrt{\lambda}|\gamma^*|)|^2}{|\gamma^*|^2}\ . \tag{4.5.25}$$

Due to the asymptotic formula

$$J_1(z) \sim \frac{C}{\sqrt{z}} \cos\left(z - \frac{3\pi}{4}\right),\ z \to \infty,$$

and leaving only two terms $\gamma_1^*, 2\gamma_1^*$ in (25) (where γ_1^* is the closest point to the origin in Γ), we easily estimate:

$$\sum_{\gamma^* \in \mathbf{Z}^2\backslash 0} \frac{|J_1(\sqrt{\lambda}|\gamma^*|)|^2}{|\gamma^*|^2} \geq C. \tag{4.5.26}$$

Combining now (21) and (23) – (26), we get

$$R_\pm(\lambda) \geq C\lambda^\alpha$$

with some $\alpha > 0$. This obviously shows that

$$\lim_{\lambda\to\infty} (N_+(\lambda-t) - N_-(\lambda+t)) = -Ct + \lim_{\lambda\to\infty} (R_+(\lambda-t) + R_-(\lambda+t)) = \infty.$$

The theorem is proved.

Corollary 4.5.5. *If $n = 2$ and $V \in L_\infty(\mathbf{R}^2)$ in (4), then the spectrum of L has only a finite number of lacunas.*

Proof. Let $||V||_\infty < t < \infty$. We choose $C > 0$ such that

$$N_+(\lambda-t) > N_-(\lambda+t)$$

for all $\lambda > C$ and denote $m_\pm = N_\pm(\lambda \mp t)$. Then $m_+ > m_-$ and both zones $S^{(0)}_{m_+}$ and $S^{(0)}_{m_-}$ for the unperturbed operator $-\Delta$ contain the interval $[\lambda - t, \lambda + t]$. The perturbation theory says [199] that the band functions $\Lambda_{m_\pm}(p)$ for the perturbed operator can not deviate for more than t from the unperturbed ones. This means that the zones S_{m_+} and S_{m_-} for the perturbed operator still contain the point λ, so $\lambda \in \sigma(L)$. The proof is completed.

As we have already mentioned before, the last statement holds also for all $n \geq 2$ and for a wider class of potentials. Let us summarize now what is the information that we have already gotten about the spectrum of the periodic Schrodinger operator (4). Its spectrum consists of the union of a finite number of allowed zones

$$\sigma(L)= [a_0,b_0]\cup[a_1,b_1]\cup \ldots\cup [a_k,\infty),$$

where $a_{i+1}>b_i$ ($i=0, \ldots, k-1$). The spectrum can be obtained as the union of (overlapping) stability zones $S_m=\{\Lambda_m(p)|\ p\in\mathbf{R}^n\}$.

The next natural question arises: what kind of spectrum is it – discrete, continuous, absolutely continuous?

Theorems 4.1.7 and 4.1.8 show that under some mild conditions on the periodic potential the operator has no discrete spectrum: $\sigma_d(L) = \emptyset$. The analogous result also holds for Dirac operator (theorem 4.1.9) and some other periodic operators (see [224, 236]). It turns out that the spectrum is even absolutely continuous both for Schrödinger and Dirac operators.

However, it is better to start with an eigenfunction expansion for the operator (4) with a continuous potential V. This expansion can be easily carried over to rather general elliptic selfadjoint operators.

Let us denote, as above, by $\Lambda_m(p)$ the m-th eigenvalue of the operator L_p ($p\in\mathbf{R}^n$). We know that $\Lambda_m(p)$ is continuous.

Lemma 4.5.6. *For any m there exists a proper analytic subset A_m in K, such that $\Lambda_m(p)$ is analytic in $K\setminus A_m$.*

Proof. The graph of $\Lambda_m(p)$ is a part of the Bloch variety $B_{\mathbf{C}}(L)$ (see section 4.4). We can observe that $B_{\mathbf{C}}(L)$ cannot coincide with $\mathbf{C}^{n+1}$, because its intersection with the line $p=p_0\in \mathrm{K}$ is the spectrum of L_p, and hence is discrete. Theorem 4.4.2 says that there exists an entire function $f(p,\lambda)$ on $\mathbf{C}^n$ such that $B_{\mathbf{C}}(L)$ is its set of zeros. Let $\partial^k f/\partial\lambda^k$ be the highest order derivative that is still zero on the graph of $\Lambda_m(p)$ (obviously, $k<\infty$). Then, apart from the zero set of $\partial^{k+1}f/\partial\lambda^{k+1}$, the implicit function theorem gives the statement of the lemma.

Let us consider the orthonormal eigenfunctions $\psi_m(x,p)$ of L_p, i.e. $L_p\psi_m(\cdot,p) = \Lambda_m(p)\ \psi_m(p)$. If the eigenvalue is not simple, the choice of ψ_m has to be specified, so we prove the following statement.

Lemma 4.5.7. *The eigenfunctions $\psi_m(\cdot,p)$ can be chosen as measurable $H^2(\mathbb{T}^n)$-valued functions.*

Proof. Outside the proper analytic (and hence of zero measure) subset A_m, the family of operators $\mathcal{A}(p) = L_p - \Lambda_m(p)$ is analytic and Fredholm. Let

$$k = \min_{p \in \mathcal{K}} \dim Ker\mathcal{A}(p).$$

Then, outside of an analytic subset Υ we have

$$\dim Ker\mathcal{A}(p) = k$$

(see theorem 1.6.11). Due to theorem 1.6.13, $Ker\mathcal{A}(p)$ is an analytic vector bundle over the set $K\backslash\Upsilon$. Let us divide $K\backslash\Upsilon$ into a finite number of contractible pieces. Then over each of these pieces the vector bundle is trivial, so we can choose a smooth orthonormal basis $\psi_{m,l}(p)$ (l=1, ..., k) of it. The lemma is proved.

Let us denote by $\psi_m(x,p)$ the corresponding Bloch eigenfunctions

(4.5.27) $$\phi_m(x,p) = exp(ip \cdot x)\ \psi_m(x,p).$$

Now we are ready to prove the following expansion into Bloch eigenfunctions.

Theorem 4.5.8. *Let $f(x) \in L_2(\mathbf{R}^n)$ and*

(4.5.28) $$\alpha_m(p) = \int_{\mathbf{R}^n} f(x)\ \overline{\phi_m(x,p)}\ dx.$$

Then

(i) $\displaystyle\int_{\mathbf{R}^n} |f(x)|^2 dx = \sum_m \int_K |\alpha_m(p)|^2 dp$

(ii) $\displaystyle f(x) = \sum_m \int_K \alpha_m(p)\ \phi_m(x,p) dp.$

Remark. The convergence of the integral in (28) and of the series in (ii) is understood in the L_2-sense.

Proof. First of all, we split the integration over $\mathbf{R}^n$ in (28) into integrations over cubes $\mathcal{K}+k$ ($k \in \mathbf{Z}^n$):

$$\alpha_m(p) = \sum_{k \in \mathbf{Z}^n} \int_{\mathcal{K}+k} f(x)\, \overline{\phi_m(x,p)}\, dx =$$

$$= \sum_{k \in \mathbf{Z}^n} \int_{\mathcal{K}} f(x+k)\, e^{-ip\cdot k}\, \overline{\phi_m(x,p)}\, dx =$$

$$= \int_{\mathcal{K}} \widehat{\mathbf{F}}f(ip,x)\, \overline{\phi_m(x,p)}\, dx.$$

The function $\widehat{\mathbf{F}}f(ip,x)$ belongs to $L_2(\mathbf{T}^n)$ for almost every $p \in K$, so

$$\| \widehat{\mathbf{F}}f(ip,\cdot)\|^2_{L_2} = \sum_m |\alpha_m(p)|^2 \tag{4.5.29}$$

Integrating both sides of (29) with respect to $p \in K$ and taking into account that $\widehat{\mathbf{F}}$ is an isometry of $L_2(\mathbf{R}^n)$ onto $L_2(K, L_2(\mathbf{T}^n))$ (theorem 2.2.5), we get the first statement of the theorem.

The right hand side of (ii) is obviously equal to

$$\int_K \widehat{\mathbf{F}}f(ip,x)dp = \sum_{k \in \mathbf{Z}^n} f(x+k) \int_K e^{-ip\cdot k} dp.$$

All terms in the last sum vanish except the one with $k=0$, which gives $f(x)$. The proof is complete.

We are able now to prove the absolute continuity of the spectrum of the Schrodinger operator L.

Theorem 4.5.9. *The spectrum of the Schrodinger operator (4) with continuous $\mathbf{Z}^n$-periodic potential is absolutely continuous, i.e. $\sigma(L) = \sigma_{ac}(L)$.*

Proof. Theorem 8 says that the mapping $f(x) \to \{\alpha_m(p)\}$ provides an isometry from $L_2(\mathbf{R}^n)$ onto $L_2(K, l^2)$. (Here l^2 is the space of square-summable sequences $\{\alpha_m\}$). This isometry transforms our operator L into the operator $\widetilde{L}$:

$$\widetilde{L}: \{\alpha_m(p)\} \to \{\Lambda_m(p)\ \alpha_m(p)\}.$$

We know that each of $\Lambda_m(p)$ is analytic outside of a closed subset Λ_m of measure zero, and none of them is constant (which according to theorem 4.1.5 is equivalent to the absence of discrete spectrum). The operator $\widetilde{L}$ is the orthogonal sum of operators $\widetilde{L}_m$ on $L_2(K)$, acting as $(\widetilde{L}_m\alpha)(p) = \Lambda_m(p)\alpha(p)$. Hence, it is sufficient to prove the absolute continuity of the spectrum of $\widetilde{L}_m$ for each m. The spectral projection of the operator $\widetilde{L}_m$ is the multiplication by the characteristic function χ_t of the set $\{p \in K | \Lambda_m(p) \leq t\}$. To prove the absolute continuity we have to show that the preimage $\Lambda_m{}^{-1}(M)$ is of measure zero whenever $M \subset \mathbb{R}$ is of measure zero. Let us recall that Λ_m is analytic on a set of complete measure and is not constant, so the set of critical points $\{p \in K | \nabla \Lambda_m(p) = 0\}$ is of zero measure. Apart of these two null sets our function $\Lambda_m(p)$ can be chosen as a local coordinate function. Now $\Lambda_m^{-1}(M)$ is of zero measure for any null set $M \subset \mathbb{R}$ due to Fubini's theorem. The theorem is proved.

4.6. Positive solutions

In this section we present some very interesting results of S. Agmon [4] that give a nice description (Floquet expansion) of all positive solutions for periodic elliptic equations of the second order.

We shall consider here the following operator in $\mathbb{R}^n$:

$$Pu = -\sum_{i,j=1}^{n} a_{i,j}(x)\partial_i\partial_j u + \sum_{i=1}^{n} b_i(x)\partial_i u + c(x)u. \tag{4.6.1}$$

The coefficients are assumed to be smooth, real and $\mathbb{Z}^n$-periodic. (The smoothness assumption is unnecessarily strong and has been chosen just to avoid non-principal technicalities. It is sufficient to assume that the coefficients are Hölder continuous, or belong to appropriate Lp-spaces [4].)

We assume that P is elliptic in the sense that

$$\sum a_{ij}(x)\xi_i\xi_j \geq C|\xi|^2 \quad \text{for } \xi \in \mathbb{R}^2, C > 0.$$

We are interested in positive solutions $u(x) \geq 0$ of the equation

(4.6.2) $$Pu = 0.$$

There is a fundamental result about such solutions, called Harnack's inequality [135]:

for any compact set $K \subset \mathbf{R}^n$ there exists a constant $c = c_K > 0$ such that

(4.6.3) $$cu(x) \le u(y)$$

for any positive solution $u(x)$ of (2) and for all points $x, y \in K$.

Let us denote by $\mathcal{S}$ the set of all solutions of equation (2), endowed with the topology of uniform convergence on compacts in $\mathbf{R}^n$. Denote also by $\mathcal{S}_+ \subset \mathcal{S}$ the set of all positive solutions of (2). Then $\mathcal{S}_+$ is a convex cone in $\mathcal{S}$. Consider now positive solutions normalized by the condition $u(0)=1$, and denote by $\mathcal{S}_1 \subset \mathcal{S}_+$ the set of all such solutions.

Lemma 4.6.1. *i) $\mathcal{S}_1$ is convex; ii) $\mathcal{S}_1$ is a base of the cone $\mathcal{S}_+$, i.e. for any $u \in \mathcal{S}_+$ there exists a constant c such that $cu \in \mathcal{S}_1$; iii) $\mathcal{S}_1$ is compact.*

Proof. The statement (*i*) is obvious. In order to prove (*ii*) we choose $c = u(0)^{-1}$, and the task is to justify that $u(0) \neq 0$. This is, however, an obvious consequence of either the maximum principle that is valid for equations of the type (1)-(2) (see [135, 338]), or of (3). The compactness follows easily from the Harnack's inequality (3) and standard Schauder's estimates. Namely, for any compact domain $\Omega \subset \mathbf{R}^n$ and a neighborhood Ω_1 of its closure, we have the estimate

$$||u||_{H^m(\overline{\Omega})} \le C_m ||u||_{L_2(\Omega_1)},$$

so the Harnack's inequality gives

(4.6.4) $$||u||_{H^m(\overline{\Omega})} \le C'_m \, |u(0)|$$

for any $m>0$. For sufficiently large m, the inequality (3) together with the standard embedding theorems gives (*iii*), so the proof is complete.

Theorem 4.6.2. *If equation (2) has a positive solution in* $\mathbf{R}^n$, *then it has a positive Bloch solution*

$$u(x) = e^{\xi \cdot x} \varphi(x), \tag{4.6.5}$$

where $\xi \in \mathbf{R}^n$ *and* $\varphi(x)>0$ *is* $\mathbf{Z}^n$*-periodic. If* u_1, u_2 *are two positive solutions of the form (5) with the same* $\xi \in \mathbf{R}^n$, *then they differ only by a constant factor.*

Proof. Let us start with the existence of a positive Bloch solution. Due to the assumption, $\mathcal{S}_1 \neq \emptyset$, so lemma 1 and Krein-Milman theorem (see [210]) show that the set $ex\ \mathcal{S}_1$ of extreme points of $\mathcal{S}_1$ is not empty. The first statement of the theorem follows now from the next result.

Lemma 4.6.3. *If* $u \in ex\ \mathcal{S}_1$, *then it can be represented in the form (5).*

Proof of the lemma. Let $u \in ex\ \mathcal{S}_1$ and $\gamma \in \mathbf{Z}^n$. Consider the family of functions $u_t(x) = tu(x) + (1-t)\ u(x+\gamma)/u(\gamma)$. This family is correctly defined because, as it has been already mentioned before, $u(x)>0$ everywhere. For any $t \in \mathbf{R}$, $u_t(x)$ satisfies the equation (2), and obviously $u_t(0)=1$. Let us prove that for t close to 1 $u_t(x)$ is positive. First of all, due to the Harnack's inequality, there exists a constant $c>0$ such that

$$\max_{y \in \mathcal{K}} u(y) \leq c\ u(x),\ x \in \mathcal{K}.$$

Here $\mathcal{K}$ is the fundamental cube of the lattice $\mathbf{Z}^n$. Due to periodicity of the equation, we can get the estimate

$$\max_{y \in \mathcal{K}+x} u(y) \leq cu(x) \tag{4.6.6}$$

uniformly with respect to $x \in \mathbf{R}^n$. Let us choose now a neighborhood $I=(1-\epsilon,\ 1+\epsilon)$ of 1 such that

$$\frac{c|1-t|}{tu(\gamma)} < 1\ ,\ t \in I.$$

Then for $t \in I$ we get $u_t(x) > 0$ for all $x \in \mathbf{R}^n$. Hence, $u_t(x) \in \mathcal{S}_1$ for all $t \in I$, and $u(x)$ is an interior point of this interval. Since $u \in ex\ \mathcal{S}_1$, this can happen only if

$u(x)=u(x+\gamma)/u(\gamma)$. In other words, $u(x+\gamma)= \alpha_\gamma u(x)$ for all $\gamma \in \mathbf{Z}^n$. Then obviously $u(x)$ is a Bloch function, and the lemma is proved.

We have already proved the existence of a Bloch solution (5). Let us prove now that there can be at most one such solution with a given quasimomentum.

Suppose that $u_1, u_2 \in \mathcal{S}_1$ have the form (5) with the same ξ, i.e.

$$u_1(x) = e^{\xi \cdot x}\varphi_1(x),\ u_2(x) = e^{\xi \cdot x}\varphi_2(x)$$

with positive periodic functions φ_j (j=1,2). Consider the operator $P_\xi = e^{-\xi \cdot x} P \circ e^{\xi \cdot x}$. This operator, as the reader can easily check, also has the form (1). The functions φ_j (j=1,2) both satisfy the equation

(4.6.7)
$$P_\xi \varphi = 0 \quad \text{in } \mathbf{R}^n.$$

The Harnack's inequality together with the periodicity of φ_j's shows the existence of a constant c such that

(4.6.8)
$$\max_{\mathbf{R}^n} \varphi(x) \le c \min_{\mathbf{R}^n} \varphi(x)$$

for all periodic positive solutions of (4.6.7). Let us consider the function

$$\varphi_\epsilon(x) = \varphi_1(x) - \epsilon\, \varphi_2(x).$$

It satisfies (7), is periodic, and due to (8) is positive for sufficiently small ϵ. Let ϵ_0 be the maximal value of ϵ such that $\varphi_\epsilon(x)$ is still non-negative. Then $\varphi_{\epsilon_0}(x) \ge 0$, and at some point x_0 we have $\varphi_{\epsilon_0}(x_0)$=0. Because φ_{ϵ_0} is non-negative and periodic, (8) shows that $\varphi_{\epsilon_0} = 0$ identically, and the functions u_1 and u_2 are linearly dependent. The proof is now complete.

Let us denote by $\Xi \subset \mathbf{R}^n$ the set of all ξ's in (5), i.e.

$$\Gamma=\{\xi \in \mathbf{R}^n\text{: equation } Pu=0 \text{ has a positive solution of the form (5)}\}.$$

For every $\xi \in \Xi$ we denote by $u_\xi(x)$ the corresponding positive Bloch solution, normalized by $u_\xi(0) = 1$. We also denote by M the set of all such Bloch solutions.

Theorem 4.6.4.

i) M is compact with respect to the topology of uniform convergence on compact subsets of $\mathbb{R}^n$.

ii) Ξ is compact.

iii) The correspondence $h: \xi \longrightarrow u_\xi$ is a homeomorphism between Ξ and M.

iv) Every positive normalized by $u(0)=1$ solution of (2) has a unique representation

$$u = \int_\Xi u_\xi \, d\mu = \int_M u_\xi \, dh_*\mu, \tag{4.6.9}$$

where μ is a probability Borel measure on Ξ, and $h_*\mu$ is its image under h on M.

v) $M = ex\, \mathcal{S}_1$.

Proof. First of all, let us show that Ξ is bounded in $\mathbb{R}^n$. Let $\xi \in \Xi$ and $u_\xi(x) \in M$. Harnack's inequality shows that

$$u_\xi(e_j) \le C u_\xi(0) = C,$$

where e_j is the j-th standard basic vector of $\mathbb{Z}^n$. On the other hand,

$$u_\xi(e_j) = e^{\xi \cdot e_j} \varphi_\xi(e_j) = e^{\xi \cdot e_j} \varphi_\xi(0) = e^{\xi \cdot e_j} = exp\ \xi_j.$$

This provides us with the estimate $exp\ \xi_j < C$. Taking $-e_j$ instead of e_j, we get the analogous estimate for $exp\ (-\xi_j)$. Hence,

$$\max_{j=1,\dots,n} |\xi_j| \le R$$

for all $\xi \in \Xi$. In other words, Ξ is a subset of the cube $B_R = \{x\colon |x_j| \le R\}$. Let us consider now the set $\widetilde{M}$ of all Bloch solutions of the equation (2) that have the form

$$u = e^{\xi \cdot x} \varphi(x)\ ,\ \xi \in B_R.$$

Then obviously $M = \widetilde{M} \cap \mathcal{S}_1$. Due to the statement (*iii*) of lemma 1, closedness of $\widetilde{M}$ would imply that M is compact, so let us prove that $\widetilde{M}$ is closed in the compact-open topology. Let $v_j \in \widetilde{M}$ be a sequence of Bloch solutions with the corresponding exponents $\xi_j \in \mathbf{R}^n$, $|\xi_j| \le R$, and $v_j \longrightarrow v$ in the compact-open topology. Then Schauder estimates show that v_j all its derivatives converge uniformly on compacts, and v is also a solution. We have to show now that v is also a Bloch solution. Choosing a subsequence, we may assume that $\xi_j \longrightarrow \xi \in \mathbf{R}^n$ (because the sequence is bounded by R). The function $v_j(x)$ satisfies the equality

$$v_j(x+\gamma) = e^{\xi_j \cdot \gamma} v_j(x), \ x \in \mathbf{R}^n, \gamma \in \mathbf{Z}^n.$$

Taking the limit, we get $v(x+\gamma) = exp\ (\xi \cdot \gamma) v(x)$, so $v(x)$ is also a Bloch solution. This implies the statement (*i*) of the theorem. The correspondence $u_\xi \longrightarrow \xi$ is a bijection between M and Ξ, due to theorem 2. This correspondence is continuous, because

$$\xi_j = log\ u_\xi(e_j).$$

Now compactness of M implies that Ξ is compact, and h is a homeomorphism.

Due to lemma 3 and the Krein-Milman theorem, we conclude that for each point $u \in \mathcal{S}_1$ there exists a probability Borel measure ν on M such that

$$u = \int_M d\nu.$$

Transforming ν into a measure μ on Ξ by means of h, we get the representation (8). Let us prove now uniqueness of the representation. Consider the closed linear subspace A in $C(\Xi)$, generated by all exponents $exp\ k \cdot \xi$, where $k \in \mathbf{Z}^n$. This is obviously an algebra that contains the unit, and separates points in Ξ, so by the Stone-Weierstrass theorem $A = C(\Xi)$. Suppose now that for a function $u \in \mathcal{S}_1$ we have two different representations.

$$u(x) = \int_\Xi u_\xi(x)\ d\mu_j(\xi) \qquad (j=1,2).$$

Then

$$\int_{\Xi} u_\xi(x)\, d(\mu_1-\mu_2) = 0\,, \quad \forall x \in \mathbf{R}^n.$$

Let us choose $x{=}k \in \mathbf{Z}^n$, then $u_\xi(k){=}e^{k\cdot\xi}$, so

$$\int_{\Xi} e^{k\cdot\xi}\, d(\mu_1-\mu_2) = 0\,, \quad \forall k \in \mathbf{Z}^n.$$

This means that $\mu_1-\mu_2$ is orthogonal to the whole space $C(\Gamma)$, so $\mu_1{=}\mu_2$.

It is easy now to prove (v). In fact, let $u \in M$, then the Dirac measure ν_{u_0}, concentrated at u_0, represents u_0:

$$u_0{=}\int_M u\ \delta_{u_0}.$$

Hence, according to the uniqueness part, we get $u_0 \in ex\ \mathcal{I}_1$, and the proof is complete.

Our last task in this section is to investigate the analytic structure of the set $\Xi \subset \mathbf{R}^n$. According to the general results (see corollary 3.1.6 and theorem 3.1.7), we expect analyticity (with possible singularities). In our case, however, much more can be proved. Let us start with some auxiliary results.

Lemma 4.6.5. *If $c(x) < 0$ in (1), then equation (2) has no positive periodic solutions.*

Proof. Suppose that $u{>}0$, and $Pu{=}0$. Consider the function

$$\tilde{u}\,(x) = u(x) - \min_{y \in \mathbf{R}^n} u(y).$$

This function is nonnegative, it does have some zeros, and $P\tilde{u} > 0$ (due to the inequality $c(x) < 0\}$. In other words, $\tilde{u}$ is a supersolution of (2). Because $\tilde{u} \geq 0$ and $\tilde{u}$ has zeros, $\tilde{u} = 0$ (apply the maximum principle [135, 338]). This means that $u(x) = const \neq 0$, which contradicts the condition $c(x) < 0$. The proof is finished.

Theorem 4.6.6. *For every $\xi \in \mathbf{R}^n$ there exists a unique $\lambda \in \mathbf{R}$ such that the equation*

$$Pu = \lambda u \tag{4.6.9}$$

has a positive Bloch solution $u(x)=e^{\xi\cdot x}\varphi(x)$. *If there is another pair* λ', u' *with the same* ξ, *then* $u'=u$, $\lambda'=\lambda$.

Proof. First of all, we can reduce the problem to the case of periodic solutions. Namely, $e^{-\xi\cdot x}Pu=P_\xi\varphi$, where $u(x)=e^{\xi\cdot x}\varphi(x)$, $\varphi(x)$ is periodic and

$$P_\xi = P-\left(\sum\gamma_j(x)\xi_j+\sum_{ij}a_{ij}(x)\xi_i\xi_j\right) \tag{4.6.10}$$

Here $\gamma_j(x)$ are some periodic functions obtained from b_j and a_{ij}. Hence, we are interested in the periodic solutions of the equation

$$P_\xi u = \lambda u. \tag{4.6.11}$$

Second, adding a constant to $c_\xi(x)$ (which obviously does not change the statement of the theorem), we may assume that $c_\xi(x)>0$. Here $c_\xi(x)$ is the zero order term

$$c_\xi(x) = c(x) -\left(\sum_{ij}a_{ij}(x)\xi_j\xi_j +\sum_i\gamma_j(x)\xi_j\right) \tag{4.6.12}$$

in (10). Now let $f\in C(\mathbb{T}^n)$ be any continuous periodic function. We shall prove that the equation

$$P_\xi u =f \tag{4.6.13}$$

can be uniquely solved in $C(\mathbb{T}^n)$. This is a standard problem about Perron solutions [135]. Let $||f||_C=\alpha>0$, $inf\ c_\xi(x) = \beta>0$, and $h_\pm(x)= \pm\alpha/\beta$. Consider the set U_f of all continuous subsolutions u of the equation $Pu=f$ that satisfy the inequality $h_-\leq u\leq h_+$. This set is nonempty, because $h_-\in U_f$. Proofs of theorems 6.11 or 2.12 from [135] show that

$$u_f(x) = \sup_{u\in U_f} u(x)$$

is a solution of (13), and obviously $||u_f||_C \leq \beta^{-1}||f||_C$. The uniqueness follows from the maximum principle in the standard way. Namely, if u_1, u_2 are two distinct bounded solutions, then $v=u_1-u_2$ is bounded and satisfies $P_\xi v=0$. Consider the function $\widetilde{v}=||v||_c-v$. Then $P_\xi \widetilde{v}\geq 0$ (due to $c_\xi(x)>0$), $\widetilde{v}\geq 0$, and $\widetilde{v}$ has zeros. Hence, $\widetilde{v}=0$. This means that the equation $P_\xi u=0$ has the non-zero constant solution v, which contradicts to the condition $c_\xi(x)>0$. Our conclusion now is that P_ξ^{-1} is a bounded operator in $C(\mathbb{T}^n)$. In fact, it is even a compact operator, due to ellipticity and embedding theorems. Indeed, P_ξ^{-1} maps $C(\mathbb{T}^n)$ into $C^2(\mathbb{T}^n)$ and the embedding $C^2(\mathbb{T}^n)\longrightarrow C(\mathbb{T})$ is compact.

We consider the cone K of positive functions in $C(\mathbb{T}^n)$:

$$K=\{f\in C(\mathbb{T}^n)|\ f(x)\geq 0,\ x\in\mathbb{T}^n\}.$$

This cone is solid [207, 210], i.e. *int* $K\neq\emptyset$. We denote by $u_0\in K$ the function $u_0(x)\equiv 1$. Let us recall [207, 210] that an operator L in a Banach space E with a cone $K(\ni u_0)$ is said to be u_0 -positive, if $LK\subset K$ and for every $u\in K$ there exist $n\in\mathbb{Z}_+,\alpha,\beta>0$ such that

$$\alpha u_0\leq L^n u\leq \beta u_0$$

with respect to the order generated by K. (In our case the inequality $u\leq v$ means $u(x)\leq v(x)$ for all $x\in\mathbb{T}^n$.) Our claim is that P_ξ^{-1} is u_0-positive. Let $f\in K$. It is obvious from the Perron's construction that the corresponding solution u_f is also nonnegative. This follows from the fact that for nonnegative $f(x)$ the set U_f contains zero function. Hence, P_ξ^{-1} maps K into itself. We can also show that u_f has no zeros. If it does have a zero, then the inequality $Pu_f=f\geq 0$ with the maximum principle show that $u_f=0$, and $f=0$. Because $u_f>0$, there are constants $\alpha,\beta>0$ such that $\alpha\leq u_f(x)\leq\beta$ for all $x\in\mathbb{T}^n$. This exactly means u_0-positivity of P_ξ^{-1}. Now we can apply the general theorem of M. Krasnoselskii [207, 210] that guarantees the existence and uniqueness of a positive eigenvalue μ_0 and corresponding positive eigenfunction $v_0\in K$ for the compact u_0-positive operator P_ξ^{-1}. Moreover, μ_0 is simple. Now the trivial correspondence between eigenvalues of P_ξ and P_ξ^{-1} finishes the proof.

The preceding theorem shows the existence of the function $\Lambda:\mathbf{R}^n \longrightarrow \mathbf{R}$ such that

(4.6.14) $$\Lambda(\xi) = \text{unique eigenvalue of } P \text{ that corresponds to a positive Bloch eigenfunction } u(x) = e^{\xi \cdot x}\varphi(x).$$

The next theorem describes the crucial properties of this function.

Theorem 4.6.7. *The function (14) has the following properties:*

i) *it is bounded from above;*

ii) *it is strictly concave;*

iii) *it is analytic;*

iv) *its gradient is nondegenerate except the maximum point.*

Proof. *i)* By the definition, $\Lambda(\xi)$ is the unique eigenvalue corresponding to a positive periodic eigenfunction of P_ξ. Taking into account (12) and the ellipticity condition, we see that $c_\xi(x)$ is uniformly bounded:

$$c_\xi(x) < M, \; x \in \mathbf{R}^n, \xi \in \mathbf{R}^n.$$

If now $\lambda \geq M$, then $c_\xi(x) - \lambda < 0$. Thus, lemma 5 shows that for this λ the equation $(P_\xi - \lambda)u = 0$ has no positive periodic solutions. Hence, $\Lambda(\xi) \leq M$.

ii) Let $\xi_1, \xi_2 \in \mathbf{R}^n$. Denote by u_j (j=1,2) the positive periodic solution of the equation $(P_{\xi_j} - \Lambda(\xi_j))u = 0$. Consider the numbers $\widetilde{\lambda} = (\Lambda(\xi_1) + \Lambda(\xi_2))/2$ and $\lambda = \Lambda((\xi_1+\xi_2)/2)$. Let also v be the positive periodic solution of the equation

(4.6.15) $$\left(P_{(\xi_1+\xi_2)/2} - \lambda\right)v = 0.$$

We have to show that $\lambda > \widetilde{\lambda}$. Consider

$$u(x) = \sqrt{u_1(x)u_2(x)}\,.$$

It is obvious that $u>0$, and

$$u(x) = e^{(\xi_1+\xi_2)\cdot x/2}\varphi(x)$$

with a periodic $\varphi(x)$. Direct substitution shows that $[P-\widetilde{\lambda}\,]u \geq 0$, but $[P-\widetilde{\lambda}\,]u \neq 0$. Suppose now that $\lambda \leq \widetilde{\lambda}$, then

$$(P-\lambda)u=(P-\widetilde{\lambda}\,)u-(\lambda-\widetilde{\lambda}\,)u \geq 0. \tag{4.6.16}$$

Let us chose $\epsilon_0 > 0$ such that the function $\widetilde{u} = u - \epsilon_0 e^{(\xi_1+\xi_2)\cdot x/2} v$ is positive, but has some zeros. Then, due to (15)-(16), $\widetilde{u}$ has some zeros and is a positive supersolution : $(P-\lambda)\widetilde{u} \geq 0$. Hence, $\widetilde{u}=0$, and u is proportional to $e^{(\xi_1+\xi_2)\cdot x/2}v(x)$. In this case

$$(P-\widetilde{\lambda}\,)u=(P-\lambda)u + (\lambda-\widetilde{\lambda}\,)u = (\lambda-\widetilde{\lambda}\,)u \leq 0.$$

This is a contradiction.

iii) The analiticity is an immediate consequence of the perturbation theory, because $\Lambda(\xi)$ is a simple eigenvalue of P_ξ^{-1}.

iv) Follows from (*ii*) and (*iii*).

The proof is complete.

An immediate consequence of the last theorem is the following result, describing the structure of Ξ.

Theorem 4.6.8. *Let the equation (2) has a positive solution. Then either $\Xi \subset \mathbf{R}^n$ consists of a single point, or Ξ is an analytic $(n-1)$-dimensional submanifold of $\mathbf{R}^n$ that is the boundary of a bounded strictly convex domain.*

Proof is immediate. Indeed,

$$\Xi=\{\xi \in \mathbf{R}^n | \Lambda(\xi) = 0\}.$$

If 0 is the maximal value of $\Lambda(\xi)$, then, due to theorem 7, Ξ consists of a single point. If $0 < max\,\Lambda(\xi)$, then the same theorem implies the statement, where the domain is

$$\{\xi \in \mathbf{R}^n |\ \Lambda(\xi)>0\}.$$

The theorem is proved.

4.7. Comments and references

Section 4.1. Theorem 4.1.2 (proved by the author [233]) has a simple matrix explanation: upper-triangular perturbations with zeros on the diagonal do not change the spectrum of a diagonal matrix. The first version of this theorem was proved by A. Miloslavskii [297] for parabolic equations that are periodic in the time variable (see theorems 5.3.7 and 5.3.8). Unfortunately, the conditions of the theorem exclude all cases of physical interest. It must be true that under the conditions of the theorem $B_{\mathbf{C}}(L_0) = B_{\mathbf{C}}(L_0 + L_1)$, but the author has not been able prove this yet.

Due to theorem 4.1.5, the statement of theorem 4.1.4 can be reformulated in the following way: if $index_{\mathbf{T}^n} L > 0$, then the equation $Lu = 0$ has a non-zero L_2-solution in $\mathbf{R}^n$. In this form the theorem was proved by M. Atiyah in the general situation of elliptic operators on non-compact manifolds, invariant with respect to an action of a discrete group, when the orbit manifold is compact [21]. Theorem 4.1.5 was proved by the author [229]. It is interesting to observe that for proper pseudo-differential elliptic periodic operators the statement c) is no longer equivalent to a) and b). (This remark has been made by M. Shubin.) Indeed, we can consider an elliptic convolution operator with a symbol that vanishes in a neighborhood of the origin. The corresponding equation has L_2-solutions, but no exponentially decreasing ones. It is also worth mentioning that theorem 4.1.5 does not say that any L_2-solution is an exponentially decreasing one (compare with results of [2]), which is not true here. Existence of an L_2-solution implies existence of another solution that decays exponentially. Theorem 4.1.6 is also due to the author. The weak uniqueness assumption of theorem 4.1.6 seems unnecessary. A related question is: Let L be an elliptic operator of zero index on a compact manifold M. Is it possible to make the operator invertible by adding a smooth zero order term? We proved this in the case of weak uniqueness (however, the proof also works, if the *adjoint* operator has weak uniqueness property). The reader can find some discussion of weak uniqueness property in [174, 302]. Theorem 4.1.7 and partly 4.1.8 in the self-adjoint case were proved by L. Thomas [388]. Theorem 4.1.8 can also be found in [341, v.4]. The first correct proof of non-existence of bounded states ($= L_2$-eigenfunctions) for the periodic Dirac operator was given by L. Danilov [83-86]. He also proved this result under some milder conditions on the potential. We follow his approach in theorems 4.1.9, 4.1.10. Theorem 4.1.11 is a simple combination of theorem

4.1.5 with a result of E. Landis [245]. There are several results that prohibit too fast decaying solutions for "nice" elliptic problems [61, 245, 246, 288-290]. Some of them can produce theorems similar to 4.1.11.

In section 4.2 we prove an analog of a known result about ordinary differential equations [177]. Some analogs of this result can be easily proved, following the same scheme, for elliptic systems, hypoelliptic equations and systems and pseudodifferential equations (modulo some remarks about the Fredholm index and about non-closed range in hypoelliptic case).

Section 4.3 is devoted to the proof of theorem 4.3.1 [223]. This theorem generalizes the well known theorems of F. Bloch [54] and E. Shnol' [356]; see also [136]. During the last couple of years several works have been devoted to this problem [203, 360]. In [203] the Bloch's theorem was in some cases generalized to Schrödinger equations that are periodic with respect to some non-commutative discrete group of transformations of the underlying manifold. M. Shubin in [360] generalized the Bloch's theorem to the case of elliptic operators on non-compact manifolds with bounded geometry. In this case no periodicity is assumed, so the statement of the theorem is slightly modified: existence of not too fast growing solutions implies that we are on the spectrum.

Theorem 4.4.2 of section 4.4 was proved by the author (it was published, except for the last statement, in [229]). It was rediscovered (without the statement about periodicity) for the case of Schrödinger equations in [202]. The analytic structure of Bloch variety (without any estimates) was proved independently by C. Wilcox [405]. Theorems 4.4.3 and 4.4.4 are just reformulations of well known properties of periodic ordinary differential equations [19, 94, 77]. The equivalence (i)$\Leftrightarrow$(iii) in theorem 4.4.4 is the famous Borg's theorem [56]. Many interesting applications of the analysis on the Bloch variety to inverse scattering can be found in [111-119, 250, 286, 287, 312, 313]. Inverse spectral problems in the periodic case are discussed in [103-105]. The directional compactification of the Bloch variety was invented, and theorems 4.4.5 and 4.4.6 were proved in [202, 39]. See also [37, 38, 134] for some further developments in this direction. There are several unsolved problems related to the analytic structure of Bloch variety (see very interesting discussion of these properties in [27, 134, 202, 314]). We would like to mention some of them. First of all, in spite of strong results of [202], the problem of irreducibility of Bloch variety remains still open in the dimensions greater than two. The next question should be not too hard to answer. "Everybody knows" that in general the components of the Bloch variety are not algebraic (however, this can happen, according to theorem 4.1.2, in some exceptional cases). This statement has never been proved rigorously, but an appropriate transversality theorem should lead to the proof. Another interesting hypothesis is [27, 314] that a selfadjoint periodic elliptic operator in "general position" satisfies the property: for any two branches $\Lambda_i(p)$ and $\Lambda_j(p)$ of eigenvalues (where p is the quasimomentum) the set A_{jk} of all real quasimomentums p such that $\Lambda_i(p) = \Lambda_j(p)$ is an analytic subset of real codimension 3.

There are several books devoted (at least partially) to spectral theory of periodic ordinary and partial differential equations [94, 177, 276, 337, 341, 372, 389]. That is, why in section 4.5 we presented only a sketch of this theory. Theorem 4.5.1 in the self-adjoint case was proved by F. Odeh and J.B. Keller [315]; cf also [126, 94, 341]. It was proved for ordinary differential operators in self-adjoint case in [389], and in non-selfadjoint one in [343, 352, 391]. Theorem 4.5.3, along with some generalizations to arbitrary dimension $n > 1$ and to some more general equations, was proved by M. Skriganov [336, 368-372]. In the two-dimensional case see also [81]. We follow Skriganov's approach in the proofs of theorem 4.5.4 and corollary 4.5.6. A different approach was developed by O. Veliev and Yu. Karpeshina [194-198, 399, 400]. They developed a perturbation theory with respect to the parameter α for the operator $(-\Delta)^m + \alpha q(x)$. This enabled them to get rid of some unnecessary arithmetic restrictions on the periods' lattice. Some closely related problems are discussed in [109, 110, 124]. Lemmas 4.5.6, 4.5.7 are due to C. Wilcox [405]; cf. also [315]. Theorem 4.5.8 was proved by I. Gelfand [126]; cf. also [94, 257, 315, 405]. Theorem 4.5.9 is a part of L. Thomas' result [388].

There is a vast literature (see, for instance, [3,4, 11-13, 19, 207, 260, 268, 285, 304, 305, 329, 330, 332-334, 364, 376]) devoted to relationships between existence of positive solutions, spectral properties of elliptic and parabolic operators and geometry of underlying manifold. The results that we present in section 4.6 were obtained by S. Agmon [4]; cf. also [260, 332-334]. The proof of lemma 4.6.3 follows an idea of I. Pinchover. Possibilities of generalizations of these results to some non-commutative groups of periods were discussed in [4, 260].

There are many problems related to the properties of periodic ordinary and partial differential equations that we have not discussed here. We can mention among them : inverse scattering [111-119, 216, 250, 276, 277, 284, 286, 287, 303, 312-314], inverse spectral problems [103-105, 337, 349, 392, 393], semi-infinite crystals [75-78], Landau singularities [132], semi-classical approximation [70, 291, 292, 403, 404], periodic point-like potentials [14, 192, 193], etc.

CHAPTER 5

EVOLUTION EQUATIONS

We consider here, in the spirit of the previous chapters, an equation $\dot{u} = -A(t)u$ with a periodic unbounded operator coefficient in a Hilbert space. In section 5.1 some kind of hypoellipticity (more general than parabolicity) is assumed. We deal here with solutions defined on the whole axis (Cauchy problem is in general ill-posed for this class of equations). The next section is devoted to some degenerate cases arising in applications. We discuss Cauchy problem for abstract parabolic problems in section 5.3 and parabolic and elliptic boundary problems in cylindrical domains in section 5.4. The last section contains some remarks and references.

5.1. Abstract hypoelliptic evolution equations on the whole axis

Let H be a complex Hilbert space with a norm $\|\cdot\|$ (all spaces under consideration are separable) and $A(t)$ $(t \in \mathbf{R})$ be a 1-periodic in t family of closed operators in H (all results can be easily reformulated for any value of the period). We will need the following conditions on $A(t)$:

i) The domain $D(A(t))$ of $A(t)$ does not depend on t and coincides with a dense linear subspace H_1 in H.

ii) Graph norms on $D(A(t))$ are equivalent (uniformly in t) to a fixed norm $\|.\|_1$ on H_1, i.e.

$$C^{-1}\|x\|_1 \leq \|x\| + \|A(t)x\| \leq C\|x\|_1$$

for all $x \in H_1$, $t \in \mathbf{R}$ and some $C > 0$.

iii) The family $A(t)$, as a function with values in the space $L(H_1,H)$ of linear bounded operators from H_1 into H, is norm-continuous.

iv) There exists a real number σ such that the line $\sigma + i\mathbf{R}$ does not intersect the spectrum of $A(t)$ for all t and the following inequality is satisfied on this line:

$$\|(A(t) - \lambda)^{-1}\|_{L(H,H)} \leq C(|\lambda| + 1)^{-1}$$

uniformly in $t \in \mathbf{R}$. We shall see below that this condition guarantees some kind of hypoellipticity of the equation. For parabolic (inverse parabolic) equations this estimate holds in a left- (correspondingly, right-) half plane.

The conditions v)-vii) repeat i)-iii) for the adjoint operator $A^*(t)$ instead of $A(t)$ and another dense subspace H_1' in H instead of H_1:

v) $D(A^*(t)) = H_1'$.

vi) $C^{-1}\|x\|_{H_1'} \leq \|x\| + \|A^*(t)\, x\| \leq C\|x\|_{H_1'}$ for all $x \in H_1'$ and $t \in \mathbf{R}$.

vii) The family $A^*(t)$, as a function with values in the space $L(H_1', H)$ of linear bounded operators from H_1' into H, is norm-continuous.

The next assumption excludes (in the infinite dimensional case) equations with bounded operator coefficients. It is commonly satisfied when the operator coefficient corresponds to an elliptic boundary value problem in a bounded domain.

viii) The embedding operator $\mathfrak{J}: H'_1 \subset H$ is compact.

A locally square integrable function $u(t)$ with values in H (i.e. $u \in L_{2,loc}(\mathbf{R}, H)$) is said to be a solution of the equation

$$\frac{du}{dt} + A(t)\, u = 0, \tag{5.1.1}$$

if

$$\int_{\mathbf{R}} (u(t),\ -\phi'(t) + A^*(t)\, \phi(t))_H\, dt = 0$$

for all H'_1-valued C_0^∞-functions $\phi(t)$ on $\mathbf{R}$.

Let $\mathbf{T} = \mathbf{R}/\mathbf{Z}$ be the circle of unit length. Due to periodicity, we may consider our operator $\frac{du}{dt} + A(t)$ also on functions defined on $\mathbf{T}$.

Let $\mathcal{H}_0 = L_2(\mathbf{T},H)$ be the Hilbert space of all strongly measurable, square norm integrable H-valued functions on $\mathbf{T}$ (we will not discuss the definition of such standard function spaces in details that you can find in chapter I of [262]). For $f_1, f_2 \in \mathcal{H}_0$ we will write $\frac{df_1}{dt} = f_2$, if for any function $\phi(t) \in C^\infty(\mathbf{T})$ we have

$$\int_{\mathbf{T}} f_2(t)\, \phi(t)\, dt = -\int_{\mathbf{T}} f_1(t)\, \phi'(t)\, dt.$$

Let us denote by $\mathcal{H}_1$ the linear subspace in $L_2(\mathbf{T}, H_1)$ consisting of all functions f such that $\frac{df}{dt} \in \mathcal{H}_0$. This space is a Hilbert space with respect to the standard norm

$$\|f\|^2_{\mathcal{H}_1} = \|f\|^2_{L_1(\mathbf{T},H_1)} + \|\frac{df}{dt}\|^2_{\mathcal{H}_0}.$$

Analogous spaces on an interval are considered in [262].

The definition of the space $\mathcal{H}'_1$ is analogous to the previous one with H'_1 instead of H_1. It is obvious that there are natural continuous embeddings $\mathcal{H}_1(\mathcal{H}'_1) \subset \mathcal{H}_0$. It is also clear from the next lemma that these embeddings are dense.

Lemma 5.1.1. *The correspondence*

$$f(t) \to \sum_k f_k \; exp\; 2\pi ikt$$

between 1-periodic functions and their Fourier series generates isomorphisms of the spaces $\mathcal{H}_1, \mathcal{H}_1', \mathcal{H}_0$ *onto the following spaces of vector sequences* $\{h_k\}$ *respectively:*

1) $\{ h_k \in H_1 \mid \sum_{-\infty}^{\infty} \|h_k\|^2_{H_1} < \infty, \sum_{-\infty}^{\infty} k^2 \|h_k\|^2_H < \infty \}$

2) $\{ h_k \in H'_1 \mid \sum_{-\infty}^{\infty} \|h_k\|^2_{H'_1} < \infty, \sum_{-\infty}^{\infty} k^2 \|h_k\|^2_H < \infty \}$

3) $\{ h_k \in H \mid \sum_{-\infty}^{\infty} \|h_k\|^2_H < \infty \}$

Proof follows immediately from the Parseval's equality.

We observe that the derivative $\frac{d}{dt}$ can be uniquely extended to a continuous operator from $\mathcal{H}_1$ (or $\mathcal{H}_1'$) into $\mathcal{H}_0$.

Let us define H_{-1} as the antidual space to H_1', i.e. the space of all antilinear continuous functionals l on H_1' :

$$l(x+y) = l(x) + l(y), \quad l(\lambda x) = \bar{\lambda}\, l(x), \quad x,y \in H_1', \quad \lambda \in \mathbf{C}.$$

There is a standard norm on H_{-1}, and the inclusion $H_1' \subset H$ generates the continuous embedding $H \subset H_{-1}$, where the functional $x(y) = (x,y)_H$ corresponds to the element $x \in H$.

Let us introduce a space $\mathcal{H}_{-1}$ of vector sequences $\{h_k\}$. This space is the linear hull of two subspaces

$$\mathcal{H}_{-1}^{(1)} = \{ \{ h_k \} \mid h_k \in H_{-1}, \ (\|h\|^{(1)})^2 = \sum_{-\infty}^{\infty} \|h_k\|^2_{H_{-1}} < \infty \} \approx L_2(\mathbf{T}, H_{-1}),$$

$$\mathcal{H}_{-1}^{(2)} = \{ \{ h_k \} \mid h_k \in H, \ (\|h\|^{(2)})^2 = \sum_{-\infty}^{\infty} \frac{\|h_k\|^2_H}{k^2+1} < \infty \}.$$

The unit ball in $\mathcal{H}_{-1}$ is the convex hull of the unit balls in $\mathcal{H}_{-1}^{(j)}$ $(j = 1, 2)$. Thus

$$\|h\|_{\mathcal{H}_{-1}} = \inf_{\substack{h = th_1 + (1-t)h_2 \\ h_j \in \mathcal{H}_{-1}^{(j)}}} \max\left(\|h_1\|^{(1)}, \|h_2\|^{(2)} \right).$$

There exists some natural duality between the spaces $\mathcal{H}_1'$ and $\mathcal{H}_{-1}$: for $x = \{x_k\} \in \mathcal{H}_1'$ and $y = \{y_k\} \in \mathcal{H}_{-1}$ we define

$$\int_{\mathbf{T}} (y(t), x(t))\, dt := \langle y, x \rangle := \sum_{-\infty}^{\infty} (y_k, x_k)_H .$$

Here $\{x_k\}$ are the Fourier coefficients of our function $x = x(t) \in \mathcal{H}_1'$.

We can define analogously (with H_1' instead of H_1) the spaces H_{-1}' and $\mathcal{H}_{-1}'$ and the duality between $\mathcal{H}_1$ and $\mathcal{H}_{-1}'$.

We notice that $\frac{d}{dt}$ can be uniquely extended from $\mathcal{H}_1$ (or from the space of smooth functions) to a continuous operator from $\mathcal{H}_0$ into $\mathcal{H}_{-1}$ and from $\mathcal{H}_0$ into $\mathcal{H}_{-1}'$:

$$\left(\frac{d}{dt}\,h(t)\right)_k := 2\pi i k h_k.$$

Here h_k are the Fourier coefficients of a function $h(t)$. It is easy to check that for $h \in \mathcal{H}_0$, $f \in \mathcal{H}_1'$ the equality

$$\int_{\mathbb{T}} (h'(t), f(t))\,dt = -\int_{\mathbb{T}} (h(t), f'(t))_H\,dt$$

holds. Here we regard the left hand side integral in the meaning of the duality between $\mathcal{H}_{-1}$ and $\mathcal{H}_1'$.

We also need some vector function spaces on an interval. Let us denote by $\mathcal{H}_0[a,b]$ the space $L_2([a,b], H)$. The space $\mathcal{H}_1\,[a,b]$ (respectively, $\mathcal{H}_1'[a,b]$) consists of H_1-valued (respectively, H_1'-valued) functions that belong to $L_2([a,b], H_1)$ (respectively, to $L_2([a,b], H_1'))$ and that their first distributional derivatives belong to $\mathcal{H}_0[a,b]$. It was proved in [262] that the space of smooth H_1-valued (respectively, H_1-valued) functions is dense in $\mathcal{H}_1[a,b]$ (respectively, in $\mathcal{H}_1'[a,b]$). An immediate consequence is (see section 2.2 of chapter I in [262]) that the operators of boundary values (i.e. values at the points a and b) are continuous, as operators from $\mathcal{H}_1[a,b]$ (respectively, $\mathcal{H}_1[a,b]$) into H. It can also be shown (section 2.2 of chapter I in [262] that if $x_1(t) \in \mathcal{H}_1[0,1]$, $x_2(t) \in \mathcal{H}_1[1,2]$ and $x_1(1) = x_2(1)$, then

$$x(t) = \begin{cases} x_1(t), & t \in [0,1] \\ x_2(t), & t \in [1,2] \end{cases}$$

belongs to the space $\mathcal{H}_1[0,2]$.

Lemma 5.1.2. *The embedding $\mathcal{H}_1' \subset \mathcal{H}_0$ is compact.*

Proof. Let us represent both spaces as spaces of sequences $\{h_k\}$ of Fourier coefficients . Let also $J : \mathcal{H}_1' \to \mathcal{H}_0$ be our embedding and

$$P_j(\{h_k\}) = \{\ldots, 0, \ldots, 0, h_{-j}, \ldots, h_j, 0, \ldots\}.$$

It is easy to check that

$$\| J - P_j \|_{L(\mathcal{H}_1', \mathcal{H}_0)} \leq j^{-1} .$$

Hence, $J = \lim_{j \to \infty} P_j$ in the operator norm. Operator P_j is the direct sum of zero operator and of a finite number of copies of the compact embedding operator $H_1' \to H$. The lemma is proved.

Let us consider the operators

$$L := \frac{d}{dt} + A(t) , \qquad L^* := - \frac{d}{dt} + A^*(t) .$$

Lemma 5.1.3. *The operator L is continuous from $\mathcal{H}_s$ into $\mathcal{H}_{s-1}$ for $s = 1, 0$. Analogously, L^* is continuous from $\mathcal{H}_1'$ into $\mathcal{H}_0$ and from $\mathcal{H}_0$ into $\mathcal{H}'_{-1}$.*

We omit the proof that is quite simple and follows immediately from the definitions of the spaces and properties of the operator function $A(t)$.

We also observe that the operators $L: \mathcal{H}_1[a,b] \to \mathcal{H}_0[a,b]$ and $L^*: \mathcal{H}_1'[a,b] \to \mathcal{H}_0[a,b]$ are continuous.

Theorem 5.1.4. *The operator $L^*: \mathcal{H}_1' \to \mathcal{H}_0$ is a Fredholm one.*

Proof. Let us fix a point $t_0 \in \mathbf{T}$ and let also $\sigma \in \mathbf{R}$ be the number from the assumption iv). We consider the operator

$$\mathcal{L}_{t_0} = -\frac{d}{dt} + A^*(t_0) - \sigma I : \mathcal{H}'_1 \to \mathcal{H}_0 \quad (\mathcal{L}_{t_0} : \mathcal{H}_0 \to \mathcal{H}'_{-1}).$$

We shall prove that $\mathcal{L}_{t_0}$ is invertible. In fact, let $x \in \mathcal{H}'_1$ and $\{x_k\}$ be its Fourier coefficients. In this case

$$(\mathcal{L}_{t_0} x)_k = (- 2\pi i k - \sigma + A^*(t_0)) \, x_k .$$

The operator $A^*(t_0) - (\sigma + i\mu)$ is invertible for all values $\mu \in \mathbf{R}$. Therefore, we can define the operator $B_{t_0} = (\mathcal{L}_{t_0})^{-1}$ in the following way:

$$(B_{t_0}x)_k = (A^*(t_0) - (\sigma + 2\pi ik))^{-1} x_k \, .$$

Let us check its boundedness. If $x \in \mathcal{H}_0$, then

$$\| B_{t_0} x \|^2_{\mathcal{H}_1'} = \sum k^2 \, \| (A^*(t_0) - (\sigma + 2\pi ik))^{-1} x_k \|^2_H +$$

$$+ \sum \| (A^*(t_0) - (\sigma + 2\pi ik))^{-1} x_k \|^2_{H_1'} \leq$$

$$\leq C \sum (k^2 + 1) \, \| (A^*(t_0) - (\sigma + 2\pi ik))^{-1} x_k \|^2_H +$$

$$+ C \sum \| A^*(t_0)(A^*(t_0) - (\sigma + 2\pi ik))^{-1} x_k \|^2_H \leq$$

$$\leq C \sum \frac{k^2 + 1}{4\pi^2 k^2 + \sigma^2 + 1} \, \| x_k \|^2_H + C \sum \| x_k \|^2_H +$$

$$+ C \sum (\sigma^2 + 4\pi^2 k^2) \, \| (A^*(t_0) - (\sigma + 2\pi ik))^{-1} x_k \|^2_H \leq$$

$$\leq C \sum \| x_k \|^2_H = C \| x \|^2_{\mathcal{H}_0} \, .$$

Here we denote by C different constants (that also do not depend on t_0) .

We shall also use compactness of B_{t_0} in $\mathcal{H}_1'$, which follows from compactness of $P_j B_{t_0}$ (where the operator P_j was introduced earlier) and from the equality

$$B_{t_0} = \lim_{j \to \infty} P_j B_{t_0}$$

(in the operator norm). Indeed, the operator $P_j B_{t_0}$ in $\mathcal{H}_1'$ is the direct sum of a finite number of operators $(A^*(t_0) - (\sigma + 2\pi ik))^{-1}$ in H_1' . Let us prove compactness of the last operator. Let $f_l \in H_1'$ ($l = 1,2,\ldots$) and $\| f_l \|_{H_1'} < C$. Due to the condition iv), this means that $\| f_l \|_H < C$ and $\| A^*(t_0) f_l \|_H < C$. According to the conditions iv) and viii), the operator $(A^*(t_0) - (\sigma + 2\pi ik))^{-1}$ is compact in H, so we can choose a subsequence f_{l_j} such that the sequences

$$(A^*(t_0)-(\sigma+2\pi ik))^{-1}f_{l_j}$$

and

$$(A^*(t_0)-(\sigma+2\pi ik))^{-1}\,A^*(t_0)\,f_{l_j}$$

are convergent in H. This implies convergence of $(A^*(t_0)-(\sigma+2\pi ik))^{-1}f_{l_j}$ in H_1'. Hence, our operator $P_jB_{t_0}$ is compact in $\mathcal{H}_1'$.

Let now $x\in\mathcal{H}_1'$. Then

$$\begin{aligned}\|(B_{t_0}-P_jB_{t_0})x\|^2_{\mathcal{H}_1'} &= \sum_{|k|>j} k^2\|(A^*(t_0)-(\sigma+2\pi ik))^{-1}x_k\|^2_H + \\ &+\sum_{|k|>j}\|(A(t_0)-(\sigma+2\pi ik))^{-1}x_k\|^2_{H_1'}\le \\ &\le C\sum_{|k|>j}(k^2+1)\|(A^*(t_0)-(\sigma+2\pi ik))^{-1}x_k\|^2_H \\ &+C\sum_{|k|>j}\|(A^*(t_0)-(\sigma+2\pi ik))^{-1}A^*(t_0)x_k\|^2_H\le \\ \le C\sum_{|k|>j}\Big(\frac{k^2+1}{4\pi^2k^2+\sigma^2+1}\|x_k\|^2_H &+\frac{1}{4\pi^2k^2+\sigma^2+1}\|A^*(t_0)x_k\|^2_H\Big)\le \\ &\le\frac{C}{4\pi^2j^2+\sigma^2+1}\sum(k^2\|x_k\|^2_H+\|A^*(t_0)x_k\|^2_H)\le \\ &\le\frac{C}{4\pi^2j^2+\sigma^2+1}\|x\|^2_{\mathcal{H}_1'}.\end{aligned}$$

Therefore, $\|B_{t_0}-P_jB_{t_0}\|_{L(\mathcal{H}_1')}\underset{j\to\infty}{\to}0.$

We are going now to construct some left and right regularizers for our operator. According to theorem 1.2.8, this will imply Fredholm property.

Let $\{\mathcal{U}_j\}$ be a covering of the circle $\mathbf{T}$ by open intervals of the length $2\lambda>0$ such that any point of $\mathbf{T}$ is covered at most twice. Let also t_j be the center of $\mathcal{U}_j$. We denote by $\phi_j(t)$ a C^∞-partition of unity subordinate to our covering. Consider also C^∞-functions $\psi_j(t)$ such that $\psi_j=1$ on the support of ϕ_j, $supp\ \psi_j\subset\mathcal{U}_j$ and $0\le\psi_j\le 1$.

We introduce the following bounded operator B that maps $\mathcal{H}_0$ into $\mathcal{H}_1'$:

$$Bf=\sum_j \phi_j B_{t_j} \psi_j f.$$

Here

$$(B_{t_j} x)_k=(A^*(t_j)-(\sigma+2\pi ik))^{-1} x_k .$$

Boundedness of B is obvious, due to the known properties of B_{t_j} and of functions ϕ_j, ψ_j. It is also easy to see that B is compact as an operator in $\mathcal{H}_1'$, due to the analogous property of B_{t_j} .

Let us calculate the composition L^*B (that is obviously a bounded operator in $\mathcal{H}_0$) :

$$L^*Bf=\sum_j L^*\phi_j B_{t_j}\psi_j f=\sum_j \phi_j L^* B_{t_j} f+\sum_j [L^*,\phi_j] B_{t_j}\psi_j f=$$

$$\sum_j \phi_j \mathcal{L}_{t_j} B_{t_j}\psi_j f-\sum_j \phi_j' B_{t_j}\psi_j f+\sigma\sum_j \phi_j B_{t_j}\psi_j f+\sum_j \phi_j (A^*(t)-A^*(t_j))B_{t_j}\psi_j f .$$

The first sum is equal to f and therefore provides us with the identity operator. The second and third ones are compact operators (because they can be factored through the compact inclusion $\mathcal{H}_1\subset\mathcal{H}_0$). We have to estimate the last term. For $t\in\mathcal{U}_j$,

$$\| A^*(t)-A^*(t_j)\|_{L(H_1',H)} \le \omega(\lambda)\underset{\lambda\to 0}{\to} 0.$$

Hence,

$$\int_{\mathbf{T}} \| \phi_j(t)[A^*(t)-A^*(t_j)]B_{t_j}\psi_j f\|_H^2 \, dt\le\omega^2(\lambda)\int_{\mathbf{T}} \| B_{t_j}\psi_j f\|_{H_1'}^2 dt\le$$

$$c\omega^2(\lambda)\int_{\mathbf{T}} \| \psi_j f\|_H^2 dt .$$

Let us notice that for any given point of $\mathbf{T}$ our sums over j contain at most two terms.

Therefore, for pointwise estimations, the expressions

$$\sum_j \| \quad \|^2$$

and

$$(\sum_j \| \quad \|)^2$$

are equivalent (with constants that do not depend on t and λ). We shall use this reasoning without additional discussions. We shall also use the letter C for different constants. Further,

$$\int_{\mathbf{T}} \| \sum \phi_j [A^*(t) - A^*(t_j)] B_{t_j} \psi_j f \|_H^2 dt \leq C \sum \int_{\mathbf{T}} \| \phi_j [A^*(t) - A^*(t_j)] B_{t_j} \psi_j f \|_H^2 \leq$$

$$\leq C\omega^2(\lambda) \sum_j \int_{\mathbf{T}} \| \psi_j(t) f(t) \|_H^2 dt \leq C\omega^2(\lambda) \int_{\mathbf{T}} \| f(t) \|_H^2 dt = C\omega^2(\lambda) \| f \|_{\mathcal{H}_0}^2.$$

Thus our last summand is a contraction for sufficiently small values of $\lambda > 0$ and

$$L^* B = I + S_1 + K_1,$$

where S_1 is a contraction and K_1 is a compact operator in $\mathcal{H}_0$. The operator $R_r = B(I + S_1)^{-1}$ is obviously a right regularizer for L^*.

Now we construct a left regularizer. We need a new set of functions $\widetilde{\psi}_j \in C^\infty(\mathbf{T})$ such that $supp\ \widetilde{\psi}_j \subset U_j$, $0 \leq \widetilde{\psi}_j \leq 1$, and $\widetilde{\psi}_j \equiv 1$ on the support of ψ_j. Let us calculate the composition

$$B_{t_j} L^* \psi_j u = B_{t_j} \mathcal{L}_{t_j} \psi_j u + \sigma B_{t_j} \psi_j u + B_{t_j} (A^*(t) - A^*(t_j)) \widetilde{\psi}_j \psi_j u =$$

$$= [I + B_{t_j} (A^*(t) - A^*(t_j)) \widetilde{\psi}_j] \psi_j u + \sigma B_{t_j} \psi_j u.$$

The norm of the operator $(A^*(t) - A^*(t_j)) \widetilde{\psi}_j$ as an operator from $\mathcal{H}_1'$ into $\mathcal{H}_0$ is obviously small (uniformly with respect to t_j) for sufficiently small values of λ. The norms of $B_{t_j}: \mathcal{H}_0 \to \mathcal{H}_1'$ are uniformly bounded. Therefore, the operator

$$V = B_{t_j}(A^*(t) - A^*(t_j))\widetilde{\psi}_j$$

is a contraction in $\mathcal{H}_1'$ for small λ. Let us denote by S_j the inverse operator to $(I+V)$. Then we get the equality

$$S_j B_{t_j} L^* \psi_j u = \psi_j u + K\psi_j u$$

where K is a compact operator in $\mathcal{H}_1'$. Denoting $R_l = \sum \phi_j S_j B_{t_j} \psi_j$, we obtain a left regularizer for L^*. The theorem is proved.

Remark. Due to compactness of the embedding $\mathcal{H}_1' \subset \mathcal{H}_0$, the operator

$$L^* - z: \mathcal{H}_1' \to \mathcal{H}_0$$

is Fredholm for any $z \in \mathbf{C}$.

The following theorem shows that under our conditions the equation is in some sense hypoelliptic..

Theorem 5.1.5. *Any solution of the equation $Lu = 0$ on $\mathbf{T}^n$ belongs to the space $\mathcal{H}_1$.*

Proof. The definition of a solution yields $u \in \mathcal{H}_0$, so $Lu \in \mathcal{H}_{-1}$. Since u is a solution, we have $Lu = 0$. Now we can construct operators C_{t_j} analogously to the construction of B_{t_j} in the proof of theorem 4, but by means of L instead of L^*. We shall prove that they can be extended to continuous operators $C_{t_j}: \mathcal{H}_{-1} \to \mathcal{H}_0$. Let $h \in \mathcal{H}_{-1}^{(1)}$, i.e. $h = \{h_k\}$, where $\sum \| h_k \|_{H_{-1}}^2 < \infty$. In this case

$$\| C_{t_j} h \|_{\mathcal{H}_0}^2 = \sum \| (A(t_j) - (\sigma - 2\pi ik))^{-1} h_k \|_H^2 \leq C \sum \| h_k \|_{H_{-1}}^2 \leq C \sum \| h \|_{\mathcal{H}_{-1}^{(1)}}^2 .$$

This means that the operator $C_{t_j}: \mathcal{H}_{-1}^{(1)} \to \mathcal{H}_0$ is continuous. Let also $h = \{h_k\} \in \mathcal{H}_{-1}^{(2)}$, i.e. $\sum (k^2 + 1)^{-1} \| h_k \|_H^2 < \infty$. In this case

$$\| C_{t_j} h \|^2_{\mathcal{H}_0} = \sum \| (A(t_j) - (\sigma - 2\pi i k))^{-1} h_k \|^2_H \leq$$

$$\leq C \sum (4\pi^2 k^2 + \sigma^2 + 1)^{-1} \| h_k \|^2_H \leq C \| h \|^2_{\mathcal{H}^{(2)}_{-1}}.$$

This means that C_{t_j} is continuous from $\mathcal{H}_{-1}$ into $\mathcal{H}_0$.

Now we need to extend operators of multiplication by smooth functions to the space $\mathcal{H}_{-1}$.

Let $\phi \in C^\infty(\mathbf{T})$ and $h \in \mathcal{H}^{(1)}_{-1} = L_2(\mathbf{T}, H_{-1})$. In this case the product $\phi h \in L_2(\mathbf{T}, H_{-1}) \subset \mathcal{H}_{-1}$ is correctly defined. Let us now pick $h \in \mathcal{H}^{(2)}_{-1}$, i.e. $h = \{h_k\}$ and

$$\sum (k^2 + 1)^{-1} \| h_k \|^2_H < \infty.$$

In this case the function $\widetilde{h}(t) = \sum_{k \neq 0} k^{-1} h_k exp(2\pi i k t)$ belongs to $L_2(\mathbf{T}, H)$. Let us define now ϕh as

$$\phi h = \phi h_0 + \frac{d}{dt}[\phi \widetilde{h}\,] - \frac{d\phi}{dt} \widetilde{h}\,.$$

It is obvious that $\phi h \in \mathcal{H}_{-1}$. Hence, the multiplication by $\phi(t)$ is a continuous operator in $\mathcal{H}_{-1}$ (we omit some simple discussion of the sum $\mathcal{H}_{-1} = \mathcal{H}^{(1)}_{-1} + \mathcal{H}^{(2)}_{-1}$ that is common in the interpolation of linear operators [210, 394]). As in the end of the proof of theorem 4, we see now that for sufficiently small $\lambda > 0$ the operator

$$C_{t_j}(A(t) - A(t_j)) \widetilde{\psi}_j$$

is a contraction; finally we construct a continuous operator $C\colon \mathcal{H}_{-1} \to \mathcal{H}_0$ such that $CL = I + K$, where K maps $\mathcal{H}_0$ into $\mathcal{H}_1$. If now $u \in \mathcal{H}_0$ and $Lu = 0$, then

$$0 = BLu = u + Ku,$$

so $u = -Ku \in \mathcal{H}_1$. The theorem is proved.

Remark. If there is a higher power of $A^*(t)$ with constant domain, then one can prove that any solution is even smoother.

Let $u(t)$ be a solution of the equation (1) on the whole axis. Suppose also that it grows exponentially in the following sense:

$$\int_0^1 \| u(t+\alpha) \|_H^2 \, dt \leq C \exp(a |\alpha|)$$

for some $C, a > 0$ and each $\alpha \in \mathbf{Z}$.

Let us introduce the following function spaces (that play the role of the spaces of section 2.1):

$$\Phi_1 = \{u(t) \in \mathcal{H}'_{1,loc}(\mathbf{R}) \colon \sup_{\alpha \in \mathbf{Z}} \| u(t) \|_{\mathcal{H}'_1[\alpha,\alpha+1]} \exp b|\alpha| < \infty \ \ \forall\, b > 0\}$$

$$\Phi_2 = \{u(t) \in L_{2,loc}(\mathbf{R}) \colon \sup_{\alpha \in \mathbf{Z}} \| u(t) \|_{L_2([\alpha,\alpha+1],H)} \exp b|\alpha| < \infty \ \ \forall\, b > 0\}.$$

Here $\mathcal{H}'_{1,loc}(\mathbf{R})$ is the space of strongly measurable H_1'-valued functions on $\mathbf{R}$ belonging to the space $\mathcal{H}'_1[a,b]$ for any $-\infty < a < b < \infty$. These spaces are naturally equipped with structures of Frechet spaces. It is obvious that our operator $L^* = -\frac{d}{dt} + A^*(t)$ is continuous as an operator from Φ_1 into Φ_2.

If u is a solution of exponential growth, the equality

$$\int_{\mathbf{R}} (u(t), L^*\phi(t))_H \, dt = 0 \tag{5.1.2}$$

holds for all $\phi \in \Phi_1$. In fact, for finite smooth functions ϕ this is just the definition of a solution. The space of such functions is dense in Φ_1 and our growth conditions on $u(t)$ imply that the expression $\int (u,\psi)_H dt$ is a continuous linear functional on Φ_2. Therefore, (2) holds also for all $\phi \in \Phi_1$. Thus, any exponentially increasing solution of (1) may be interpreted as a linear functional on Φ_2 that annihilates the image of the operator $L^*\colon \Phi_1 \to \Phi_2$. We shall determine below how all such functionals look like.

Definition 5.1.6. A nonzero solution $u(t)$ of the equation (1) is said to be a *Floquet solution* if it has the form

$$u(t) = (\exp ipt) \sum_{l=0}^{n} (t^l g_l(t)) ,$$

where the functions $g_l(t)$ are 1-periodic and $p \in \mathbf{C}$.

The number $p \in \mathbf{C}$ is said to be the ***quasimomentum*** of the solution and the nonzero number $z = \exp ip \in (\mathbf{C}\backslash 0)$ is said to be the ***Floquet exponent*** (or ***Floquet multiplier***) of the solution.

Theorem 5.1.7. *Under the assumptions* i) – viii)

A. The set N of Floquet exponents of all Floquet solutions to (1) *is either discrete in* $\mathbf{C}|0$, *or coincides with the whole* $\mathbf{C}|0$.

B. If N is discrete, i.e. $N = \{z_j\}$, *then for any* z_j *there exists a finite collection* $f_{j,k}(t)$ *of Floquet solutions with the Floquet exponent* z_j *such that*

a) any exponentially increasing solution $u(t)$ *of* (1) *can be uniquely represented as a finite linear combination*

(5.1.3)
$$u(t) = \sum_{j,k} \alpha_{j,k} f_{j,k}(t),$$

where $\alpha_{j,k} \in \mathbf{C}$;

c) for any choice of a finite collection of numbers $\alpha_{j,k} \in \mathbf{C}$ *the sum* (3) *represents an exponentially increasing solution of* (1).

C. If $N = \mathbf{C}\backslash 0$, *then there exist:*

a) a discrete subset $\{z_k\} \subset \mathbf{C}\backslash 0$;

b) a finite collection of functions

$$F_j(z,t) : \mathbf{C}|0 \times \mathbf{R} \to H ;$$

d) for any z_k *a finite collection* $f_{k,l}(t)$ *of functions from* $\mathbf{R}$ *into H;*
such that

1) any function $F_j(z,t)$ *is analytic with respect to* $z \in \mathbf{C}|0$, *as a function with values in* $L_{2,loc}(\mathbf{R},H)$; *for any fixed z it is a Floquet solution of* (1) *with the Floquet exponent z;*

2) any function $f_{k,l}$ *is a Floquet solution of* (1) *with the Floquet exponent* z_k;

3) any exponentially increasing solution $u(t)$ *of the equation* (1) *on* $\mathbf{R}$ *can be (not uniquely in general) represented as a finite sum*

$$u(t) = \sum_{k,l} \alpha_{k,l} f_{k,l}(t) + \sum_j \int F_j(z,t)\, \mu_j(z)\, dz_1 \wedge dz_2 , \tag{5.1.4}$$

where $\mu_j \in C_0^\infty(\mathbf{R}^2 \backslash 0)$ *and* $\alpha_{k,l} \in \mathbf{C}$;

4) for any choice of such $\alpha_{k,l}$ *and* μ_j *the expression* (4) *represents an exponentially increasing solution of* (1).

Proof. Let us consider a function $\phi \in \Phi_2$ and introduce (formally at the moment) the function

$$[F\phi(z)]\,(t) = \sum_{k \in \mathbf{Z}} \phi(t-k)\, z^k \tag{5.1.5}$$

for $t \in [0,1]$ and $z \in \mathbf{C} \backslash 0$. The estimate

$$\| \phi(t-k) \|_{L_2([0,1],H)} \leq C_b\, exp(-b\,|\,k\,|) \qquad \forall b > 0$$

implies the convergence of the series for $z \neq 0$ as a series with terms from $L_2([0,1],H)$. Its sum, $F\phi(z)$, is a holomorphic in $\mathbf{C}\backslash 0$ function with values in $L_2([0,1],H)$. And conversely, any such holomorphic function $f(z)$ can be represented in this way. Indeed, let $f(z) = \sum f_k z^k$ be the Laurent series for $f(z)$. It follows from analiticity in $\mathbf{C}\backslash 0$ and standard estimates of Laurent series coefficients that

$$\| f_k \|_{L_2([0,1],H)} \leq C_a\, exp(-a\,|\,k\,|) \qquad \forall a > 0.$$

We define $\phi(t) = f_{-k}(t-k)$ for $t \in [k,k+1]$. It is easy to see that $\phi \in \Phi_2$ and $F\phi(z) = f(z)$. It is also obvious that the topology of Φ_2 is equivalent to the topology of the convergence of $F\phi$ on compact subsets in $\mathbf{C}\backslash 0$.

Now we are interested in the range of Φ_1 under this transform. As in the case of Φ_2, the image $F\phi(z)$ of a function $\phi \in \Phi_1$ is an $\mathcal{H}_1'[0,1]$ – valued, analytic in $\mathbf{C}\backslash 0$

function. Not any such analytic function, however, can be represented as $\boldsymbol{F}\phi$ with $\phi \in \Phi_1$. In fact, if $f(z) = \boldsymbol{F}\phi(z)$, we get

$$[f(z)](1) = \sum \phi(1-k)z^k = z\sum \phi(0-k)z^k,$$

$$[f(z)](0) = \sum \phi(0-k)z^k.$$

Therefore, $[f(z)](1) = z[f(z)](0)$. Let us denote by $E_z \subset \mathcal{H}_1'[0,1]$ the subspace consisting of all functions g satisfying the condition $g(1) = zg(0)$. This subspace is closed, due to continuity of the boundary value operators (see the discussion preceding lemma 2). We can notice that $\mathcal{E}_1 = \bigcup_z E_z$ is an analytic subbundle of the trivial bundle $(\mathbb{C}\backslash 0) \times \mathcal{H}_1'[0,1]$. This means the existence of a projector $P(z)$: $\mathcal{H}_1'[0,1] \to E_z$ that depends analytically upon $z \in \mathbb{C}\,|\,0$. It is very easy to construct such a projector. Moreover, it is also easy to find explicitly a local trivialization of $\mathcal{E}_1$. Let $z_0 = exp\ ip_0 \in \mathbb{C}\,|\,0$. We can choose a holomorphic coordinate p in a neighborhood U of z_0 in such a way that $z = exp\ ip$ for $z \in U$. Let us consider the transform that multiplies any function of the fiber of $\mathcal{E}_1$ over any point $z \in U$ by the function $h_z(t) = exp(-ipt)$ $(t \in [0,1])$. Under this transform we obtain the trivial bundle $U \times \mathcal{H}_1'$ over U (let us recall that $\mathcal{H}_1'$ consists of periodic functions). Hence, $\mathcal{E}_1$ is a locally trivial holomorphic Hilbert bundle over $\mathbb{C}\backslash 0$. Due to theorems 1.3.2, 1.3.3 and 1.5.23; $\mathcal{E}_1$ is trivial (but we can not provide any explicit global trivialization). As we discussed before, $\boldsymbol{F}\phi(z)$ is a holomorphic section of $\mathcal{E}_1$ for any $\phi \in \Phi_1$. Let us also denote by $\mathcal{E}_2$ the trivial bundle $(\mathbb{C}\backslash 0) \times L_2([0,1],H)$.

Lemma 5.1.8. *The transform $\phi \to \boldsymbol{F}\phi(z)$ isomorphically maps the space Φ_j onto the space $\Gamma(\mathbb{C}\,|\,0, \mathcal{E}_j)$ of all holomorphic sections of the bundle $\mathcal{E}_j$ $(j = 1,2)$, equipped with the topology of uniform convergence on compact subsets of $\mathbb{C}\,|\,0$.*

Proof is completely analogous to the proof of theorem 2.2.3.

We shall also need the following version of transform $\widehat{\boldsymbol{F}}$ of section 2.2:

$$\widehat{\boldsymbol{F}}\phi(\lambda) = \boldsymbol{F}\phi(e^{\lambda}). \tag{5.1.6}$$

Let us return to the proof of the theorem. Our operator L^* commutes with the translation onto the period of the equation. Hence, $F(L^*\phi)(z) = L^*F\phi(z)$. We consider the operator L^* in the right hand side of this equality as the same operator on $\mathcal{H}_1'[0,1]$ for any $z \in \mathbb{C}\setminus 0$. But we are really interested only in the restriction of L^* to the subspace $E_z \subset \mathcal{H}_1'[0,1]$. Let us investigate the local (in z) structure of this restriction. In a neighborhood U of any point $z_0 = exp\ ip_0$ we can introduce the coordinate p and trivialization of $\mathcal{E}_2|_U$ as before. After this trivialization we have trivial bundles $U \times \mathcal{H}_1'$ and $U \times \mathcal{H}_0$. Our operator L^* has the form

$$L^* - ip = -\frac{d}{dt} + A^*(t) - ip : \mathcal{H}_1' \to \mathcal{H}_0$$

on the fiber over the point p. As we have already proved, this operator is a Fredholm one. Now, after the global trivialization we have the following situation: there are two Hilbert spaces G_1 $(= \mathcal{H}_1')$ and G_2 $(= \mathcal{H}_0)$ and a holomorphic Fredholm operator function $L(z)$: $G_1 \to G_2$, $z \in \mathbb{C}\setminus 0$ $(L(z) = L^* - ip,\ z = exp\ ip)$. We consider the spaces $\Gamma(\mathbb{C}\setminus 0, G_j)$ of holomorphic G_j-valued functions on $\mathbb{C}\setminus 0$ equipped with the topology of the uniform convergence on compact subsets. Our operator function $L(z)$ defines the operator $\mathcal{L}$ of multiplication: $x(z) \in \Gamma(\mathbb{C}\setminus 0, G_1) \to L(z)x(z) \in \Gamma(\mathbb{C}\setminus 0, G_2)$. According to the discussion preceding definition 6, we are interested in the cokernel of the operator

$$\mathcal{L}: \Gamma(\mathbb{C}\setminus 0, G_1) \to \Gamma(\mathbb{C}\setminus 0, G_2)$$

(i.e., in the set of all continuous functionals that annihilate the image of $\mathcal{L}$). According to 1.6.11, this gives the statement A of the theorem. Concerning Floquet expansion of solutions of exponential growth, we are exactly in the situation of the theorem 3.2.1. Now the rest of the proof follows the proof of theorem 3.2.1, if we take into account that analytic subsets in $\mathbb{C}\setminus 0$ are either discrete, or equal to $\mathbb{C}\setminus 0$.

Remark. The statement A of the theorem also holds under conditions i) – iv) and under the assumption of compactness of the inclusion $H_1 \subset H$.

The situation, when the Floquet exponents set for an equation (1) coincides with the entire space $\mathbb{C}\setminus 0$, is apparently a pathology. If we deal, for instance, with a parabolic problem, then the monodromy operator is defined and compact (see section 5.3). The set of Floquet exponents coincides with the spectrum of this operator and is, therefore, discrete. As we know (see section 4.1), in the case of spatially periodic elliptic equations this pathology does occur. As far as it is known to the author, the corresponding question for evolution equations has not been answered yet.

Theorem 5.1.9. *Let us assume that only conditions i) – iv) are fulfilled and the embedding $H_1 \subset H$ is compact. In this case the following assertions are equivalent:*

i) the set of Floquet exponents for the equation $Lu = 0$ coincides with $\mathbf{C} \setminus 0$;

ii) there exists a nonzero solution of this equation that belongs to $L_2(\mathbf{R}, H)$;

iii) there exists a nonzero solution of this equation that satisfies the estimate:

$$\int_k^{k+1} \| u(t) \|_H^2 \, dt \leq C_a \exp(-a|k|) \qquad \forall a > 0,\ k \in \mathbf{Z}.$$

Proof is completely analogous to the proof of theorem 4.1.5.

Now we shall discuss solutions of more rapid growth. Let s_j be the s-numbers of the embedding operator $\mathfrak{z}$: $H_1' \subset H$.

Theorem 5.1.10. *Suppose that there exists a number $p > 1$ such that the series*

$$\sum j\, s_j^p\, s_{j+1}^{-1}$$

converges. Let also the Fredholm index of the operator $L^*: \mathcal{H}_1' \to \mathcal{H}_0$ be equal to zero and the equation $Lu = 0$ has no nonzero subexponential (in the meaning of the part iii) of theorem 9) solutions. Then the linear hull of all Floquet solutions is $L_{2,loc}(\mathbf{R}, H)$-dense in the class of all solutions defined on the entire axis and satisfying the estimate

$$\int_0^1 \| u(t+k) \|_H^2 dt \leq C \exp a|k|^{1+\gamma}$$

for some $a > 0$ and $\gamma < (p-1)^{-1}$.

Proof. Let us prove first an auxiliary result.

Lemma 5.1.11. *If $\sum j\, s_j^p\, s_{j+1}^{-1} < \infty$, then the embedding operator $\Lambda: \mathcal{H}_1' \subset \mathcal{H}_0$*

belongs to the class $\mathfrak{I}_p$, i.e. the sequence of its s-numbers $s_j(\Lambda)$ belongs to l_p :

$$\sum s_j(\Lambda)^p < \infty$$

Proof of the lemma. Consider the spaces $\mathcal{H}_1'$ and $\mathcal{H}_0$ as corresponding spaces of sequences $\{h_k\}$ (see lemma 1). We denote by R_j a $(j-1)$-dimensional operator from H_1' into H that provides the j-th s-number of $\mathfrak{I}$ (see section 1.2). We introduce the operator $\tilde{R}_j : \mathcal{H}_1' \to \mathcal{H}_0$ as follows :

$$\tilde{R}_j\{h_k\} = P_{k_j}\{R_j h_k\},$$

where $k_j = [s_j^{-1}]$, and the operators P_j were introduced in the proof of lemma 2. Then,

$$\| h - \tilde{R}_j h \|^2_{\mathcal{H}_0} = \sum_{-k_j}^{k_j} \| h_k - R_j h_k \|^2_H + \sum_{|k| > k_j} \| h_k \|^2_H \leq$$

$$\leq s_j^2 \sum_{-k_j}^{k_j} \| h_k \|^2_{H_1'} + (k_j + 1)^{-2} \sum_{|k| > k_j} k^2 \| h_k \|^2_H \leq$$

$$\leq max \{s_j^2, (k_j + 1)^{-2}\} \| h \|_{\mathcal{H}_1'} = s_j^2 \| h \|^2_{\mathcal{H}_1'}.$$

Let us mention that the rank of $\tilde{R}_j$ is equal to $(2k_j + 1)(j-1) = (2[s_j^{-1}] + 1)(j-1)$, so

$$s_{(2k_j + 1)j - 2k_j}(\Lambda) \leq s_j .$$

Due to the monotonicity of s-numbers , $s_k(\Lambda) \leq s_j$ for

$$j(2k_j + 1) - 2k_j \leq k.$$

Thus,

$$\sum s_k^p(\Delta) \leq \sum \{(j+1)(2k_{j+1} + 1) - 2k_{j+1} - j(2k_j + 1) + 2k_j\} \ s_j^p.$$

Due to the definition of k_j, in order to get convergence of the latter series, we need convergence of the series $\sum j\, s_j^p s_{j+1}^{-1}$, $\sum j\, s_j^p$, $\sum s_j^p s_{j+1}^{-1}$ and $\sum j\, s_j^{p-1}$. The worst of them is the first one, and its convergence is guaranteed by the assumption. The lemma is proved.

Now the proof of the theorem is completely analogous to the proof of theorem 3.3.1.

For $\gamma < p^{-1}$ we can obtain some expansion into Floquet solutions.

Theorem 5.1.12. *Suppose that the Fredholm index of the operator* $L:\mathcal{H}_1 \to \mathcal{H}_0$ *is zero, and the set* N *of Floquet exponents for* (1) *is discrete (i.e., it does not coincide with* $\mathbf{C}\,|\,0$*). Denote the collection of all quasimomentums of* (1) *with their multiplicities by* $\{\lambda_n, p_n\}$. *Define* $\beta_{i,k,m,n}$ *as in section 1.5. Suppose also that* $\sum j\, s_j^p s_{j+1}^{-1} < \infty$ *for some* $p > 1$. *Then for each triple* (m,n,k) *there are finitely many Floquet solutions* $h_{k,l,m,n}(t)$ *of the problem* (1) *(with quasimomentum* λ_m *of multiplicity* k*) such that*

i) *Any solution of* (1) *satisfying*

$$\| u(t) \| \le C(u)\; exp\; a(u)\; |t|^{1+\gamma} \quad \textit{for some } \gamma < p^{-1} \tag{5.1.7}$$

can be represented as the sum

$$u(t) = \sum_{i,k,l,m,n} \phi_{i,n,l}\, \beta_{i,k,n,m}\, h_{k,l,m,n}(t), \tag{5.1.8}$$

where after summation (finite for each n) over i,k,l,m *the series converges absolutely in* $L_{2,\,loc}(\mathbf{R}, H)$ *and*

$$\sum_{i,n} \Big(\sum_{l} |\phi_{i,n,l}|^2\Big)^{\frac{1}{2}} exp\,(a\,|\lambda_n|^{p+1}) < \infty. \tag{5.1.9}$$

ii) *For any collection* $\{\phi_{i,n,l}\}$ *of complex numbers, satisfying* (9), *the series* (8) *converges in the sense indicated above, and defines a solution of* (1) *satisfying* (7).

Proof. First of all, we introduce the following functional spaces:

$$\Phi_{1,p} = \{f \in \mathcal{H}'_{1,loc}(\mathbf{R}) \mid \sup_{k \in \mathbf{Z}} \|f\|_{H_1'[k,k+1]} \exp(b|k|^p) < \infty \text{ for all } b > 0\}$$

$$\Phi_{2,p} = \{f \in L_{2,\,loc}(\mathbf{R},H) \mid \sup_{k \in \mathbf{Z}} \|f\|_{L_2[k,k+1]} \exp(b|k|^p) < \infty \text{ for all } b > 0\}.$$

Operator L^* is obviously continuous from $\Phi_{1,p}$ into $\Phi_{2,p}$, and we can regard any solution of (1) that satisfies (7) as a functional on $\Phi_{2,p}$ that annihilates the image of the operator. After the transformation $\widehat{F}$ this operator becomes multiplication by the Fredholm morphism $L^* - z$ on the spaces $A_{p/(p-1)}(\mathcal{E}^1) \to A_{p/(p-1)}(\mathcal{E}^2)$ (where the bundles $\mathcal{E}^1$ and $\mathcal{E}^2$ were introduced before lemma 8 and the latter spaces were introduced in section 1.5).

Lemma 5.1.13. *There is a holomorphic projection P from the trivial bundle $\mathbf{C} \times \mathcal{H}_1'$ onto the bundle $\widehat{\mathcal{E}}^1$. This projection is generated by the multiplication by an entire $2\pi i$-periodic operator function $P(\lambda)$ of order 1 in $\mathcal{H}_1'$.*

Proof. Let us recall that the fiber of $\widehat{\mathcal{E}}^1$ over a point $\lambda \in \mathbf{C}$ consists of all functions $f \in \mathcal{H}_1'[0,1]$ satisfying the condition $f(1) = e^{\lambda} f(0)$. Consider the restriction operator $f \to f(1)$. As it has been mentioned before, it is a continuous linear operator from $\mathcal{H}_1'$ into H. Denote its kernel (which is a closed subspace of $\mathcal{H}_1'$) by M_1. We have then the natural surjective "trace" operator $f \to f(1)$ from $\mathcal{H}_1'$ onto $F_1 := \mathcal{H}_1'/M_1$. Let T: $F_1 \to \mathcal{H}_1'$ be its right inverse. We can always assume that $(Tu)(0) = 0$ for all u, otherwise we can multiply T by a fixed smooth function ϕ such that $\phi(0) = 0$, $\phi(1) = 1$. We can introduce the spaces M_0, F_0 in the similar way, considering the point 0 instead of 1. The change of variables $t \to 1 - t$ switches the points 1 and 0 and enables us to identify the spaces F_0 and F_1 in the natural way. Consider now the operator $P(z)f = f - T(f(1) - e^{\lambda} f(0))$. It obviously satisfies the properties we need. This proves the lemma.

Now we are able to finish the proof of the theorem. By the last lemma, after the transform $\widehat{F}$ we are in the situation of theorem 1.8.2, which gives all the statements that we need. The proof is complete.

The condition iv) on the operator A, as we have already mentioned before, means some kind of hypoellipticity of the problem. For instance, parabolic and inverse parabolic problems belong to this class (since the inequality iv) is satisfied in these cases in a half-plane instead of a single vertical line). Uncoupled systems of such equations also belong to the class. We shall mention a class of problems that can be considered as "weakly coupled" parabolic and inverse parabolic problems.

Consider so called anti-canonical equation [88]:

$$G\frac{du}{dt} = H(t)u. \tag{5.1.10}$$

We assume here that $D(H(t)) \equiv H_1 \subset H$, operator G is selfadjoint, bounded and invertible, $GH_1 \subset H_1$, $H(t)$ is selfadjoint and positive definite (i.e. $H(t) \geq cI,\ c > 0$). We also assume that the commutator $[G,H(t)]$ is bounded in H. In this case we can prove the estimate iv). We shall prove that the operator $i\beta - G^{-1}H(t)$ is invertible for $\beta \in \mathbf{R}$ and

$$\| (i\beta - G^{-1}H(t))^{-1} \|_{H\to H} \leq \frac{C}{1+|\beta|}\,. \tag{5.1.11}$$

Indeed, we can consider the operator $i\beta G - H(t)$ instead of $i\beta - G^{-1}H(t)$. Then

$$\| (i\beta G - H(t))\, x\| \, \|x\| \geq |((i\beta G - H(t))\, x,\, x)| = |\, i\beta(Gx,x) - (H(t)x,x)\,|\,.$$

Taking into account selfadjointness of G and *of* H and the inequality $H \geq cI$, we get

$$\| (i\beta G - H(t))\, x\| \geq c\, \|x\|,\ \beta \in \mathbf{R}.$$

The analogous estimate holds also for the adjoint operator. Hence, the operator $(i\beta G - H(t))^{-1}$ does exist. Now,

$$((i\beta G - H(t))\, x,\ Gx) = i\beta\, \|Gx\|^2 - (H(t)x,\ Gx) =$$

$$i(\beta\, \|Gx\|^2 + \tfrac{1}{2}i([G,H(t)]x,x)) - Re\ (H(t)x,\ Gx).$$

Since the commutator $[G,H(t)]$ is bounded and G is bounded and has bounded inverse, we get for sufficiently large values of $|\beta|$:

$$(5.1.12)\qquad \|(i\beta G - H(t))\,x\|\;\|x\| \geq |\,Im\,((i\beta G - H(t))\,x,\ Gx)| \geq C\,(1 + |\beta|)\,\|x\|^{2}.$$

This provides the estimate (11) for sufficiently large $|\beta|$. Since the operator is invertible for all real β, we get the estimate on the whole real line of values of β.

The simplest example of a problem of this type is the following one. Suppose that our operator G has the form:

$$G = \begin{bmatrix} I_{H_1} & 0 \\ 0 & -I_{H_2} \end{bmatrix},$$

where $H = H_1 \oplus H_2$ is an orthogonal decomposition of the space. Then $H(t)$ has a block structure with respect to this decomposition:

$$H(t) = \begin{bmatrix} H_{11}(t) & H_{12}(t) \\ H_{21}(t) & H_{22}(t) \end{bmatrix}.$$

The condition that we require is that the operators $H_{12}(t)$ and $H_{21}(t)$ are bounded (uniformly with respect to $t \in \mathbf{R}$) and the diagonal blocks $H_{11}(t)$ and $H_{22}(t)$ give rise to a parabolic and an inverse parabolic problems. The out-of-diagonal blocks provide a "weak coupling" of these two problems. The Cauchy problem is incorrect in this case, so the approaches based on the monodromy operator can not be applied.

5.2. Some degenerate cases

The aim of this section is to consider periodic in time evolution PDE systems of the type:

$$(5.2.1)\qquad \begin{cases} u_t(t,x) = A(t,x,D)\ a(t)u(t,x) + b(t)u(t,x), & t>0,\ \ x\in\Omega \\ B(x,D)a(t)u(t,x)\,|_{x\in\partial\Omega} = 0, & t>0, \end{cases}$$

where Ω is a bounded region in $\mathbf{R}^l$ with sufficiently smooth boundary, $u(t,x)$ is a function on $\mathbf{R}\times\bar{\Omega}$ with values in $\mathbf{C}^n$, $a(t)$ and $b(t)$ are periodic $n\times n$-matrix-functions, differential operators A and B have scalar coefficients and the scalar problem

$$(5.2.2)\qquad \begin{cases} v_t = Av \\ Bu|_{x\in\partial\Omega} = 0 \end{cases}$$

is in some sense hypoelliptic (for instance, parabolic, or inverse parabolic). The matrix $a(t)$ is supposed to have real spectrum without adjoint vectors corresponding to the zero eigenvalue. We prove that the set of Floquet exponents of (1) is either discrete with only possible accumulation points 0, ∞, $\chi_1, \ldots, \chi_m$ (where $\dim \operatorname{Ker} a(t) \equiv m$), or contains $\mathbf{C}\backslash\{0, \chi_1, \ldots, \chi_m\}$. Here $\chi_1, \ldots, \chi_m$ are the Floquet exponents for an m-dimensional system of ordinary differential equations constructed by means of (1). If we have an exponential bound for solutions of (1) (for example, if (1) is parabolic), the set is discrete with the set of accumulation points $\{0, \chi_1, \ldots, \chi_m\}$. Let us mention that without the degeneration caused by the non-zero kernel of $a(t)$ (i.e. in the situation of the section 5.1) the Floquet exponents do not have finite accumulation points except $\{0\}$, provided the set of all exponents is not the whole $\mathbf{C}\backslash\{0\}$.

We shall consider our problem in an abstract operator setting (the PDE case is a particular one). Consider first the case of a nondegenerate matrix $a(t)$. It will be convenient to denote it for some time by $q(t)$.

Let H be a separable Hilbert space and $A(t)$ be a 1-periodic family of (unbounded) linear operators in H such that

i) The domain $\mathfrak{D}(A(t))$ coincides with a fixed dense linear subspace $H_1\subset H$ and the graph norms on $D(A(t))$ are uniformly (with respect to t) equivalent to a norm $\|\cdot\|_1$ on H_1. In this norm H_1 is a Hilbert space and the embedding $H_1\subset H$ is compact.

ii) The operator function $A(t)$ is continuous in the operator norm of the space $L(H_1, H)$.

iii) There exists a continuous (may be supposed to be real-valued) function $\delta(t)$ such that the line $l(t) = \delta(t) + i\mathbf{R}$ lies in the resolvent set of $A(t)$ for each $t \in \mathbf{R}$ and on this line

$$\|(\lambda I - A(t))^{-1}\|_{H \to H} \le C(|\lambda| + 1)^{-1}$$

uniformly with respect to t.

We also consider a 1-periodic invertible continuous $n \times n$ matrix-function $q(t)$ and assume that :

iv) The spectrum of $q(t)$ is real for any $t \in \mathbf{R}$.

The spaces $\mathcal{H}_1$ and $\mathcal{H}_0$ are defined as in section 5.1.

Theorem 5.2.1. *Under the conditions* i-iv *the operator*

$$L = d/dt - A(t)q(t) : [\mathcal{H}_1]^n \to [\mathcal{H}_0]^n \tag{5.2.3}$$

is Fredholm.

Proof. Let us fix a point $t_0 \in \mathbf{T}$ and choose a basis e_{jl} of eigen and adjoint vectors for $q(t_0)$; its biorthogonal sequence we denote by f_{jl}. Here e_{j0} is an eigenvector corresponding to the eigenvalue λ_j ($\neq 0$) and $e_{j1}, \ldots, e_{jr_j}$ are the corresponding adjoint vectors. The norm of the vector

$$a = \sum a_{jl}\, e_{jl} \in \mathbf{C}^n \quad (\text{ where } a_{jl} = \langle f_{jl}, a \rangle)$$

is equivalent to $(\sum |a_{jl}|^2)^{1/2}$. (Let us mention that we do not need here any continuity in t_0). Consider the operator

$$L_{t_0} = d/dt + \delta(t_0)\, q(t_0) - A(t_0)q(t_0) : [\mathcal{H}_1]^n \to [\mathcal{H}_0]^n .$$

Due to our assumptions i-iv, it is bounded. This operator function with values in

$L(\mathcal{H}_1^n, \mathcal{H}_0^n)$ is norm continuous with respect to t_0. We are going to prove invertibility of L_{t_0}. Let $u \in \mathcal{H}_0{}^n$ and

$$u = \sum_0^{\infty} u_k \, exp(2\pi ikt)$$

be its Fourier expansion, where $u_k = (u_k^{(1)}, \ldots, u_k^{(n)}) \in H^n$. We can expand each u_k into the finite sum $u_k = \sum_{j,\, l} u_{kjl} \, e_{jl}$, where $u_{kjl} \in H$ (we just put

$$u_{kjl} = < f_{jl}, u_k > = \sum_k u_k^{(r)} f_{jl}^{(r)} \, ,$$

where

$$f_{jl} = (f_{jl}^{(1)}, \ldots, f_{jl}^{(n)})$$

is the biorthogonal sequence introduced above). Therefore,

$$u = \sum_{k=-\infty}^{\infty} (\sum_{j,\, l} u_{kjl} \, e_{jl}) \, exp(2\pi ikt) \, ,$$

where $u_{jkl} \in H$. Then

$$(L_{t_0} u)_{kjl} = (2\pi ik + \sigma(t_0) \, \lambda_j - A(t_0) \, \lambda_j) \, u_{kjl} +$$

$$(\sigma(t_0) - A(t_0)) \, u_{kjl-1} \qquad (\text{ for } l > 0), \tag{5.2.4}$$

$$(L_{t_0} u)_{kj0} = (2\pi ik + \sigma(t_0) \, \lambda_j - A(t_0) \, \lambda_j) \, u_{kj0} \, .$$

Notice that the operator

$$(2\pi ik + \sigma(t_0) \, \lambda_j - A(t_0)\lambda_j) = \lambda_j (2\pi ik\lambda_j^{-1} + \sigma(t_0) - A(t_0) \,)$$

is invertible in H, due to conditions iii) and iv) , and

$$\| (2\pi ik + \sigma(t_0)\lambda_j - A(t_0)\lambda_j)^{-1} v \|_{H_1} \leq C \| v \|_{H},$$

$$\|(2\pi ik+\sigma(t_0)\lambda_j-A(t_0)\lambda_j)^{-1}v\|_H\le C(|k|+1)^{-1}\|v\|_H.$$

Hence,

$$(L_{t_0}^{-1}u)_{kj0}=(2\pi ik+\sigma(t_0)\lambda_j-A(t_0)\lambda_j)^{-1}u_{kj0},$$

$$(L_{t_0}^{-1}u)_{kjl}=(2\pi ik+\sigma(t_0)\lambda_j-A(t_0)\lambda_j)^{-1}(u_{kjl}-(\sigma(t_0)-A(t_0))(L_{t_0}^{-1}u)_{kjl-1})$$

for $l>0$, and the following estimates hold:

$$\|(L_{t_0}^{-1}u)_{kj0}\|_H\le\frac{C}{|k|+1}\|u_{kj0}\|_H,$$

$$\|(L_{t_0}^{-1}u)_{kjl}\|_H\le\frac{C\max\limits_{0\le l\le r_j}\|u_{kjl}\|_H}{|k|+1},\qquad l>0,$$

and

$$\sum_{l=0}^{r_j}\|(L_{t_0}^{-1}u)_{kjl}\|_H^2\le\frac{C}{(|k|+1)^2}\sum_{l=0}^{r_j}\|u_{kjl}\|_H^2.$$

Besides,

$$\|(A(t_0)(L_{t_0}^{-1}u)_{kjl}\|_H^2\le C\sum_{l=0}^{r}\|u_{kjl}\|_H^2.$$

Therefore,

$$\|L_{t_0}^{-1}u\|^2_{[\mathcal{H}_1]^m}\le C(\sum_{k,j,l}(|k|+1)^2\|(L_{t_0}^{-1}u)_{kjl}\|_H^2+$$
$$\|A(t_0)(L_{t_0}^{-1}u)_{kjl}\|_H^2\le C\sum_{k,j,l}\|u_{kjl}\|_H^2\le C\|u\|^2_{[\mathcal{H}_0]^n}.$$

(here we denote by C various constants) .

We observe that the operator $L_{t_0}^{-1}$ is compact in $[\mathcal{H}_1]^n$. Indeed, consider the operator

$$(L_p u)_{k\,j\,l} = \begin{cases} (L_{t_0}^{-1} u)_{k\,j\,l} & \text{for } |k| \le p \\ 0 & \text{for } |k| > p,\ p = 1, 2, \dots \end{cases}$$

We will prove compactness of L_p in $[\mathcal{H}_1]^n$ and the equality

$$\lim_{p \to \infty} \| L_p - L_{t_0}^{-1} \|_{[\mathcal{H}_1]^n} = 0.$$

This will imply compactness of $L_{t_0}^{-1}$. Let us consider the norm $\| (L_{t_0}^{-1} - L_p) u \|^2_{[\mathcal{H}_1]^n}$:

$$\| (L_{t_0}^{-1} - L_p) u \|^2_{[\mathcal{H}_1]^n} \le C \Big(\sum_{|k| > p} \sum_{j\,l} (|k| + 1)^2 \, \| (L_{t_0}^{-1} u)_{k\,j\,l} \|^2_H + \\ + \| A(t_0)(L_{t_0}^{-1} u)_{j\,k\,l} \|^2_H \Big) \le C (p+1)^{-2} \sum_{k\,j\,l} \big((|k| + 1)^2 \| u_{k\,j\,l} \|^2_H + \\ + \| A(t_0) u_{k\,j\,l} \|^2_H \big) \le C (p+1)^{-2} \| u \|^2_{[\mathcal{H}_1]^n} .$$

This estimate shows that

$$\| (L_{t_0}^{-1} - L_p) \|_{[\mathcal{H}_1]^n \to [\mathcal{H}_1]^n} \ \to \ 0 \text{ when } p \to \infty.$$

According to (4), operator L_p is a finite direct sum of operators T in $[H_1]^r$ that are inverse to operator matrices

$$\begin{bmatrix} G & (\sigma(t_0) - A(t_0)) & 0 & \dots & 0 \\ 0 & G & (\sigma(t_0) - A(t_0)) & 0 & \dots \\ \dots & \dots & \dots & \dots & \dots \\ 0 & \dots & 0 & G & (\sigma(t_0) - A(t_0)) \\ 0 & \dots & \dots & \dots & G \end{bmatrix},$$

where $G = (2\pi ik + \sigma(t_0)\lambda_j - A(t_0)\lambda_j)$. Therefore, denoting by Λ the operator $(2\pi ik + \sigma(t_0)\lambda - A(t_0)\lambda)^{-1}$, we obtain that

$$T = \Lambda \begin{bmatrix} I & -\Phi & \Phi^2 & -\Phi^3 & \dots \\ 0 & I & -\Phi & \Phi^2 & \dots \\ \dots & \dots & \dots & \dots & \dots \\ \dots & \dots & \dots & \dots & \dots \\ 0 & 0 & \dots & 0 & I \end{bmatrix},$$

where $\Phi = \Lambda\,(\sigma(t_0) - A(t_0))$. The operator $\Lambda = (2\pi ik + \sigma(t_0)\lambda - A(t_0)\lambda)^{-1}$ is compact in H_1. Indeed, take a bounded sequence $\{f_k\} \subset H_1$. Together with i), the boundedness means that $\{f_k\}$ and $\{A(t_0)f_k\}$ are bounded in H. Since the sequences Λf_k and $\Lambda A(t_0)f_k$ are both bounded in H_1, and the embedding of H_1 into H is compact, we can choose a subsequence f_{k_j} that is convergent in H. This implies convergence of Λf_k in H_1. Hence, Λ is compact. The operator Φ is obviously bounded in H_1. Consequently, the operators T and L_p are compact . This means compactness of $L_{t_0}^{-1}$ in $[\mathcal{H}_1]^n$.

Let us mention that L_{t_0} is a norm continuous invertible operator function, so the norm $\| L_{t_0}^{-1} \|_{[\mathcal{H}_0]^n \to [\mathcal{H}_1]^n}$ is uniformly bounded with respect to $t_0 \in \mathbf{T}$.

Now, in order to prove the statement of the theorem, we have to construct left and right regularizers. This can be easily done exactly as in the proof of theorem 5.1.4. The theorem is proved.

Now we consider some more general situation. Let $a(t)$ be a continuously differentiable $n \times n$ matrix function such that

v) $a(t+1) = a(t)$,

vi) $dim\ Ker\ a(t) = const\ (= m \geq 0)$ (here we consider $a(t)$ as an operator in $\mathbf{C}^n$),

vii) the spectrum of $a(t)$ is real,

viii) $Ker\ a(t) \cap Im\ a(t) = \{0\} \qquad \forall t \in \mathbf{R}$.

The last condition means the absence of Jordan blocks that correspond to the zero eigenvalue. Thus, we have the direct decomposition of $\mathbf{C}^n$ into the sum of subspaces $Im\ a(t) \oplus Ker\ a(t)$. Let $P(t)$ be the projector on $Im\ a(t)$ according to the decomposition, then $P(t)$ is a C^1-smooth operator function. Let us denote by $Q(t) := I\text{-}P(t)$ the corresponding projector onto $Ker\ a(t)$. We also denote by $\mathcal{E} = \bigcup_{t \in \mathbf{T}} Im\ a(t)$ the vector bundle of images of $a(t)$. Now we can define some suitable Sobolev type spaces.

Definition 5.2.2. The space $\mathcal{H}_j(\mathcal{E})$ $(j=0,1)$ consists of all functions $f \in \mathcal{H}_j^n$ such that $P(t)f(t)=f(t)$ almost everywhere.

Let

$$\mathcal{F}=\bigcup_{t\in \mathbf{T}} Ker\ a(t)$$

be the kernel bundle. We introduce the space $\mathcal{H}_j(\mathcal{F})$ in the analogous way. Due to C^1-smoothness of $P(t)$, we get the following decomposition

$$\mathcal{H}_j^n=\mathcal{H}_j(\mathcal{E})\oplus\mathcal{H}_j(\mathcal{F}) \qquad (j=0,1).$$

It is also clear that the operator

$$\mathcal{A} = P(t)\,\frac{d}{dt} - A(t)\ a(t)$$

maps $\mathcal{H}_1(\mathcal{E})$ into $\mathcal{H}_0(\mathcal{E})$.

Theorem 5.2.3. *Operator $\mathcal{A}$ is a Fredholm operator from $\mathcal{H}_1(\mathcal{E})$ into $\mathcal{H}_0(\mathcal{E})$.*

Proof. Let us consider the matrix function $q(t)=a(t)+Q(t)$. It is evident that the conditions of theorem 1 are satisfied. Hence, the operator $S=\frac{d}{dt}-A(t)\ q(t)$ is Fredholm from $\mathcal{H}_1{}^n$ into $\mathcal{H}_0{}^n$. In our direct decomposition it has the following block structure

$$S=\begin{bmatrix} P\frac{d}{dt}-Aa & -P\frac{dP}{dt} \\ -Q\frac{dQ}{dt} & Q\frac{d}{dt}-AQ \end{bmatrix}$$

According to lemma 5.1.2, the operator

$$\begin{bmatrix} 0 & -P\frac{dP}{dt} \\ -Q\frac{dQ}{dt} & 0 \end{bmatrix}$$

is compact, since it factorizes through the compact embedding $\mathcal{H}_1^n \subset \mathcal{H}_0^n$. This means (see corollary 1.2.5) that operator

$$S^0 = \begin{bmatrix} P\frac{d}{dt} - Aa & 0 \\ 0 & Q\frac{d}{dt} - AQ \end{bmatrix}$$

is Fredholm. Then obviously $\mathcal{A} = P\frac{d}{dt} - Aa$ is also Fredholm. The theorem is proved.

Let now $b(t)$ be a continuous 1-periodic $n \times n$-matrix function. We consider the following main equation

$$\frac{du}{dt} = A(t)\ a(t)\ u(t) + b(t)\ u(t) \tag{5.2.5}$$

Let us assume that conditions i)-iii) and v)-viii) are satisfied. We are interested in the operator

$$\mathcal{B} = Q(t)\ [\ \frac{d}{dt}\ - A(t)\ a(t) - b(t)] = Q\frac{d}{dt} - Qb$$

on the space of sections of the bundle $\mathcal{F}$. This bundle over $\mathbf{T}$ is complex, and thus trivial. There exists a C^1-basis $e_j(t)$ of this bundle. Every section $u \in \mathcal{H}_0(\mathcal{F})$ may be interpreted as a collection $\{u_j(t)\}$ of H-valued functions on $\mathbf{T}$ such that

$$u(t) = \sum_j u_j(t)\ e_j(t).$$

In terms of these functions we can describe the operator $\mathcal{B}$ in the following way: for each m-tuple u of H-valued functions $u(t) = \{u_1(t), \ldots, u_m(t)\}$

$$\mathfrak{B}u = \frac{du}{dt} + \widetilde{\mathfrak{B}}(t)\, u,$$

where $\widetilde{\mathfrak{B}}(t)$ is the 1-periodic $m \times m$ matrix function with the entries

$$\widetilde{b_{ij}}(t) = \mu_{ij}(t) - \gamma_{ij}(t).$$

Here $Q\dfrac{de_j}{dt} = \sum \mu_{ij}\, e_i$ and $Q\, b\, e_j = \sum \gamma_{ij}\, e_i$. Consider the m-dimensional system

$$\frac{dv}{dt} + \widetilde{\mathfrak{B}}(t) v = 0, \tag{5.2.6}$$

where $v(t) \in \mathbf{C}^n$. That system has m Floquet exponents $\chi_1, \dots \chi_m$ which will play the crucial role below.

Let $\mathcal{H}_1(\mathbf{T})$ be the standard Sobolev space of complex-valued functions on the circle $\mathbf{T}$ that belong to $L_2(\mathbf{T})$ together with their first distributional derivatives. We also consider the Sobolev space $\mathcal{H}_1(\mathbf{T},H)$ of H-valued functions. Let us introduce the space $\widetilde{\mathcal{H}}_1(\mathcal{F})$ of sections $u(t) \in \mathcal{H}_0(\mathcal{F})$, such that $u = \sum u_j\, e_j$, where $u_j \in \mathcal{H}_1(\mathbf{T},H)$. It is obvious that $\widetilde{\mathcal{H}}_1(\mathcal{F}) \approx H \otimes \mathcal{H}_1(\mathbf{T})^m$ in the sense of Hilbert spaces tensor products [341]. Our operator $\mathfrak{B}$ is equal to $I \otimes (\frac{d}{dt} + \widetilde{\mathfrak{B}})$. It is useful now to consider the following function space:

$$\widetilde{\mathcal{H}}_1 := \mathcal{H}_1(\mathcal{E}) \oplus \widetilde{\mathcal{H}}_1(\mathcal{F}). \tag{5.2.7}$$

Let us mention that a function u belongs to $\widetilde{\mathcal{H}}_1$ if and only if it belongs to $\mathcal{H}_1(\mathbf{T},H^n)$, and simultaneously $a(t)u(t)$ belongs to $L_2(\mathbf{T},H_1{}^n)$. It is easy to see that the operator $(\frac{d}{dt} - Aa - b)$ is bounded as an operator from $\widetilde{\mathcal{H}}_1$ into $\mathcal{H}_0^n$. With respect to the direct decompositions (7) and $\mathcal{H}_0^n = \mathcal{H}_0(\mathcal{E}) \oplus \mathcal{H}_0(\mathcal{F})$, this operator has the following block structure:

$$\begin{bmatrix} \mathcal{A} & \Gamma_1 \\ \Gamma_2 & I \otimes (\frac{d}{dt} + \widetilde{\mathfrak{B}}) \end{bmatrix},$$

where $\mathcal{A} = P(t)\frac{d}{dt} - A(t)a(t) - P(t)b(t)$, operator $\widetilde{\mathfrak{B}}$ was defined above, and $\Gamma_2 = Q\frac{dP}{dt} - Qb$. Since Γ_2 is an operator that can be factorized through the compact embedding of $\mathcal{H}_1^n$ into $\mathcal{H}_0^n$, it is compact . Now let us try to find a Floquet solution to the equation (5) of the form $u(t) = exp(i\lambda t)g(t)$ with 1-periodic $g(t)$. It is the same as to find a nonzero solution of the system

$$(5.2.8) \qquad \begin{bmatrix} \mathcal{A} + i\lambda & \Gamma_1 \\ \Gamma_2 & I \otimes (\frac{d}{dt} + \widetilde{\mathfrak{B}} + i\lambda) \end{bmatrix} \begin{bmatrix} g_1 \\ g_2 \end{bmatrix} = \begin{bmatrix} 0 \\ 0 \end{bmatrix}.$$

Theorem 5.2.4. *The operator in the left side of* (8) *is Fredholm for each* λ *except the quasimomentums of the system* (6).

Remark. The operator is considered as an operator from $\widetilde{\mathcal{H}}_1$ into $\mathcal{H}^n$.

Proof. If we are interested in the Fredholm property, we may omit any compact summand, so we may assume that $\Gamma_2 = 0$. Thus, instead of the operator matrix (8) we get the matrix:

$$(5.2.9) \qquad \begin{bmatrix} \mathcal{A} + i\lambda & \Gamma_1 \\ 0 & I \otimes (\frac{d}{dt} + \widetilde{\mathfrak{B}} + i\lambda) \end{bmatrix}.$$

Here operator Γ_1 is bounded, $\mathcal{A}+i\lambda$ is Fredholm for any λ (due to theorem 3 and lemma 5.1.2) and $\frac{d}{dt} + \widetilde{\mathfrak{B}} + i\lambda$ is the operator from $\mathcal{H}_1(\mathbb{T})^m$ into $L_2(\mathbb{T})^m$ that corresponds to the m-dimensional system (6). If λ is not a quasimomentum of (6), the latter operator is invertible, so $I \otimes (\frac{d}{dt} + \widetilde{\mathfrak{B}} + i\lambda)$ is invertible too. It is evident now that the operator matrix (9), which has Fredholm and invertible diagonal entries, is Fredholm for such λ. The theorem is proved.

The main result of the section is the following

Theorem 5.2.5. *Under the conditions* i)-iii) *and* v)-viii), *the intersection of the set* *of all Floquet exponents of equation* (5) *with the domain* $\Omega = \mathbf{C} \setminus \{0, \chi_1, \ldots, \chi_m\}$ *is an analytic subset in* Ω (*where* χ_j *are the Floquet exponents of system* (6)) .

Remark. We consider solutions of (5) that belong to the space $\widetilde{\mathcal{H}}_1$. This means that $u \in \mathcal{H}_1(\mathbf{T}, H^n)$ and $a(t)u(t) \in L_2(\mathbf{T}, H_1^n)$. This class is obviously the natural one.

Proof. Let us consider the set of quasimomentums λ instead of Floquet exponents χ (where $\chi = exp\ i\lambda$). It is sufficient to prove that the set of quasimomentums of (5) is analytic in the domain $\Omega_1 = \mathbf{C} \setminus$ {quasi-momentums of (6)}. But, as we have already mentioned above, this set is equal to the spectrum of the operator function

$$\begin{bmatrix} \mathcal{A} + i\lambda & \Gamma_1 \\ \Gamma_2 & I \otimes (\frac{d}{dt} + \widetilde{\mathfrak{B}} + i\lambda) \end{bmatrix} : \widetilde{\mathcal{H}}_1 \to \mathcal{H}_0{}^n.$$

Theorem 4 implies that this operator function is Fredholm and analytic in Ω_1. Now theorem 1.6.16 states that the spectrum is an analytic subset in Ω_1. The theorem is proved.

Corollary 5.2.6. *Under the conditions of theorem* 5, *there are only two possibilities:*

1) the set of Floquet exponents of (5) *contains* $\mathbf{C} \setminus \{\chi_1, \ldots, \chi_m\}$;

2) the set is discrete with only possible accumulation points 0, ∞ and $\chi_1, \ldots, \chi_m$.

If our equation (5) is of parabolic type, then we have an exponential bound

$$\| u(t) \| \leq C\ exp \gamma t$$

for all solutions. In this case all Floquet exponents must be contained in the disk $|z| \leq \exp\gamma$. Thus, corollary 6 implies that the set of these exponents is discrete with the accumulation points at $0, \chi_1, \ldots, \chi_m$ (but not at ∞). Nevertheless, even in the case when the equation $u_t + A(t)u = 0$ is parabolic, the equation (5) is not necessarily parabolic without additional restrictions on $a(t)$ (like those of [80] or of section 5.3). For example, if $a(t)$ is the 2×2-matrix of the form

$$\begin{bmatrix} 1 & 0 \\ 0 & -1 \end{bmatrix},$$

the equation (4) is the direct sum of parabolic and inverse parabolic problems and Floquet exponents have accumulation points both at 0 and ∞. If $n=1$ and $a=-1$, we have an inverse parabolic problem with accumulation of Floquet exponents at infinity, but not at zero. Hence, under our conditions all situations with accumulation points at 0 and ∞ are possible.

Let us describe some cases when we can guarantee an exponential bound for solutions.

Theorem 5.2.7. *Let, in addition to the conditions of theorem* 5, *one of the following conditions be satisfied:*

i) for some Λ *and for each* t *the operator* $A(t)-\Lambda$ *generates a bounded holomorphic semigroup (see* [341], *chapter X), and the nonzero spectrum of* $a(t)$ *is positive (i.e., it is contained in* $\mathbf{R}_+$ *);*

ii) *for some* Λ *and for each* t *the operator* $A(t)-\Lambda$ *generates a semigroup of contractions (see* [341], *chapter X), and the spectrum of* $a(t)$ *is positive and without Jordan blocks.*

Then the only possible accumulation points for the set of Floquet exponents of the problem (5) *are* $\{ 0, \chi_1, \dots \chi_m \}$.

Proof. Due to theorem 5, it is sufficient to prove that any solution of (5) satisfies the estimate $\| u(t) \|_H \leq C \exp \gamma t$, where γ does not depend on the solution.

i) By adding and subtracting (and changing the matrix $b(t)$) we may assume that the operator $A(t)$ generates a bounded (and even exponentially decreasing) holomorphic semigroup for each t. Consider the equation

$$\frac{du}{dt} = A(t)a(t) + b(t) - \lambda \tag{5.2.10}$$

for a large positive λ. We assume that λ is so large that $b(t)-\lambda$ is accretive (see [341], chapter X) for any t. If we prove the boundedness of all solutions of (10), we shall obviously have the $exp\lambda t$-bound for all solutions of (5). Therefore, we will consider (10) instead of (5). Now, let us prove that for any t the operator $A(t)a(t)+(b(t)-\lambda)$ with the natural domain $\mathfrak{D}=\{u\in H^n \mid au\in H_1^n\}$ generates a semigroup of contractions. We may decompose H^n according to Jordan structure of $a(t)$, as in the proof of theorem 1: $H^n = M_0 \oplus M_1 \oplus \ldots \oplus M_k$. Here $M_0 = H\otimes Ker\ a(t)$ and $M_j = H\otimes V_j$ $(j\neq 0)$, where V_j is the subspace corresponding to the Jordan block

$$\begin{bmatrix} \lambda_j & 1 & 0 & \ldots & 0 \\ 0 & \lambda_j & 1 & \ldots & 0 \\ 0 & 0 & \ldots & \ldots & \ldots \\ \ldots & \ldots & \ldots & \ldots & 1 \\ 0 & 0 & \ldots & 0 & \lambda_j \end{bmatrix}$$

of $a(t)$. If we prove that on each M_j our operator $A(t)a(t)$ generates a semigroup of contractions, then it will generate a bounded semigroup on the whole space H^n, thus $A(t)a(t)-\varepsilon$ will generate a semigroup of contractions for any $\varepsilon>0$.

The case of M_0 is trivial, because $A(t)a(t)\mid_{M_0}=0$. Let us consider the case of M_j $(j\neq 0)$. In an appropriate block structure the operator $A(t)a(t)$ has on M_j the following operator matrix :

$$\begin{bmatrix} \lambda_j A(t) & A(t) & 0 & \ldots & 0 \\ 0 & \lambda_j A(t) & A(t) & \ldots & 0 \\ 0 & 0 & \ldots & \ldots & 0 \\ \ldots & \ldots & \ldots & \ldots & A(t) \\ 0 & 0 & \ldots & 0 & \lambda_j A(t) \end{bmatrix}.$$

Here $\lambda_j>0$. Hence, $\exp\tau A(t)a(t)$ has the matrix

$$\begin{bmatrix} e^{\lambda_j \tau A(t)} & \tau A(t)\, e^{\lambda_j \tau A(t)} & -\tau^2 A(t)^2 e^{\lambda_j \tau A(t)} & \cdots & \cdots \\ 0 & e^{\lambda_j A(t)} & \tau A(t)\, e^{\lambda_j \tau A(t)} & \cdots & \cdots \\ 0 & 0 & \cdots & \cdots & \cdots \\ \cdots & \cdots & \cdots & \cdots & \cdots \\ 0 & 0 & \cdots & 0 & e^{\lambda_j A(t)} \end{bmatrix}$$

Condition i) and corollary 2 to theorem X.52 in [341] imply that all the entries of this matrix are operators, uniformly bounded from above.

Now let us consider the operator

$$(A(t)a(t) - \varepsilon) + (b(t) - \lambda) = A_0 + B.$$

The operator A_0, as we have proved earlier, generates a semigroup of contractions, and B is a bounded accretive operator. Hence, according to the lemma after theorem X.49 in [341], the whole operator is an accretive operator that generates a semigroup of contractions in H^n. It is evident that any solution of (10) is bounded (we can just differentiate the value $\| u(t) \|_H^2$, and use the accretivity).

ii) The considerations in this case differ only slightly from the ones above. On the space M_j our operator is just $\lambda_j A(t)$, so $exp A(t)a(t)\tau = exp \lambda_j \tau A(t)$. In this case we do not have expressions like $(\tau A(t))^k\, exp \lambda \tau A(t)$ and it is sufficient to know that $\lambda_j > 0$ and that $A(t)$ generates a semigroup of contractions. The theorem is proved.

It is easy to see that in the case of negative spectrum of $a(t)$ the set of accumulation points is $\{\infty, \chi_1, \ldots \chi_m\}$. It is also obvious that in the case of nonreal spectrum of $a(t)$ (even if $n = 1$) the set of Floquet exponents is not necessarily analytic.

Some routine considerations enable us to apply our results to the case of PDEs.

Let us consider the following 1-periodic in t parabolic problem:

$$\text{(5.2.11)} \qquad \begin{cases} v_t = A(t,x,D)v & t > 0,\ x \in \Omega \\ B_j(x,D)v\,|_{\partial\Omega} = 0, & t>0,\ j = 1, \ldots, l, \end{cases}$$

where Ω is a bounded domain in $\mathbf{R}^p$ with smooth boundary, the operator A is properly elliptic of order $2l$ with the coefficients that belong to $C(\mathbf{T}, C^\infty(\bar\Omega))$, the coefficients of B_j belong to $C^\infty(\bar\Omega)$. Moreover, suppose that the following condition is satisfied: for any $\theta \in [-\pi/2, \pi/2]$ the operator $\Lambda_\theta = A(t,x,D_x) + e^{i\theta}(-1)^l D_y^{2l}$ is properly elliptic in $\bar\Omega \times \mathbf{R}_y$ and the system $\{B_j\}$ of boundary operators on $\partial\Omega \times \mathbf{R}_y$ satisfies covering conditions (see section 5.4 or [5, 262]) with respect to Λ_θ. (In fact, smoothness conditions on the boundary and coefficients can be reduced (see [5] for details)). Let also $a(t)$ and $b(t)$ be some 1-periodic $n \times n$-matrices, $b(t)$ be continuous, $a(t)$ be of class C^1, the spectrum of $a(t)$ be real and positive, $dim\ Ker\ a(t) \equiv m < n$ and $Im\ a(t) \cap Ker\ a(t) = \{0\}$ for any t.

Now we consider the $n \times n$-system

$$\text{(5.2.11)} \qquad \begin{cases} \dfrac{\partial u(t,x)}{\partial t} = A(t,x,D)a(t)u(t,x) + b(t)u(t,x), & t > 0,\ x \in \Omega \\ B(x,D)a(t)u(t,x)\,|_{\partial\Omega} = 0, & t > 0. \end{cases}$$

We consider only solutions u of (11) such that $u \in H^1_{loc}(\mathbf{R}_+, L_2(\Omega))$ and $a(t)u \in L_{2,loc}(\mathbf{R}_+, H^{2l}(\bar\Omega))$.

Theorem 5.2.8. *The set of Floquet exponents of* (11) *is discrete, with possible accumulation points at* $0, \chi_1, \ldots, \chi_m$. *Here* χ_j *are Floquet exponents of the finite dimensional system* (6), *constructed as above.*

This is an obvious consequence of theorem 7 and of standard estimates for the parabolic problem (11) (see, for instance, [5, 262]).

5.3. Cauchy problem for abstract parabolic equations

We considered in previous two sections evolution equations on the whole axis. For parabolic equations, however, this is not natural. The reasonable problem here is to investigate completeness of Floquet solutions in the class of all solutions of the Cauchy problem.

All equations that we shall consider here will have the special form:

$$du/dt + Au = B(t)A^{\alpha}u. \tag{5.3.1}$$

Here A is a positive definite self-adjoint operator in a Hilbert space H, A^{α} ($\alpha<1$) is its power and $B(t)$ is a bounded operator function. The conditions will be specified when necessary.

We start with a result that shows how the general methods of chapter 3 and of section 5.1 can be transformed for the case of Cauchy problem.

We shall assume the following conditions:

(5.3.2) A is a selfadjoint, bounded from below operator in H;

(5.3.3) The domain $\mathfrak{D}(A)$ of A is a dense subspace $H_1 \subset H$;

(5.3.4) Equipped with the graph norm of $\mathfrak{D}(A)$, H_1 is a Hilbert space, compactly embedded into H, and singular numbers of the embedding operator $J: H_1 \subset H$ satisfy for some $p \in [1, \infty)$ the condition:

$$\sum_j s_j^p s_{j+1}^{-1} < \infty .$$

(5.3.5) The operator-function $B(t)$ is 1-periodic and norm continuous with respect to t, as function with values in $L(H)$.

In fact, our next result holds under the following weaker condition:

(5.3.6) $B(t)$ is a bounded 1-periodic measurable function on $\mathbb{R}$ with values in $L(H)$.

Theorem 5.3.1. *Let operators A and $B(t)$ in the equation*

(5.3.7) $$\frac{du}{dt} + Au + B(t)u = 0$$

satisfy the conditions (5.3.2) - (5.3.5) *and, additionally,*

(5.3.8) $$\|B(t)\| \leq C < \pi,\ t \in \mathbf{R}.$$

Then the set of all Floquet solutions of (5.3.7) *is complete in the space of solutions of Cauchy problem*

(5.3.9) $$\begin{cases} du/dt + Au + B(t)u = 0 \\ u(0) = u_0 \end{cases},$$

equipped with topology of $L_{2,comp}$ $(\mathbf{R}^+, H)$. *(Here* u_0 *is not fixed.)*

Proof. Suppose this is not true. Then there is a non-trivial function $\varphi \in L_{2,comp}(\mathbf{R}^+, H)$ such that

(5.3.10) $$\int_0^\infty \big(u(t), \varphi(t)\big)_H dt = 0$$

for all Floquet solutions of (7) and at the same time

(5.3.11) $$\int_0^\infty \big(u_0(t), \varphi(t)\big)_H\, dt \neq 0$$

for some solution $u_0(t)$ of the Cauchy problem. Suppose that *supp* $\varphi \in [0, a]$. Let us try to draw some conclusions from condition (10). Consider the following inverse time abstract parabolic Cauchy problem:

(5.3.12) $$\begin{cases} -\frac{d\psi}{dt} + A\psi(t) + B^*(t)\,\psi(t) = \varphi(t)\,, \quad t \leq a + 1 \\ \psi(a + 1) = 0. \end{cases}$$

It is standard that under our conditions both problems (9) and (12) are parabolic (see [108, 209, 341]), so the solution ψ does exist. (This can be extracted also from the construction that follows below.) We can extend $\psi(t)$ for $t \geq a + 1$ as zero and get a solution on **R** with its support in $(-\infty, a+1]$.

The interesting conclusion that can be drawn from (10) is that

$$\text{supp}\ \psi \subset [0, a+1], \tag{5.3.13}$$

i.e. $\psi(0)=0$. Let us assume for a moment that this has been proved already. Then the end of the proof is trivial: just plug $(-\dot{\psi}+A\psi+B^*(t)\psi)$ instead of $\varphi(t)$ into (11). The integration by parts shows that

$$\int_0^\infty (u_0(t), -\dot{\psi}+A\psi+B^*(t)\psi)_H dt = \int_0^{a+1} (\dot{u}_0 + Au_0+B(t)u_0, \psi) = 0.$$

We employed here the facts that $\psi(0)=0$, A is self-adjoint and u_0 satisfies equation (7). It is very easy to justify the integration by parts here.

Now our goal is to prove (13). Let us consider the following family of operators depending on $p \in \mathbf{C}$:

$$\mathcal{L}(p) = -\frac{d}{dt} + A+B^*(t) - ip. \tag{5.3.14}$$

We regard $\mathcal{L}$ as an operator between the spaces $\mathcal{H}_1$ and $\mathcal{H}_0$ of periodic vector-functions (see section 5.1). The definitions of all pertinent spaces are the same as in section 5.1; we only drop "primes", since $H_1' = \mathcal{D}(A^*) = \mathcal{D}(A) = H_1$. As we know from the remark that follows the proof of theorem 5.1.4, $\mathcal{L}(p)$ is Fredholm for all values of p. Consider the real number $\nu=\sigma_0-\pi$, where σ_0 is the bottom of the spectrum of A. Let us introduce the following "comb" Γ, which is the subset of $\mathbf{C}$ consisting of the line $\Lambda=\{\lambda \in \mathbf{C} \mid Im\ \lambda = -\nu\}$ and of the rays $\mathcal{R}_n = \{\lambda \in \mathbf{C} \mid Re\ \lambda = (2n+1)\pi,\ Im\ \lambda \le -\nu\}$.

The crucial step of the proof is the following simple lemma.

Lemma 5.3.2. *If $||B(t)|| \le C < \pi$, then for all $p \in \Gamma$ the operator $\mathcal{L}(p)$ is invertible in $\mathcal{H}_0$ and*

$$\max_{p \in \Gamma \cup \{Im\ p \ge -\nu\}} ||\mathcal{L}^{-1}(p)||_{\mathcal{H}_0 \to \mathcal{H}_0} \le const < \infty. \tag{5.3.15}$$

Proof. We shall prove that the operator

$$\mathcal{L}_0(p) = -\frac{d}{dt}+A-ip$$

satisfies the estimate

$$\|\mathcal{L}_0(p)v\|_{\mathcal{H}_0} \geq \pi\|v\|_{\mathcal{H}_0},\ p\in\Gamma\cup\{p\mid\ Im\ p\geq -\nu\}. \tag{5.3.16}$$

This will imply

$$\|\mathcal{L}(p)v\|_{\mathcal{H}_0} = \|\mathcal{L}_0(p)v+B^*(t)v\|_{\mathcal{H}_0} \geq (\pi - \max_t \|B(t)\|)\|v\|_{\mathcal{H}_0}$$

and the proof will be complete. Let us expand $u \in \mathcal{H}_0$ into Fourier series:

$$u(t) = \sum_{K\in\mathbb{Z}} u_k e^{2\pi ikt}.$$

Then $\|u\|^2_{\mathcal{H}_0} = \sum\|u_k\|^2_H$. The operator $\mathcal{L}_0(p)$ has constant coefficients, so we get

$$\mathcal{L}_0(p)\ u = \sum_{k\in\mathbb{Z}}(-2\pi ik-ip+A)u_k e^{2\pi ikt}\ .$$

We shall show that $\|(A-i(2\pi k+p))^{-1}\|_H \leq \pi$ for all $p\in\Gamma\cup\{p\mid\ Im\ p\geq -\nu\}$ and the lemma will be proved. Let us consider first $p\in\Lambda$, i.e. $p=\alpha-i(\sigma_0-\pi)$. Then *dist* $(ip+2\pi ik,\sigma(A))\geq\pi$, so the standard resolvent estimate for selfadjoint operators provide the latter inequality. The same consideration works also if $Im\ p\geq -\nu$. Let us introduce the new "comb" Γ^1 that consists of the line

$$\Lambda^1=\{\lambda\in\mathbb{C}\mid\ Re\ \lambda=\nu\}$$

and of the rays

$$\mathcal{R}^1_n=\{\lambda\in\mathbb{C}\mid\ Im\ \lambda=(2n+1)\pi,\ Re\lambda\geq\nu\},\quad n\in\mathbb{Z}.$$

Then, for $p\in\Gamma^1$ and $k\in\mathbb{Z}$, the point $\lambda = 2\pi k+p$ belongs to Γ^1. Hence, we have to prove that

$$||(A-\lambda)^{-1}v|| \leq \pi^{-1}||v||$$

for all $v \in \mathcal{H}_0$ and $\lambda \in \Gamma^1$. It is obvious, however, that *dist* $(\lambda, \sigma(A)) \geq \pi$ for all $\lambda \in \Gamma^1$. Since A is self-adjoint, the same standard resolvent estimate shows that $||(A-\lambda)^{-1}|| \leq \pi^{-1}$ and the lemma is proved.

Now we apply the transform $\mathfrak{F}$ (see sections 5.1 and 2.2) to both functions φ and ψ, and denote the results by $\Phi(p)=\mathfrak{F}\varphi, \Psi(p)=\mathfrak{F}\psi$. As we know, $\Phi(p)$ is an entire $\mathcal{H}_0$-valued function $\Phi(p,t) = \sum_k \varphi(t-k)e^{-ip(t-k)}$, $t \in [0, 1]$. Since *supp* $\varphi \in [0, a]$, we have $-a \leq k \leq 0$, so

$$||\Phi(p)||_{\mathcal{H}_0} \leq C exp\ C\ Im\ p \quad \text{for } Im\ p > 0,$$

$$||\Phi(p)||_{\mathcal{H}_0} \leq C \quad \text{for } Im\ p < 0.$$

The function $\Psi(p)$ cannot be defined a priori for all $p \in \mathbf{C}$ (because $\psi(t)$ could grow as $t \longrightarrow -\infty$). It can be defined, however, as an analytic function in the half-plane $Im\ p > N$ for sufficiently large $N > 0$. This follows from the standard parabolic estimate for the Cauchy problem (12):

$$||\psi(t)|| \leq C exp\,(-Ct)\,, \quad t \longrightarrow -\infty$$

(see, for instance, [209]).

Now equality (12) can be written as

$$\mathcal{L}(p)\Psi(p)=\Phi(p), \quad Im\ p > N. \tag{5.3.17}$$

Theorem 3.1.3 and corollary 1.7.2 imply that orthogonality (10) of $\varphi(t)$ to Floquet solutions guarantees (unique) solvability of (17) in entire functions. For $Im\ p > N$ this solution must coincide with $\Psi(p)$, so we conclude that $\Psi(p)$ is an entire function. (This means, in particular, that $\psi(t)$ decays for $t \longrightarrow -\infty$. We will prove, however, that $\psi(t)$

is even zero for $t \le 0$.)

Due to (4), lemma 5.1.11 and lemma 1.2.21, $\mathcal{L}^{-1}(p)$ can be represented as the ratio $U(p)/\Delta(p)$, where $U(p)$ is an entire operator-valued function of finite order, and $\Delta(p)$ is an entire scalar function of finite order. Hence, (17) can be written as

$$\Psi(p) = U(p)\Phi(p)/\Delta(p). \tag{5.3.18}$$

The estimates of $\Delta(p)$ from below (theorem 1.5.6) show that $\Psi(p)$ has a finite order. The last important conclusion that we need to make here is the following:

Lemma 5.3.3. *For any real α the function $\Psi(p)$ is bounded in the half-plane $Im\ p \le \alpha$.*

Proof. Consider the representation

$$\Psi(p)=\mathcal{L}^{-1}(p)\Phi(p).$$

Since $\Phi(p)$ is bounded in any half-plane $Im\ p \le \alpha$ and due to lemma 2, we have:

$$||\Psi(p)||_{\mathcal{H}_0} \le C<\infty \quad \text{for } p \in \Gamma \bigcup \{Im\ p \in [-\nu, 0]\}. \tag{5.3.19}$$

Let us apply the Phragmen-Lindelöf principle ([55], theorem 1.4.2, or [256], section I.14) to each strip between two adjacent "teeth" of Γ. Take into account here that the border rays of the strip are parallel, and $\Psi(p)$ is an entire function of a finite order. This gives us the estimate

$$||\Psi(p)||_{\mathcal{H}_0} \le const < \infty$$

in any half-plane $Im\ p \le \alpha$. The lemma is proved.

We return to the proof of the theorem. Consider the representation

$$\Psi(p,t) = \sum_{k \in \mathbb{Z}} \psi(t-k)e^{-ip(t-k)}, \quad t \in [0,1].$$

We claim that only the terms with $k \leq 0$ are present. Indeed, due to lemma 3, we have the following estimate for the function $\Psi_1(p,t) = e^{ipt}\,\Psi(p,t)$:

$$\|\Psi_1(p,\ \cdot)\|_{\mathcal{H}_0} \leq C\, e^{-Im\, p}$$

for $Im\ p \leq 0$. Thus the function

$$\Psi_2(z,t) = \sum \psi(t-k)\, z^k$$

can be estimated as

$$\|\Psi_2(z,\cdot)\|_{\mathcal{H}_0} \leq C\,|z|$$

for $|z| \geq 1$. Hence, this Laurent's polynomial may contain only one positive order term, namely $\psi(t-1)\,z$. In fact, this term is not present either, otherwise the sum

$$\Psi(p,t) = \sum_{k=-a}^{1} \psi(t-k)\, e^{-ip(t-k)}$$

would grow in $\mathcal{H}_0$-norm as $Im\ p \to -\infty$. We conclude that $supp\ \psi \in [0,\ \infty)$. Due to continuity of $\psi(t)$, $\psi(0) = 0$. We proved inclusion (13), so the proof is complete.

This theorem gives, for instance, the completeness of Floquet solutions for the Cauchy problem

$$\text{(5.3.19)} \qquad \begin{cases} u_t = \Delta\, u(t) + b(t,x)u(t,x), & x \in \Omega,\ t > 0 \\ u(0,x) = \varphi_0(x), & x \in \Omega \end{cases}$$

with any boundary conditions on $\partial\Omega$ that do not depend on t and generate a self-adjoint elliptic boundary problem for the Laplacian in Ω. We have to assume here, however, that

$$max\,|b(t,x)| < \pi.$$

Condition $max\ ||B(t)|| < \pi$ of theorem 1 seems to be strange. However, we shall present an example showing that if, under the rest of the conditions of theorem 1, we have $max\ ||B(t)|| > \pi$, then even the existence of Floquet solutions cannot be guaranteed.

Theorem 5.3.4. *For every $\epsilon>0$ there exists an equation* (7) *satisfying conditions* (2)-(4) *and* (6) *such that* $max\ ||B(t)|| \leq \pi + \epsilon$ *and the equation has no Floquet solutions at all.*

Proof. Let $H=l_2$, and A be the diagonal operator $diag\ (2n\delta)$, i.e.

$$Ae_n = 2n\delta e_n,\quad n = 1, 2, \ldots$$

for some fixed $\delta > 0$ and for the standard basis e_n in l_2. The operator function $B(t)$ will be piecewise constant:

$$B(t) = \begin{cases} -B_1, t \in [0, 1/2) \\ -B_2, t \in [1/2, 1) . \end{cases}$$

Here B_1 has the block-diagonal structure

$$\begin{bmatrix} B & 0 & \cdots & \cdots \\ 0 & B & \cdots & \cdots \\ \cdots & \cdots & \cdots & 0 \\ \cdots & \cdots & 0 & B \end{bmatrix},$$

where B is the following two-dimensional matrix:

$$B = \begin{bmatrix} -\delta & -\pi \\ \pi & \delta \end{bmatrix}.$$

The operator B_2 also has block-diagonal structure:

$$\begin{bmatrix} 0 & 0 & \dots & \dots \\ 0 & B & \dots & \dots \\ \dots & \dots & B & 0 \\ \dots & \dots & 0 & B \end{bmatrix}$$

with the same B and with one-dimensional first block on the main diagonal. It is obvious that for sufficiently small δ we have $||B(t)|| \leq \pi + \epsilon$.

Consider the equation

$$\frac{du}{dt} = -Au + B(t)u.$$

If there is a Floquet solution, then the monodromy operator U: $u(0) \longrightarrow u(1)$ has an eigenvector. We shall calculate U explicitly and show that it has no eigenvalues at all.

Due to the simple (piecewise constant) structure of the coefficient $B(t)$, we observe that

$$U = exp\left(\frac{1}{2}(-A+B_2)\right) exp\left(\frac{1}{2}(-A+B_1)\right).$$

Let us compute both factors of this representation. Consider first $exp\,\frac{1}{2}(-A+B_1)$. This operator is obviously of block-diagonal structure. In order to find the n-th block of $exp\,\frac{1}{2}(-A+B_1)$, we have to solve the system of differential equations:

$$\begin{cases} \dot{x} = (-4n+1)\delta x - \pi y \\ \dot{y} = (-4n+1)\delta y + \pi x, \end{cases}$$

$$x(0) = x_0,\ y(0) = y_0.$$

The solution is

$$x = exp\,(-4n+1)\,t\delta(x_0\,cos\,\pi t - y_0\,sin\,\pi t)$$

$$y = exp\,(-4n+1)\,t\delta(y_0\,cos\,\pi t + x_0\,sin\,\pi t).$$

We plug $t=1/2$ and get the n-th block of $exp\,\frac{1}{2}(-A+B_1)$, so

$$exp\,\tfrac{1}{2}(-A+B_1) = diag\begin{bmatrix} 0 & -e^{(-4n+1)\,\delta/2} \\ e^{(-4n+1)\,\delta/2} & 0 \end{bmatrix}.$$

The analogous calculation gives us

$$exp\,\tfrac{1}{2}(-A+B_2) = \begin{bmatrix} exp(-\delta) & 0 \\ 0 & diag\,M_n \end{bmatrix},$$

where

$$M_n = \begin{bmatrix} 0 & -e^{(-4n-1)\,\delta/2} \\ e^{(-4n-1)\,\delta/2} & 0 \end{bmatrix}.$$

Now the matrix of U can be easily computed. It has zeros on the main diagonal, the sequence $(-c, 0, 0, \ldots)$ on the upper diagonal, the sequence $(0, a_1, 0, a_2, 0, a_2, 0)$ on the second upper diagonal, the sequence $(b_1, 0, b_2, 0, b_3, 0, \ldots)$ on the second lower diagonal and zeros elsewhere:

$$\begin{bmatrix} 0 & -c & 0 & 0 & \cdot \\ 0 & 0 & 0 & a_1 & \cdot \\ b_1 & 0 & 0 & 0 & \cdot \\ 0 & 0 & \cdot & \cdot & \cdot \\ \cdot & \cdot & \cdot & \cdot & \cdot \end{bmatrix}.$$

Here all numbers c, a_n, b_n are not equal to zero (they are some exponents) and $a_n = exp\,(-d_n)$, where d_n is an increasing arithmetic progression. The first conclusion is that zero is not an eigenvalue of U. Let us assume now that $\lambda \neq 0$ is an eigenvalue, and $x=\{x_n\}_{n=1}^{\infty} \in l_2$ is the corresponding eigenvector:

$$Ux=\lambda x.$$

Considering only components x_{2n} with even numbers, we get:

$$\lambda x_{2n} = a_n x_{2n+2}.$$

From this equality we conclude that

$$x_{2n+2} = \lambda^n exp\,(\sum_{k=1}^{n} d) x_2.$$

Since $\lambda \neq 0$ and $\sum_{k=1}^{n} d_k$ behaves like n^2, the only way to avoid the divergence of x_{2n+2} is to assume that $x_2 = 0$. On the other hand, $\lambda x_1 = -Cx_2$, so $x_1 = 0$. Taking into account other equations:

$$\lambda x_{2n+1} = b_n x_{2n-1},$$

we get $x=0$. Hence, U has no eigenvectors. This means that the equation has no Floquet solutions, so the proof is complete.

Let us return to the consideration of a more general Cauchy problem

$$\text{(5.3.20)} \qquad \begin{cases} \dfrac{du}{dt} + Au = B(t)\, A^{\alpha} u, & t>0 \\ u(0) = u_0 \,. \end{cases}$$

We shall assume that

(5.3.21) A is a positive definite operator.

(5.3.22) The embedding operator $J : H_1 \subset H$ is compact.

Theorem 5.3.5. *Let the conditions* (3), (6), (21) *and* (22) *be satisfied. Denote by* $\lambda_n < \lambda_{n+1} < \ldots$ *the eigenvalues of A without taking into account their multiplicities. If*

$$\text{(5.3.23)} \qquad \liminf_{n \to \infty} \; 2 \max_t \; \|B(t)\|\, \lambda^{\alpha}_{n+1}\, (\lambda_{n+1} - \lambda_n)^{-1} < 1,$$

then the statement of theorem 1 holds.

Proof. We follow the same scheme as in the proof of theorem 1. The only difference is in the estimates of $\Psi(p)$ in the half-plane $Im\ p<0$. We have to prove that $\Psi(p)$ is bounded in this half-plane. Let us observe that $\Psi(p)$ is an entire function, but not necessarily of a finite order, because we do not make any assumptions on singular numbers of the operator $J: H_1 \subset H$.

There exists a sequence of integers $m_n \longrightarrow \infty$ such that

$$2 \max_t \|B(t)\| \lambda^{\alpha}_{m_n+1} (\lambda_{m_n+1} - \lambda_{m_n})^{-1} \le q < 1. \tag{5.3.24}$$

Let us denote by μ_n the point $(\lambda_{m_n+1} - \lambda_{m_n})/2$ (the middle point of a "gap" in the spectrum of A) and by $2\Delta_n = \lambda_{m_n+1} - \lambda_{m_n}$ the length of the gap. Consider the lines $\mathcal{L}_n$ in the p-plane:

$$\mathcal{L}_n = \{p \in \mathbb{C} \mid Im\ p = -\mu_n\}.$$

We claim that $\Psi(p)$ is bounded between each two of these lines. Let us estimate the operator $\mathcal{L}(p)^{-1}$ on $\mathcal{L}_n$:

$$\mathcal{L}(p)^{-1} = \left(-\frac{d}{dt} + A - B(t)A^{\alpha} - ip\right)^{-1}$$

$$= \left(-\frac{d}{dt} + A - ip\right)^{-1}\left(1 - B(t)A^{\alpha}\left(-\frac{d}{dt} + A - ip\right)^{-1}\right)^{-1}.$$

First of all, the operator $\left(-\frac{d}{dt} + A - ip\right)^{-1}$ is bounded for $p \in \mathcal{L}_n$ uniformly with respect to n, because

$$\left(-\frac{d}{dt} + A - ip\right)^{-1} \sum u_k\, e^{2\pi ikt} = \sum \big(A - i(2\pi k + p)\big)^{-1} u_k\, e^{2\pi ikt},$$

and $\|(A - i(2\pi k + p))^{-1}\| \le \frac{1}{\Delta_n} \le$ *const.* (Note that (24) implies the estimate $\Delta_n \ge const > 0$).

Consider now the operator

$$B(t)A^{\alpha}\left(-\frac{d}{dt}+A-ip\right)^{-1}, \quad p \in \mathcal{L}_n$$

and show that its norm is less than a constant $c < 1$, uniformly with respect to p and n. Let us apply $A^{\alpha}(-\frac{d}{dt}+A-ip)^{-1}$ to an element $u=\sum u_k\, e^{2\pi ikt} \in \mathcal{H}_0$:

$$A^{\alpha}\left(-\frac{d}{dt}+A-ip\right)^{-1} u = \sum A^{\alpha}(A-i(2\pi k+p))^{-1}u_k e^{2\pi ikt}.$$

It can be easily shown that for $p \in \mathcal{L}_n$

$$\|A^{\alpha}(A-i(2\pi k+p))^{-1}\| \leq \frac{\lambda^{\alpha}_{m_n+1}}{\Delta_n}.$$

Therefore,

$$\|B(t)\, A^{\alpha}\left(-\frac{d}{dt}+A-ip\right)^{-1}\| \leq \max \|B(t)\|\, \lambda^{\alpha}_{m_n+1}\, \Delta_n^{-1} \leq C < 1.$$

Due to the boundedness of $\Phi(p)$ in the half-plane $Im\ p<0$, we get the boundedness of $\Psi(p)$ on $\mathcal{L}_n$ (uniformly with respect to n). Let us show now that $\Psi(p)$ is bounded in each strip U between two of the lines $\mathcal{L}_n$. We know that the spectrum of operator function $\mathcal{L}(p)$ is discrete, so there exists an interval $I=\{p \in U|\ Re\ p = \alpha\}$ such that $\mathcal{L}(p)$ is invertible for $p \in I$. Hence,

$$\sup_{p \in I} \|\mathcal{L}(p)^{-1}\| \leq const.$$

Consider now the intervals $I_k = \{p \in U|\ p-2\pi k \in I\}$, $k \in \mathbb{Z}$. Since operators $\mathcal{L}(p)$ and $\mathcal{L}(p+2\pi k)$ are unitarily equivalent in $\mathcal{H}_0$ for $k \in \mathbb{Z}$, we conclude that

$$\sup \|\mathcal{L}(p)^{-1}\| \leq const < \infty,$$

where the supremum is taken over all p from the union of intervals $I_k, k \in \mathbb{Z}$. Since $\Psi(p) = \mathcal{L}(p)^{-1}\Phi(p)$, we get the boundedness of Ψ in U.

Hence, we obtained the boundedness of $\Psi(p)$ on the union of lines $\mathcal{L}_n$ and in each strip between these lines. The Phragmen-Lindelöf principle (theorem 20 of chapter I in

[256]) shows the boundedness of $\Psi(p)$ in the whole half-plane $Im\ p<0$. This finishes the proof of the theorem.

Let us note that for $\alpha=0$ the theorem guarantees completeness of Floquet solutions if there is a sequence of sufficiently large gaps in $\sigma(A)$:

$$\lambda_{n+1}-\lambda_n > (2+\epsilon)\ max\ \|B(t)\|. \tag{5.3.25}$$

The example of theorem 4 can be modified to satisfy the inequality

$$\lambda_{n+1}-\lambda_n > 2\ max\ \|B(t)\| - \epsilon.$$

The corresponding problem (20) has no Floquet solutions at all (see [297]). It is possible, however, to reduce condition (25), allowing the length of the gap to approach $2max\ \|B(t)\|$ sufficiently slowly.

Theorem 5.3.6. *Let conditions* (3), (5), (21) *and* (22) *be satisfied and*

$$1-2b(\lambda_{n+1}-\lambda_n)^{-1} \geq C\ exp\left[-\frac{1}{2}\left(\frac{\lambda_{n+1}+\lambda_n}{2K}\right)\right]^2,$$

where $c>0$, $K>b$ and $b\ =\ max\ \|B(t)\|$. *Then the statement of theorem 1 holds.*

We shall not prove this theorem here.

The problem of completeness of Floquet solutions can be treated in a different way in the parabolic case. Due to correctness of the Cauchy problem for parabolic equations, the monodromy operator

$$U: u(0) \longrightarrow u(1)$$

is well defined and bounded in H. It can be even shown that compactness condition (22) implies compactness of U. (Namely, $U=A^{-\alpha}(A^{\alpha}U)$, $A^{-\alpha}$ is compact, and $A^{\alpha}U$ is bounded.) This enables us to adjust the approach from the ordinary differential equations theory. In other words, if $u(t)$ is a Floquet solution, then obviously $u_0=u(0)$ is either an eigenvector, or an adjoint vector for the operator U. Hence, we arrive to the problem of completeness of eigenvectors of the operator U. If U can be represented as the exponent $exp\ R$ of some operator R, which is in some sense a "weak" perturbation of a self-adjoint operator, then a rich theory (that goes back to the famous Keldysh's theorem) can be applied (see [280]).

In order to obtain the exponential representation of U, however, we need an analog of the Floquet-Lyapunov reduction to constant coefficients. In some cases it can be obtained.

Let us start with some "non-physical" situation, similar to one considered in theorem 4.1.2.

Theorem 5.3.7. *Let conditions* (2), (6) *and* (22) *be satisfied. Assume also that*

(5.3.26)
$$B(t) = \sum_{k=0}^{\infty} B_k \, exp\ 2\pi ik.$$

Then for every $\lambda \in \sigma(A)$ *the number* $exp\ (-\lambda)$ *is a Floquet exponent of problem* (20).

We shall not present the proof here. It follows in all details the scheme of the proof of theorem 4.1.2; the situation now is even simpler, due to one-dimensionality.

Condition (26) is a very strong one, and it can not be satisfied when (20) is generated by a parabolic partial differential equation with real coefficients. The condition means roughly that we deal with a "triangular" perturbation of a constant coefficient problem. This perturbation does not disturb the spectrum of the "diagonal" constant coefficient part.

Theorem 5.3.8. *Let* $\alpha=0$ *and the conditions of theorem 7 be satisfied. Then the equations*

(5.3.27)
$$\frac{du}{dt} + Au = B(t)u$$

and

(5.3.28)
$$\frac{du}{dt} + Av = 0$$

are kinematically equivalent, i.e. there exists a bounded, norm continuous, 1-periodic, invertible operator function $Q(t)$ *such that*

$$U(t)Q(0) = Q(t)V(t)\ ,\ t \geq 0\ , \tag{5.3.29}$$

where $U(t)$ *and* $V(t)$ *are evolution operators for equations* (27) *and* (28).

We shall present here a sketch of the proof. Some technical details can be recovered by the reader.

Let us introduce the following transformator $\mathcal{A}$ on the space $L(H)$ of bounded operators in H:

$$\mathcal{A}Q = AQ - QA,$$

with the domain

$$P_1 = \mathfrak{D}(A) = \{Q \in L(H) |\ QH_1 \subset H_1,\ AQ - QA \in L(H)\}.$$

Lemma 5.3.9. *If* $Re\ p \neq 0$, *then the transformator* $(\mathcal{A}+ip)$ *is invertible, and*

$$\|\ (\mathcal{A}+ip)^{-1}\| \leq \frac{C}{|\ Re\ p\ |}\ . \tag{5.3.30}$$

The proof follows from the explicit formulas for the resolvent

$$(\mathcal{A}+ip)^{-1}Y = \pm i\int_0^\infty e^{\pm iAt}Ye^{\mp iAt}\ e^{\mp pt}dt\ ,\ Y \in L(H), \tag{5.3.31}$$

where we choose the top signs for $Re\ p>0$, and the bottom ones for $Re\ p<0$. It is easy to verify that (31) really provides the resolvent. Estimate (30) follows from the straightforward norm estimates in (31) , if we take into account that $\|e^{\pm iAt}\| = 1$.

It is easy to check that if $Q(t)$ with the required properties exists and has appropriate smoothness, then it satisfies the differential equation

(5.3.32) $$\frac{dQ}{dt} + \mathcal{A}Q = B(t)Q,\ t \geq 0\ .$$

The idea is to find a periodic solution of (32) that satisfies the initial condition $Q(0)=I$ and to prove that this is the operator function we are looking for. In order to do this, we use the approach of theorems 7 and 4.1.2.

The restrictive "non-physical" condition (26) can be avoided when the "perturbation" term $B(t)$ is small in an appropriate sense.

Theorem 5.3.10. *Let $\alpha=0$ and conditions (2), (5) and (22) be satisfied. Then there exists a constant $c>0$ such that if*

(5.3.33) $$\sum_{-\infty}^{\infty} \|B_k\| \leq c,$$

where

$$B(t) = \sum B_k\, exp\, 2\pi ikt,$$

then the evolution operator $U(t)$ can be represented in the form

$$U(t) = Q(t)\ exp\ (-A+C)t.$$

Here C is a bounded operator in H, and $Q(t)$ is a 1-periodic, continuous, bounded, invertible operator function.

The theorem can be proved using perturbation methods. Namely, we consider the following equation with parameter $\mu \in \mathbf{C}$:

(5.3.34) $$\dot{u}+Au=\mu B(t)u.$$

It can be shown that its evolution operator $U(t,\mu)$ is an entire function of μ. We can look for Q and C in the form

(5.3.35) $$Q(t,\mu) = I + \mu Q_1(t) + \mu^2 Q_2(t) + \ldots$$

$$Q_k(t+1) = Q_k(t)$$

$$C(\mu) = \mu C_1 + \mu^2 C_2 + \dots .$$

Plugging Q and C into the equation (34), we get

$$\frac{dQ(t,\mu)}{dt} + AQ(t,\mu) = \mu B(t,\mu)Q(t,\mu) - C(\mu)Q(t,\mu).$$

This equation leads to some recurrent equations for C_n and Q_n. It can be shown that for an appropriate constant c (that can be estimated explicitly) we have convergence of this procedure for $|\mu| \leq 1$. We shall not go into further details of the proof here.

The last two theorems (that are actually Floquet-Lyapunov type results about the reduction to the constant coefficient case) suggest another way of proving the completeness of Floquet solutions. Namely, in these results we reduce the original equation to an equation with a constant coefficient of the type $A + C$, where C is bounded. There are many results about completeness of eigen- and adjoint vectors for such operators [280]. Taking these vectors as the initial data for the Cauchy problem (27) and employing correctness of the problem, we get the completeness of Floquet solutions in a stronger sense than in theorems 1, 5 and 6. The following theorem is the direct combination of the last two ones with some results of V. Matsaev [283], M. Keldysh [201], and V. Katsnelson [200].

Theorem 5.3.11. *Under conditions of either of theorems 8 and 10 we have:*

i) If $A^{-1} \in \mathfrak{S}_\omega$, then the system of kernel vectors of the monodromy operator U is complete in H, and Floquet exponents ρ_k are of the form

$$\rho_k = exp\,(-2\pi\lambda_k), \quad |arg\,\lambda_k| < \epsilon, \quad k \geq k_0(\epsilon),$$

where ϵ is an arbitrary positive number.

ii) If $A^{-1} \in \mathfrak{S}_p$, and there exists a non-decreasing function $\varphi(r)$ $(0 \leq r < \infty)$ such that $\varphi(r) \underset{r\to\infty}{\longrightarrow} \infty$, $\varphi(s)/\varphi(r) \leq (s/r)^\gamma$, $\lim_{r\to\infty} \nu(r,A)/\varphi(r) = 1$, then

$$\lim_{r\to\infty} \frac{\nu(r,U)}{\nu(r,A)} = 1,$$

where $\nu(r,U)$ is the distribution function of the numbers $\{ln(|\rho_k|^{-1})\}$ and $\nu(r,A)$ is the *distribution function of the eigenvalues of the operator A.*

iii) If

$$\overline{\lim_{n\to\infty}}\, n\lambda_n^{-1} < \infty,$$

where λ_n are the eigenvalues of A (counted with their multiplicities), then the kernel vectors of the monodromy operator U constitute an unconditional block-basis in H.

We shall also present without proofs some results of A. Miloslavskii [297] concerning the basis property of kernel vectors of the monodromy operator.

Theorem 5.3.12. *Let conditions* (3), (5) *and* (21) *be satisfied and eigenvalues λ_n of A satisfy the inequality*

$$\liminf_{n\to\infty} 4 \max \|B(t)\|\, \lambda_{n+1}^{\alpha}(\lambda_{n+1}-\lambda_n)^{-1} < 1.$$

Then the kernel vectors of U constitute a block-basis in H.

Theorem 5.3.13. *If, additionally,*

$$\liminf_{n\to\infty} \lambda_{n+1}^{\alpha}(\lambda_{n+1}-\lambda_n)^{-1} = 0\ ,$$

then the kernel vectors constitute a Bari block-basis in H.

We remind the reader that a basis N_k of subspaces of a Hilbert space is said to be a *Bari basis*, if there exists an orthogonal basis M_k of subspaces such that

$$\sum \| P_k - Q_k \|^2 < \infty,$$

where P_k and Q_k are the orthogonal projections onto N_k and M_k correspondingly.

Theorem 5.3.14. *If, in addition to conditions of theorem 12, we have*

$$\lim_{n\to\infty} 4 \max \|B(t)\|\, \lambda_{n+1}^{\alpha}(\lambda_{n+1}-\lambda_n)^{-1} < 1,$$

then

$$| \ln(|\rho_n|^{-1}) - \lambda_n| \leq \max \|B(t)\| \lambda_n^{\alpha} (n = n_0, n_0 + 1, \ldots)$$

for a sufficiently large n_0. Here λ_n are the eigenvalues of A, counted with thei multiplicities, and ρ_n are the Floquet exponents of the problem.

These results can be immediately applied to some parabolic boundary valu problems.

For instance, theorem 11 together with the known spectral asymptotics (se [394]) and A. Markus' theorem [279] give the following result:

Theorem 5.3.15. *Consider the problem*

$$\text{(5.3.36.)} \quad \begin{cases} \frac{\partial u}{\partial t} + (-\Delta)^m u = \sum\limits_{|\gamma| \leq r} b_j(t,x) D^{\alpha} u, x \in \Omega \\ u|_{\partial\Omega} = \frac{\partial u}{\partial \nu}|_{\partial\Omega} = \ldots = \frac{\partial^{m-1} u}{\partial \nu^{m-1}}|_{\partial\Omega} = 0 \\ u|_{t=0} = u_0(x) \end{cases},$$

where $\partial/\partial\nu$ is the normal derivative on the (smooth) boundary of a bounded domair $\Omega \subset \mathbf{R}$. Assume that (complex) functions $b_\gamma(t,x)$ are Hölder-continuous, periodic witl respect to t and measurable with respect to x and that

$$2m > n, \ 0 \leq r < 2m - n.$$

Then the kernel vectors of the monodromy operator of (36) constitute an unconditiona block-basis in $L_2(\Omega)$.

5.4. Elliptic and parabolic boundary value problems in a cylinder

A. Elliptic problems. Let $Q = \mathbf{R}^{n_1} \times G$ be a cylindric domain in $\mathbf{R}^n$ $(n = n_1 + n_2)$, where G is a bounded domain with C^∞ – boundary in $\mathbf{R}^{n_2}$. The points of G are denoted by $x = (x_1, \ldots, x_{n_2})$ and those of $\mathbf{R}^{n_1}$ by $t = (t_1, \ldots, t_{n_1})$. We consider the following boundary value problem in Q:

$$(5.4.1)\qquad \begin{cases} L(t, x, D_t, D_x)\, u(t, x) = 0 & \text{in } Q \\ B_j(t, x, D_t, D_x)\, u|_{\partial Q} = 0 & (j = 1,\dots,m-1), \end{cases}$$

where $2m$ is the order of L, B_j is an operator of order $m_j \le 2m-1$, and the coefficients of L and of B_j are 1-periodic in all t_j $(j=1,\dots,n_1)$ functions from $C^\infty(\overline{Q})$. Suppose that the problem is regularly elliptic in the following sense:

i) *Ellipticity condition.* For all $(t,x) \in \overline{Q}$ and linearly independent vectors $\xi_1, \xi_2 \in \mathbf{R}^n$ the polynomial $L_0(t, x, \xi_1 + \tau\xi_2)$ of τ has exactly m roots $\tau_i^{\pm}(t, x, \xi_1, \xi_2)$ $(i = 1,\dots,m)$ in the open upper and lower half-planes. Here $L_0(t, x, \xi)$ is the principal symbol of L.

ii) *Normality condition.* $B_{j0}(t, x, \xi') \neq 0$ for all $(t,x) \in \Gamma := \partial Q$ and all non-zero vectors $\xi' \in \mathbf{R}^n$ that are normal to Γ at (t,x). Here B_{j0} is the principal symbol of B_j.

iii) *Covering condition.* Polynomials $B_{j0}(t, x, \xi + \tau\xi')$ $(j = 0,\dots,m-1)$ of the complex variable τ are linearly independent modulo the polynomial $\prod(\tau - \tau_i^{+}(t,x,\xi,\xi'))$ for all $(t,x) \in \Gamma$ and non-zero vectors $\xi, \xi' \in \mathbf{R}^n$, where ξ is tangent and ξ' is normal to Γ at (t,x).

In order to apply the abstract results of sections 1.7 and 1.8 as well as the approach of chapter 2, we need to set some "transposed" problem and to construct suitable function spaces, where our operators are Fredholm.

The standard Green's formula approach (see section 2.2 in chapter 2 of [262]) enables us to create a "transposed" problem. We only have to modify known results [262] to our periodic situation.

Let us recall that a system $F_0, \dots, F_{\nu-1}$ of boundary operators is said to be a *Dirichlet system of order* ν on Γ, if it is normal (i.e. satisfies ii)) and the set of orders of $\{F_j\}$ is $\{0, 1, \dots, \nu - 1\}$.

The next result (which is a simple modification of results of [262]) provides the Green formula we need.

Proposition 5.4.1. *Let* (1) *be regularly elliptic. Suppose that* L_1 *is the operator transposed to* L. *Then there exist collections of boundary operators* $\{S_j\}_{j=0}^{m-1}$, $\{C_j\}_{j=0}^{m-1}$, $\{T_j\}_{j=0}^{m-1}$, *such that*

1) all these operators are 1-periodic in each t_j $(j=1,\dots,n_1)$;

2) $\{B_j, S_j\}$ *and* $\{C_j, T_j\}$ *are Dirichlet systems of order* $2m$;

3) let $\mu_j \le 2m-1$ *be the order of* S_j, *then the order of* C_j *is* $(2m-1-\mu_j)$ *and the order of* T_j is $(2m-1-m_j)$;

4) for all $u, v \in D(\overline{Q})$ *the following equality holds:*

$$(5.4.2)\quad \int_Q (Lu)\, v\; dtdx - \int_Q u\,(L_1 v)\; dtdx = \sum_{j=0}^{m-1}\int_\Gamma (S_j u)(C_j v)\,d\sigma - \sum_{j=0}^{m-1}\int_\Gamma (B_j u)(T_j v)\; d\sigma.$$

(*Here* $d\sigma$ *is the volume element on* Γ).

Proof. Consider the compact manifold $\tilde{Q} = \overline{Q}/\mathbf{Z}^{n_1}$, where $\mathbf{Z}^{n_1}$ is the integer lattice in $\mathbf{R}^{n_1}$. Periodicity enables us to consider our problem also on $\tilde{Q}$. Let $\tilde{\Gamma}$ be the boundary of $\tilde{Q}$. By ([262], chapter 2, theorem 2.1), there exist boundary operators $\{\tilde{S}_j\}$, $\{\tilde{C}_j\}$, $\{\tilde{T}_j\}$ on $\tilde{Q}$ satisfying conditions 2) and 3) and identity

$$(5.4.3)\quad \int_{\tilde{Q}} (Lu)\, \bar{v}\; dtdx - \int_{\tilde{Q}} u\,\overline{(L^* v)}\; dtdx = \sum \int_{\tilde{\Gamma}} (\tilde{S}_j u)\overline{(\tilde{C}_j v)}\; d\sigma - \sum \int_{\tilde{\Gamma}} (\tilde{B}_j u)\overline{(\tilde{T}_j v)}\; d\sigma,$$

where L^* is the formal adjoint operator to L and u,v are arbitrary functions from $C^\infty(\tilde{Q})$. In fact, the result of [262] was proved only for regions in $\mathbf{R}^n$, not for arbitrary manifolds. The proof, however, is based on local considerations and on unity decompositions, so it can be easily carried out for smooth compact manifolds.

Denote by S_j the lifting of $\tilde{S}_j$ from $\tilde{Q}$ onto Q, and by C_j, T_j the liftings of operators

$$C_j{}'\, v = \overline{\tilde{C}_j \bar{v}} \text{ and } T_j{}'\, v = \overline{\tilde{T}_j \bar{v}}.$$

By the construction, these operators are periodic in t-variables. We have to check only the identity (2). It is an immediate consequence of (3). First of all, for functions u and v with sufficiently small non-disjoint supports (2) is equivalent to (3), because we can push u and v down to $\tilde{Q}$. The case of disjoint supports is trivial. For general u and v we apply a unity decomposition. This finishes the proof.

Using Green's formula (2), we can regard solutions of (1) in a weak sense, which is, as in chapter 2, one of the main steps of our considerations.

Proposition 5.4.2. *A function* $u(t, x) \in C^\infty(\overline{Q})$ *satisfies* (1) *if and only if*

$$\int_Q u\, L_1 v\, dt dx + \sum \int_\Gamma S_j u\, C_j v\, d\sigma = 0 \tag{5.4.4}$$

for any $v \in D(\overline{Q})$.

Proof. Necessity of (4) immediately follows from (2) and from compactness of *supp v*. Let us now prove the sufficiency. If we apply (4) to a function $v \in C_0^\infty(Q)$, we obtain

$$\int_Q u\, L_1 v\, dt dx = 0.$$

This means that $u(t, x)$ is a solution of the equation $Lu = 0$ in Q (in the distributional sense). Due to ellipticity, it is a smooth classical solution. From this equation and from (4) we obtain the equality

$$\sum \int_\Gamma B_j u\, T_j v\, d\sigma = 0 \tag{5.4.5}$$

for all $v \in D(\overline{Q})$. By lemma 2.2 of chapter 2 in [262], the space of restrictions $\{T_j v|_\Gamma\}_{j=0}^{m-1}$ for $v \in D(\overline{Q})$ coincides with the whole space $[C_0^\infty(\Gamma)]^m$. Now identity (5) shows that $B_j u|_\Gamma = 0$ for $j = 0, \dots, m-1$. The proposition is proved.

Now we construct the following Fréchet spaces:

1) $\Phi_p^s(Q) = \{f \in H_{loc}^s(\overline{Q}) \mid \gamma_a(f) = \sup_{k \in \mathbf{Z}^{n_1}} \|f\|_{H^s(\{x+k\} \times G)} \exp a|k|^p < \infty,\ \forall a > 0\}$

2) The space $\Psi_p(Q)$ consists of $(m+1)$-tuples

$$\overline{g} = (g, g_1, \dots, g_{m-1}) \in L_{2,loc}(Q) \times \prod_j H_{loc}^{\mu_j + 1/2}(\Gamma)$$

such that the semi-norm

$$\xi_a(\overline{g}) = \sup_{k \in \mathbb{Z}^{n_1}} \left(\|g\|_{L_2(\{\mathcal{K}+k\}\times G)} + \sum \|g_j\|_{H^{\mu_j+1/2}(\{\mathcal{K}+k\}\times \partial G)} \right) \exp a|k|^p$$

is finite for all $a>0$.

We define now the operator $\mathcal{L}$: $\Phi_p^{2m}(Q) \to \Psi_p(Q)$ as

$$\mathcal{L}f = \{L_1 f, C_0 f|_\Gamma, \ldots, C_{m-1} f|_\Gamma\}.$$

The continuity of this operator follows from standard embedding theorems and periodicity of operators L_1 and C_j.

We skip the proof of the next statement, which is similar to the one of lemma 3.1.2.

Proposition 5.4.3. *A function $u(t,x)$ is a solution of* (1) *that satisfies the estimate*

$$(5.4.6) \qquad |u(t,x)| \le C \exp a|t|^p \quad (p \ge 1)$$

if and only if the functional

$$\overline{\phi} \to \int_Q u\phi \, dt dx + \sum \int_\Gamma (S_j u)\phi_j d\sigma$$

can be extended to an element of $(Im\ \mathcal{L})^\perp \subset \Psi_p(Q)^*$.

These considerations enable us to treat our problems in the dual setting, as in section 3.7. The natural analogs of the transforms $\boldsymbol{F}$, $\hat{\boldsymbol{F}}$ and $\mathfrak{F}$ (with respect to t-variables) can be defined. Under the transform $\boldsymbol{F}$ our operator $\mathcal{L}$ becomes a Fredholm morphism $\mathcal{L}$: $\mathcal{E}^{2m} \to \mathcal{E}^0$ of appropriate holomorphic Banach bundles over Ω_{n_1}. The details are the same, as in section 2.2. Let us just make one remark about Fredholm property. The morphism $\mathcal{L}$ has the form $\mathcal{L}_\lambda = \mathcal{L}_0 + T(\lambda)$ on the fiber over λ, where $\lambda \in \mathbb{C}^{n_1}$ (and $z = \exp\lambda \in \Omega_{n_1}$). Here

$$\mathcal{L}_\lambda f = \mathcal{L}(x, D-i\lambda)f = \{L_1(x,D-i\lambda)f,\ C_0(x,D-i\lambda)f|_\Gamma, \ldots,\ C_{m-1}(x,D-i\lambda)f|_\Gamma\},$$

$$\mathcal{L}_0 = \{L_1 f,\ C_0 f|_\Gamma, \ldots, C_{m-1} f|_\Gamma\},$$

$$T(\lambda)f = \{M(\lambda)f,\ T_0(\lambda)f|_\Gamma, \ldots, T_{m-1}(\lambda)f|_\Gamma\}.$$

The order of $M(\lambda)$ is strictly less than the order of L_1, and the order of each of $T_j(\lambda)$ is less then the order of the corresponding C_j. All the operators are considered as operators from $H^{2m}(\tilde{Q})$ into $L_2(\tilde{Q}) \times \prod H^{\mu_j + 1/2}(\partial\tilde{Q})$. Operator $\mathcal{L}_0$ is Fredholm, due to known theorems for elliptic boundary value problems (see [171], theorem 10.6.1, or [262], chapter 2, theorem 5.3). Since the orders of operators involved into $T(\lambda)$ are smaller than of ones in $\mathcal{L}_0$, operator $T(\lambda)$ can be factorized through the compact embedding

$$H^{\alpha}(\tilde{Q}) \times \prod_j H^{\gamma_j}(\partial\tilde{Q}) \rightarrow L_2(\tilde{Q}) \times \prod_j H^{\mu_j + 1/2}(\partial\tilde{Q})$$

for some $\alpha > 0$, $\gamma_j > \mu_j + 1/2$. This means that $T(\lambda)$ is compact, being considered as an operator acting in the spaces under consideration. So we obtain the Fredholm property of morphism $\mathcal{L}$.

Floquet solutions are the ones of the form

$$u(t, x) = e^{i\lambda \cdot t}\left(\sum t^k g_k(t, x)\right),$$

where g_k are 1-periodic in t. The set N of Floquet exponents is a subset in Ω_{n_1}.

When $n_1 = 1$, then the largest possible value of k in the last sum is called the multiplicity of the quasi-momentum λ (or of the Fredholm exponent $z = exp\ i\lambda$).

This discussion easily leads to the natural analogs of main results of chapters 2 – 4. We formulate some of them.

Theorem 5.4.4. *Under conditions i)-iii) the following assertions hold:*

1) The set $N \subset \Omega_{n_1}$ of Floquet multiplies of (1) *is analytic.*

2) The expansion of the type (3.2.1) *is valid for all solutions of* (1) *of exponential growth:* $|u(x,y)| \leq C\ exp\ a|x|$.

Following the same idea as in section 3.3, we can prove the following result:

Theorem 5.4.5. *Let the Fredholm index of operator $\mathcal{L}$ on $\tilde{Q}$ be zero and $N \neq \Omega_{n_1}$.*

Then, for any $k \in \mathbb{Z}^+$ *and* $\gamma < (n_1+n_2-1)^{-1}$ *the linear hull of Floquet solutions is dense in* $C^k_{loc}(\overline{Q})$ *– topology in the space of all solutions of* (1) *that satisfy the estimate*

$$|u(t, x)| \le C_u \exp a_u |t|^{1+\gamma}. \tag{5.4.7}$$

When $n_1 = 1$, we can obtain not only completeness of Floquet solutions, but even a decomposition into them; the condition on the exponent γ, however, must be more restrictive.

Theorem 5.4.6. *Suppose that the index of the problem* (1), *considered as a problem on* $\tilde{Q}$, *is zero and that* $N \ne \Omega_1$ $(=\mathbb{C}\backslash 0)$. *Denote by* $\{\lambda_n, p_n\}$ *the collection of quasi-momentums* λ_n *of* (1) *with their multiplicities* p_n . *Define* $\beta_{i,k,m,n}$ *as in section 1.5. Then, for any triple* (m, n, k) *there is a finite collection of functions* $h_{k,l,m,n}(t, x)$ *such that*

i) $h_{k,l,m,n}(t, x)$ *is a Floquet solutions of the problem* (1) *with quasimomentum* λ_m *of multiplicity* k;

ii) any solution of (1) *satisfying*

$$|u(t, x)| \le C \exp a|t|^{1+\gamma}, \quad \gamma < \frac{1}{n_2+1} \tag{5.4.8}$$

can be represented as

$$u(t, x) = \sum_{j,k,l,m,n} \phi_{i,n,l}\, \beta_{i,k,m,n}\, h_{k,l,m,n}(x,y), \tag{5.4.9}$$

where after summation (finite for each n) over i,k,l,m the series converges absolutely in $L_{2,loc}(Q)$ *and*

$$\sum_{i,n} \Big(\sum_{l} |\phi_{i,n,l}|^2\Big)^{1/2} \exp(a|\lambda_n|^p) < \infty \tag{5.4.10}$$

for some $a>0$, $p>n_2+2$.

iii) any collection of complex numbers $\phi_{i,n,l}$ *satisfying* (10) *gives rise to the series* (9) *that converges in the sense indicated and defines a solution of* (1) *satisfying* (8).

Proof. Applying the transform $\hat{\mathcal{F}}$ to $\mathcal{L}:\Phi_p^{2m}(Q) \to \Psi_p(Q)$, we obtain some Fredholm morphism $\hat{\mathcal{L}}$: $\hat{\mathcal{E}}^{2m} \to \hat{\mathcal{E}}^0$. Multiplication by this morphism is continuous as the linear operator from $\tilde{A}_{p/(p-1)}(\mathbf{C}, \hat{\mathcal{E}}^{2m})$ into $\tilde{A}_{p/(p-1)}(\mathbf{C}, \hat{\mathcal{E}}^{0})$.

We are in the situation of section 1.8 now, so we need to prepare some preliminary results for theorem 1.8.2.

Lemma 5.4.7. *There exists an entire projector valued function of order 1* P: $\mathbf{C} \to L(H^{2m}([0,1]\times G))$ *such that its image is the subbundle* $\hat{\mathcal{E}}^{2m}$ *in the trivial bundle* $\mathbf{C}\times H^{2m}([0,1]\times G)$.

Proof. Let us recall that the fiber of $\hat{\mathcal{E}}^{2m}$ over the point $\lambda \in \mathbf{C}^1$ consists of all functions $f \in H^{2m}([0,1]\times G)$ satisfying the conditions

$$\frac{\partial^k f(t, x)}{\partial t^k}\Big|_{t=1} = e^{\lambda} \frac{\partial^k f(t, x)}{\partial t^k}\Big|_{t=0} \quad (k = 0, ..., 2m\text{-}1).$$

Consider the operator that restricts f and all its t-derivatives up to the order $2m-1$ from $G\times[0, 1]$ onto $G\times\{1\}$:

$$H^{2m}(G\times[0, 1]) \longrightarrow \prod_{j=0}^{2m-1} H^{2m-j-1/2}(G\times\{1\}).$$

This operator is surjective, so there exists a continuous right inverse operator, i.e. an extension operator

$$T: \prod_j H^{2m-j-1/2}(G\times\{1\}) \longrightarrow H^{2m}(G\times[0,1])$$

We may assume that all functions in the image of T are equal to zero on $G\times\{0\}$. If T does not satisfy this condition, we introduce the new extension operator $T_1=\varphi T$, where $\varphi(t)$ is any smooth function which is equal to one in a neighborhood of $\{1\}$ and is equal to zero in a neighborhood of 0. Consider the operator

(5.4.11) $$P(\lambda)f = f - TQ(\lambda)f,$$

where

$$(5.4.12)\qquad Q(\lambda)f = \{f|_{t=1} - e^{\lambda} f|_{t=0},\ (\frac{\partial f}{\partial t}|_{t=1} - e^{\lambda}\frac{\partial f}{\partial t}|_{t=0}), \dots\}.$$

On the fiber of $\widehat{\mathcal{E}}^{2m}$ over the point λ the operator $P(\lambda)$ is equal to the identity. Moreover, the image of $P(\lambda)$ is contained in the fiber, so $P(\lambda)$ is a projection onto the fiber. Holomorphity and estimates for $P(\lambda)$ are obvious, due to (11) and (12). The lemma is proved.

To satisfy the conditions of theorem 1.8.2, we have to study the structure of the inverse operator $\widehat{\mathcal{L}}^{-1}$. Let us make the standard trivialization (see the proof of theorem 2.2.1) and reduce $\widehat{\mathcal{L}}$ to the following operator polynomial acting on functions on the "torus" $\widetilde{Q}$:

$$\mathcal{L}(\lambda) = \{L_1(t, x, D_t - i\lambda, D_x), \dots, C_j(t, x,, D_t - i\lambda, D_x), \dots\}.$$

(see theorem 2.2.3). Expanding into powers of λ, we get

$$\mathcal{L}(\lambda) = \mathcal{L}_0 + \sum_{|\alpha|>0} \lambda^{\alpha}\mathcal{L}_{\alpha}.$$

By the conditions of the theorem, *ind* $\mathcal{L}(\lambda) = 0$ and cospectrum of $\mathcal{L}(\lambda)$ is discrete, so $\mathcal{L}(\lambda)$ is invertible for all points λ outside of a proper analytic subset. We shall assume that $\mathcal{L}_0$ itself is invertible, otherwise we can make a simple translation with respect to λ and achieve this. The order of $\mathcal{L}_\alpha$ is less than the order of $\mathcal{L}_0$ by $|\alpha|$. Hence, $\mathcal{L}_\alpha$ belongs to the class $\mathcal{S}_{q/|\alpha|}$ for any $q>n_2+1$, as an operator from $H^{2m}(\tilde{Q})$ into

$$L_2(\widetilde{Q}) \times \prod_{j=0}^{2m-1} H^{2m-j-1/2}(\partial\widetilde{Q}).$$

(see lemma 1.2.18). Due to lemma 1.2.21, $\mathcal{L}^{-1}(\lambda) = B(\lambda)/\Delta(\lambda)$, where Δ and B are scalar and operator valued entire functions of order $q>(n_2+1)$ of minimal type. Let us observe that functions $\Delta(\lambda)$ and $\Delta(\lambda+2\pi i)$ have the same sets of zeros (including multiplicities). In fact, the polynomial $\mathcal{L}(\lambda)$ remains equivalent to itself after the $2\pi i$-translation with respect to λ (the equivalence is provided by the operator of multiplication by $e^{2\pi i x}$). Therefore, the collection $\{\lambda_k, p_k\}$ of the eigenvalues (with their multiplicities) of the polynomial is $2\pi i$-periodic. This means that the function

$\Delta(\lambda+2\pi i)/\Delta(\lambda)$ is entire of order n_2+1 and without zeros. The Hadamard's theorem (see theorem 13 of chapter 1 of [256], or [390]) shows that $\Delta(\lambda+2\pi i)/\Delta(\lambda) = \exp g(\lambda)$ for some polynomial $g(\lambda)$ of order not exceeding n_2+1. Now we apply lemma 1.5.30 and find a polynomial $\tilde{g}(\lambda)$ such that

$$\tilde{g}(\lambda+2\pi i)-\tilde{g}(\lambda)=g(\lambda).$$

Let us define now $\Delta_1(\lambda) = \Delta(\lambda) \exp(-\tilde{g}(\lambda))$, $B_1(\lambda) = \exp(-\tilde{g}(\lambda))B(\lambda)$. A trivial calculation shows that Δ_1 is $2\pi i$-periodic. Besides, both B_1 and Δ_1 are entire functions of order at most n_2+2 and $\mathcal{L}^{-1}(\lambda) = B_1(\lambda)/\Delta_1(\lambda)$. This provides representation $\hat{\mathcal{L}}^{-1}=\hat{B}/\Delta_1$ for morphism $\hat{\mathcal{L}}$, where $\hat{B}$ maps $\tilde{A}_{p/(p-1)}(\mathbb{C},\hat{\mathcal{E}}^0)$ into $\tilde{A}_{p/(p-1)}(\mathbb{C},\hat{\mathcal{E}}^{2m})$, due to the conditions on p and on the order of B_1.

Now we are under the conditions of theorem 1.8.2. Hence, any element $\varphi\in\tilde{A}_{p/(p-1)}(\mathbb{C},\hat{\mathcal{E}}^0)^*$ that is orthogonal to $Im\,\hat{\mathcal{L}}$ can be represented as

$$y\to\sum_{i,k,l,m,m}\varphi_{i,n,l}\,\beta_{i,k,n,m}\left\langle\frac{\partial^k y}{\partial\lambda^k}\Big|_{\lambda=\lambda_m},h_{n,l}\right\rangle,$$

where the estimate (10) holds. Taking into account that functionals of the type

$$\left\langle\frac{\partial^k y}{\partial\lambda^k}\Big|_{\lambda=\lambda_m},h_{n,l}\right\rangle$$

correspond to Floquet functions, we get the statements of the theorem.

Let us note some disturbing circumstances concerning this theorem. The regions U_n (see the discussion preceding theorem 1.5.3) strongly overlap for different values of n, so we cannot obtain block-basis property of Floquet solutions. Thus, the following theorem, which is an analog of the preceding one, up to a different choice of the interpolation theorem, might be sometimes convenient.

Theorem 5.4.8. *The statement of theorem 6 holds, if we replace coefficients* $\beta_{i,k,n,m}$ *by* $\gamma_{i,k,\alpha,m}$ *(see the definitions before theorem 1.5.5) and* (10) *by the estimate*

$$\sum_{i,\alpha}\Big(\sum_{l}|\varphi_{i,\alpha,l}|^2\Big)^{1/2}d_\alpha^{-i}\exp|z_\alpha|^q<\infty$$

for some $q > n_2 + 1$.

We shall not prove this theorem here. The proof follows the same scheme, as the proof of theorem 6, we only apply corollary 1.8.5 instead of theorem 1.8.2.

It is easy to prove the obvious analog of theorem 4.1.5 for problems of the type (1). This enables us to obtain some results of importance for some boundary value problems arising in the wave guides theory.

Let us consider the system (1) of the following special type:

$$\text{(5.4.13)} \qquad \begin{cases} (L_1(D_x) + \Delta_t + q(t, x))\, u(t, x) = 0 \ \text{ in } Q \\ \\ B(x, D_x)\, u\,|_{\partial Q} = 0. \end{cases}$$

Here Δ_t is the Laplace operator in t-variables. We assume that conditions i)-iii) are satisfied and $q \in L_\infty(Q)$.

Theorem 5.4.9. *Let the operator* $\mathcal{L}$ *in* $L_2(G)$ *generated by* $L_1(D_x)$ *with the boundary conditions* $Bu|_{\partial Q} = 0$ *be selfadjoint. Then the following (equivalent) assertions hold:*

a) the set of Floquet exponents of the problem (13) *does not coincide with* Ω_m;

b) problem (13) *has no non-zero solutions in* $L_2(Q)$.

Proof. Due to the analog of theorem 4.1.5, it is sufficient to find a vector $\lambda \in \mathbf{C}^m$ such that the problem

$$\text{(5.4.14)} \qquad \begin{cases} (L_1(D_x) + \sum \left(i \frac{\partial}{\partial t_j} + \lambda_j \right)^2 + q)\, u = 0 \\ \\ Bu\,|_{\partial Q} = 0 \end{cases}$$

has no 1-periodic solutions. Let us choose a number $N > ||q||_\infty$. As it was shown during the proof of 4.1.7, there is a vector $\lambda \in \mathbf{C}^{n_1}$ such that $|\, Im \sum (2\pi\gamma_j + \lambda_j)^2 \,| \geq N$ for all $\gamma \in \mathbf{Z}^{n_1}$. This vector λ is the one we need. In fact, suppose that (14) has a 1-periodic in t solution $u(t, x)$ and expand this solution into the Fourier series

$$u(t, x) = \sum_{\gamma \in \mathbf{Z}^{n_1}} u_\gamma(x) e^{2\pi i \gamma t}.$$

Since $u(t, x)$ satisfies (14), all functions $u_\gamma(x)$ satisfy the boundary conditions of (14). Moreover,

$$Lu = \sum_{\gamma \in \mathbf{Z}^m} [\, \mathcal{L} + \sum_j (2\pi\gamma_j + \lambda_j)^2 \,] \, u_\gamma \, e^{i\gamma x} + qu = 0 \,. \tag{5.4.15}$$

Taking into account that $\mathcal{L}$ is selfadjoint and $|\, Im \sum_j (2\pi\gamma_j + \lambda_j)^2 \,| \geq N$, we obtain the following resolvent estimate:

$$\| \, [\mathcal{L} + \sum_j (2\pi\gamma_j + \lambda_j)^2]^{-1} \| \, _{L_2(G) \to L_2(G)} \leq \frac{1}{N} \cdot$$

This means that

$$N \, \| \, u_\gamma \, \| \, \leq \| \, (\mathcal{L} + \sum_j (2\pi\gamma_j + \lambda_j)^2) u_\gamma \| \, .$$

Together with Parseval's equality this excludes (14):

$$\| \, (L_1(D_x) + \sum \, (\, i \frac{\partial}{\partial t_j} + \lambda_j)^2 \, + q) \, u \, \| \, _{L_2} \geq (N - \| \, q \, \| \, _{L_\infty}) \, \| \, u \, \| \, _{L_2} > 0$$

for $u \neq 0$. The proof is completed.

The conditions of the last theorem are satisfied, for instance, for Dirichlet and Neumann problems in Q for the equation

$$\Delta u + q(t, x) u = 0 \,. \tag{5.4.16}$$

These problems describe electromagnetic and sound waves propagation in periodic wave guides (n_1=1) or periodic structures (n_1>1) (see, for instance, [208, 300]).

B. Parabolic Problems. In this subsection we shall not go into most of the technical details of the proofs, because they coincide with those contained in subsection *A*.

Let $G \subset \mathbf{R}^{n-1}$ be a bounded region with a smooth boundary and $Q = \mathbf{R} \times G \subset \mathbf{R}^n$. We shall denote points of G by x, and points of $\mathbf{R}$ by t. Consider the operators

$$(5.4.17)\qquad \begin{aligned} A(t) &= \frac{\partial}{\partial t} + L(t,x,D_x) = \frac{\partial}{\partial t} + \sum_{|\alpha| \leq 2m} a_\alpha(t,x) D_x^\alpha, \\ B_j(t) &= B_j(t,x,D_x) = \sum_{|\alpha| \leq m_j} b_{j,\alpha}(t,x) D_x^\alpha, \quad j = 1, \dots, m \end{aligned}$$

with smooth 1-periodic coefficients.

We shall assume that for every $t_0 \in \mathbf{R}$ the system $\{A(t_0), B_j(t_0)\}$ satisfies the following conditions:

1) for every $\theta \in [-\frac{\pi}{2}, \frac{\pi}{2}]$ the operator $\Lambda_\theta = A(t_0) + e^{i\theta}(-1)^m D_t^{2m}$ is properly elliptic in $\overline{G} \times \mathbf{R}$;

2) $\{B_j\}_{j=1}^m$ is normal on ∂G and of order m_j with $0 \leq m_j \leq 2m-1$;

3) $\{B_j\}_{j=1}^m$ covers Λ_θ on $\partial G \times \mathbf{R}$, $\forall \theta \in [-\frac{\pi}{2}, \frac{\pi}{2}]$.

Let $H_{loc}^{(1,2m)}(\overline{Q})$ be the anisotropic Sobolev space of functions whose first t-derivative and all x-derivatives up to the order $2m$ belong to $L_{2,loc}(\overline{Q})$. We define the space

$$\Phi_p^{(1,2m)}(\overline{Q}) = \{f \in H_{loc}^{(1,2m)}(\overline{Q}) \mid \gamma_a(f) = \sup_k \|f\|_{H^{(1,2m)}(G \times \{k,k+1\})} \exp a|k|^p < \infty, \ \forall a > 0\}.$$

If we replace in this definition the space $H^{(1,2m)}$ by L_2, we get the space that we denote by $\overset{0}{\Phi}_p$. We also denote by $\widetilde{Q}$ the quotient space $Q/\mathbf{Z}$.

We can follow the same scheme, as in subsection A. This leads to the operators

$$\mathcal{L}_\lambda = \{(\frac{\partial}{\partial t} + L) - i\lambda, B_j\} : H^{(1,2m)}(\widetilde{Q}) \longrightarrow H_2 := L_2(\widetilde{Q}) \times \prod_{j=1}^m H^{\mu_j + 1/2}(\partial \widetilde{Q}).$$

Here $\mu_j = 2m - m_j - 1$. For $\lambda = 0$ this operator is Fredholm. This can be easily shown by a parametrix construction on $\widetilde{Q}$. Let us sketch this construction. We shall identify functions on $\widetilde{Q}$ with 1-periodic (with respect to t) functions on Q. Consider a C^∞-function $\varphi(t)$ on $\mathbf{R}$ such that

1) $supp\ \varphi(t) \subset (0, 2)$;

2) $\sum_{k\in\mathbf{Z}} \varphi(t-k) \equiv 1.$

We shall also need a function $\psi \in C_0^\infty$ such that $\psi \equiv 1$ on $supp\ \varphi$ and $\psi \equiv 0$ for $t \leq 0$ and $t \geq 2$. Consider the subspace M of the space

$$L_2(G\times[0,2]) \times \prod_{j=1}^{m} H^{2m-m_j-1/2}(\partial G\times[0,2])$$

consisting of all function tuples $\{g_0, g_1, \ldots, g_m\}$ that have zero values at t=0 (when the restrictions are defined, i.e. when $2m-m_j>1$). We can define the operator $S: M \longrightarrow H^{(1,2m)}(G\times[0,2])$ that maps $\{g_0, g_1, \ldots, g_m\}$ into the solution f of the following Cauchy problem:

$$\text{(5.4.18)} \qquad \begin{cases} (\frac{\partial}{\partial t} + L)\ f = g_0 \text{ in } G\times(0,2) \\ B_j\ f|_{\partial G\times(0,2)} = g_j \\ f|_{t=0} = 0. \end{cases}$$

Due to conditions 1)-3), the solution exists and is unique; the operator S is continuous (see, for instance, [5]). Let T be the translation operator

$$Ts(t, x) \longrightarrow s(t+1, x).$$

We define the operator U by

$$U(\{g_0, g_1, \ldots, g_m\}) = \sum_{k\in\mathbf{Z}} T^k \psi S(\varphi\{g_0, \ldots, g_m\}).$$

We identify here the functions φ and ψ with the operators of multiplication by these functions. The operator U is continuous from H_2 into $H_{loc}^{(1,2m)}(Q)$ and its image consists of 1-periodic functions, so U maps H_2 into $H^{(1,2m)}(\tilde{Q})$. A simple calculation shows that U is a regularizer to $\mathcal{L}_0$. Due to embedding theorems, the term $-i\lambda$ in $\mathcal{L}_\lambda$ corresponds to a compact operator, so $\mathcal{L}_\lambda$ is Fredholm. As we observed in lemma 1.2.19, the embedding operator $H^{(1,2m)}(\tilde{Q}) \longrightarrow L_2(\tilde{Q})$ belongs to the class $\mathcal{S}_p$ for $p>1+(n-1)/2m$.

Integrating with respect to t the corresponding formula for the elliptic case, we can easily obtain the following Green's formula for (17):

$$\int\limits_Q (\tfrac{\partial u}{\partial t}+Lu)\, v\, dtdx - \int\limits_Q u(-\tfrac{\partial v}{dt}+L_1 v)\, dtdx = \sum_j \int\limits_{\partial Q} S_j u\, C_j v\, d\sigma - \sum_j B_j u\, T_j v\, d\sigma.$$

Taking into account these remarks and following the considerations of subsection A, we can obtain, for instance, the following results:

Theorem 5.4.10. *The set N of Floquet exponents of the problem*

$$\text{(5.4.19)} \qquad \begin{cases} u_t + Lu = 0 \text{ in } Q \\ B_j u|_{\partial Q} = 0 \end{cases}$$

is discrete in $\mathbb{C}\setminus 0$.

There exist a discrete subset $\{z_j\} \subset \mathbb{C}\setminus 0$ *and for every* j *a finite collection of functions* $F_{j,k}(t,x)$ *such that*

1) $F_{j,k}$ *is a Floquet solution of* (19) *with Floquet exponent* z_j;

2) every solution of (19) *that satisfies the estimate*

$$\text{(5.4.20)} \qquad |u(x,t)| \le C \exp a|t| \quad \text{for some } C,\ a>0$$

can be represented as the finite sum

$$\text{(5.4.21)} \qquad u(t,x) = \sum_{j,k} \alpha_{j,k} F_{j,k}(t,x), \quad \alpha_{j,k} \in \mathbb{C};$$

3) Every finite sum (21) *gives rise to a solution with estimate* (20).

Remark. The difference between this theorem and theorem 4 is that the possibility $N=\mathbb{C}\setminus 0$ is now excluded. The reason is that all solutions of parabolic problem (19) have limited rate of exponential growth: there exists $\sigma>0$ such that $u(t,x)$ can not grow faster that $exp\ \gamma t$ in the sense that $e^{-\gamma t}u$ must belong to $H^{1,2m}(G\times\mathbf{R}_+)$ (see [5]). This means that all Floquet multipliers z must satisfy the condition $|z|\leq e^{\gamma}$.

Theorem 5.4.11. *Let $\{exp\ \lambda_n, p_n\}$ be the Floquet exponents of* (19) *with their multiplicities. Let also $\beta_{i,k,u,m}$ be defined as in section 1.5. Then, for every tuple (k,n,m) such that $|\lambda_n-\lambda_m|<\delta|\lambda_n|$, $0\leq k\leq p_m-1$ there exists a finite collection of Floquet solutions $h_{k,l,m,n}(t, x)$ (with multiplier $exp\ \lambda_m$ of multiplicity k) such that every solution of* (19) *with*

$$|u(t, x)|\leq C\ exp\ a|t|^{1+\gamma},\ \gamma<\frac{2m}{2m+n-1} \tag{5.4.22}$$

can be represented as

$$u(t, x) = \sum_{i,k,l,m,n}\varphi_{i,n,l}\,\beta_{i,k,m,n}h_{k,l,m,n}(t, x), \tag{5.4.23}$$

where the complex numbers $\varphi_{i,n,l}$ satisfy the estimate

$$\sum_{i,n}\Big(\sum_{l}|\varphi_{i,n,l}|^2\Big)^{1/2}\ exp(a|\lambda_n|^p)<\infty \tag{5.4.24}$$

for some $a>0$, $p>2+\frac{n-1}{2m}$.

And conversely, any choice of $\varphi_{i,n,l}$ with (24) *provides by* (23) *a solution that satisfies* (22).

5.5. Comments and references

Some analogs of Floquet theory for the equation

$$\frac{du}{dt}=A(t)u$$

with a *bounded* periodic operator coefficient were developed in [82, 282]. Our goal is to study the equations that arise from boundary value problems for partial differential equations, so the operator $A(t)$ must be unbounded. These two cases differ significantly with regard both to the methods and to the results. The general approach adopted in this book works for equations that are hypoelliptic in some sense (for instance, for abstract parabolic equations). The situation for periodic in time hyperbolic equations, or non-stationary Schrödinger equations differs almost completely from the one for hypoelliptic equations. We shall provide the reader with some references later in this section.

Writing this chapter, we discovered that there are too many results about hypoelliptic evolution equations to be covered. Thus, we provide just a sketch of main results. We plan to cover all known results and approaches in a joint publication with A. Miloslavskii.

The results of section 5.1 were obtained by the author (see the survey [229] and references in it). Several papers of V. Dergusov [87-89] were devoted to periodic anti-canonical, canonical and Hamiltonian abstract evolution equations, mainly in relation with wave-guides theory. He has not obtained, however, any Floquet expansions. He proved the analogs of some results of sections 4.1 and 4.2 for these equations, and investigated some other problems that we do not touch here (for instance, problems related to joining of two semi-infinite wave guides).

Section 5.2 is based upon results of G. Da Prato and A. Lunardi [80] and upon some generalizations by the author [235]. In the form presented here the results were proved in [235], however the effect itself was discovered first in [80]. In [80] the authors were concerned with controllability and stabilizability of problems of the type (5.2.1). Their interest to these problems was fueled by the fact that several important systems arising in applications can be reduced (5.2.1). Among them are [80]:

i) The strongly damped wave equation

$$\begin{cases} u_{tt} = \Delta_x u + \rho(t)\Delta_x u_t, \ t > 0, \ x \in \Omega \\ B(x, D)u \mid_{\partial\Omega} = 0, \end{cases}$$

where $\rho(t) > 0$.

ii) The reaction-diffusion systems

$$\begin{cases} u_t = d_1(t)\Delta_x u + b_{11}(t)u + b_{12}(t)v \\ v_t = d_2(t)\Delta_x v + b_{21}(t)u + b_{22}(t)v \\ Bu|_{\partial\Omega} = 0, \; Bv|_{\partial\Omega} = 0 \end{cases},$$

where $d_i > 0$.

iii) The nerve axon systems

$$\begin{cases} u_t = \alpha(t)\Delta_x u + b_{11}(t)u + b_{12}(t)v \\ v_t = b_{21}(t)u + b_{22}(t)v \\ Bu|_{\partial\Omega} = 0, \end{cases}$$

where $\alpha(t) > 0$.

G. Da Prato and A. Lunardi proved analogs of corollary 5.2.6 and of theorem 5.2.7 under more restrictive conditions. They assumed that $n=2$, A is time-independent and generates an analytic semigroup, $Tr\ a(t) > 0$, $Det\ a(t) \geq 0$ and the *Coker* bundle for $a(t)$ is trivial. It was also supposed in the main theorem that $Im\ a(t)$ is constant. It is worth mentioning that in the case when $n=2$ (and $m=1$, otherwise the problem is trivial) the unique non-zero accumulation point χ can be found rather explicitly [80]. The approach of [80] is different from the one adopted in section 5.2 and in [235].

The reader can find some general information about abstract parabolic equations in the books [108, 168, 209, 210]. Section 5.3 is based on results of A. Miloslavskii [296, 297] and of the author (cf. also [298, 299, 397, 398]). Theorems 5.3.1 and 5.3.2 have been proved by the author. The example constructed in theorem 5.3.4 is based on an example of A. Miloslavskii [297] (see the remark after the proof of theorem 5.3.5), which in turn was based on an example by P. Lax [249]. Theorem 5.3.5 is the author's version of a theorem of A. Miloslavskii [297]. All theorems 5.3.6-5.3.15 were proved by A. Miloslavskii [297]. Due to space limitations, we do not discuss here the questions analogous to those considered in sections 4.1-4.5 (including solvability of non-homogeneous equations, stability, dichotomy, etc.). Results of this type can be easily obtained along the path adopted in chapter 4. Some results about solvability of non-homogeneous equations (analogous to the ones of section 4.2) were obtained in [125, 263, 264]. An important goal is to get rid of the condition that the domain of the operator coefficient is constant. This is the standard trouble one encounters, when regarding a parabolic problem for partial differential equations as an abstract evolution equation. When one works with PDEs directly (see, for instance, section 5.4), one usually does not have to worry about boundary conditions that depend on time, but when one considers the problem in the abstract setting, this condition arises almost automatically. M. Fuhrman managed to work with variable domains in [125], and we hope that it can be done for most of the results of chapter 5.

The results of section 5.4 were obtained by the author [225, 229]. There is no doubt that these results can be easily carried over to general Douglis-Nirenberg elliptic systems and general parabolic systems, using results of [265, 266, 344] in the elliptic case and of [373] in the parabolic case. Some problems for evolution equations related to the theory of periodic wave guides were treated in [87-89, 208, 212, 213, 281]. In particular, theorem 5.4.9 was proved in [89] for the particular case of Helmholtz equation and $n=1$.

As we have already mentioned, the cases of hyperbolic and non-stationary Schrödinger equations are completely different from the hypoelliptic one. Indeed, if the monodromy operator is unitary (which is rather common situation for such equations), its spectrum is confined to the unit circle, so we do not have any analyticity (like in the part A of theorem 5.1.7). This analyticity, however, is the basis for the whole our approach. On the other hand, if the monodromy operator is unitary, we can expect some kind of expansion into generalized eigenvectors, which in turn provides some Floquet expansion. This type of arguments was successfully implemented in some cases by V. Sviridov [384-387]. The reader can also look at the papers [175, 206, 406-408] devoted to the quantum scattering theory for Schrödinger equations with time-periodic potentials (cf. also sections 7.4, 7.5, 8.5 and 8.6 in [79]). This Howland-Yajima method presented there is actually based on the transform $\mathfrak{F}$.

CHAPTER 6

OTHER CLASSES OF PROBLEMS

In this chapter we briefly discuss some possible generalizations of Floquet theory. Section 6.1 deals with equations with deviating arguments. Sections 6.2-6.4 are devoted to generalizations of Ehrenpreis-Malgrange-Palamodov theory [99, 273, 274, 324] and Floquet theory to equations that are invariant with respect to more general transformation groups.

6.1. Equations with deviating arguments

The situation with Floquet theory is quite complicated even for ordinary differential equations with deviating argument. As was shown in [416], Floquet solutions need not be complete even for the equation $y'(t) = a(t)y(t-1)$ with 1-periodic coefficient $a(t)$. On the other hand, it was proved in [416] that any solution of exponential growth on the whole axis for an ordinary periodic equation with deviation in only the lower order terms is a finite linear combination of Floquet solutions. It is worth mentioning that the theorem fails for solutions defined on the whole axis but with the exponential bound on only one half-axis [416]. We shall present here some positive results that can be obtained for partial differential equations with deviations.

We introduce the class of operators that we shall consider. Let h be a C^∞-diffeomorphism of $\mathbf{R}^n$ that commutes with translations by elements of $\mathbf{Z}^n$. In this case h leads to some torus diffeomorphism $\tilde{h}:\mathbf{T}^n \to \mathbf{T}^n$. Denote by T_h the operator that acts on functions as $T_h f(x) = f(h(x))$. We consider an operator on $\mathbf{R}^n$ of the form

$$L(x, D) = \sum_{i=1}^{m} L_i(x,D)\, T_{h_i}\,. \tag{6.1.1}$$

Here h_i are diffeomorphisms of the class that we described above and $L_i(x,D)$ are linear partial differential operators with C^∞-smooth $\mathbf{Z}^n$-periodic coefficients. This operator L obviously commutes with $\mathbf{Z}^n$-translations.

We denote by $J_h(x)$ the Jacobian determinant of $h(x)$. For any two smooth functions $\phi(x)$, $\psi(x)$ on $\mathbf{R}^n$, one of which is compactly supported, the following identity holds:

$$\int (L(x,D)\ \phi(x))\ \psi(x)\ dx = \int \phi(x)\ {}^tL(x,D)\ \psi(x)\ dx, \tag{6.1.2}$$

where

$$ {}^tL(x,D) = \sum_{i=1}^{m} T_{h_i}{}^{-1}\ J_{h_i}{}^{-1}\ {}^tL_i(x,D) \tag{6.1.3}$$

and ${}^tL_i(x,D)$ is the formally transposed operator to L_i. Commuting $T_{h_i}{}^{-1}$ with differential operators, we can reduce ${}^tL$ to an operator L_1 of the form (1) with $h_i{}^{-1}$ instead of h_i. Equality (3) holds even for functions ϕ, $\psi \in H^{2m}_{loc}(\mathbf{R}^n)$ (one of which is compactly supported), where $2m$ is the order of L.

Solutions of the equation $Lu = 0$ can be considered as distributions that annihilate the range of the continuous operator

$$L_1\colon \mathfrak{D}(\mathbf{R}^n) \to \mathfrak{D}(\mathbf{R}^n).$$

This operator can be extended continuously to an operator

$$L_1\colon \Phi^{2m,a}_{n,p} \to \Phi^{0,a}_{n,p}$$

(in particular, $L_1\colon \Phi^{2m}_{n,p} \to \Phi^{0}_{n,p}$). In fact, this is already known for purely differential operators $L_i(x,D)$, and multiplication by the smooth periodic function $J_h{}^{-1}$ is obviously continuous in $\Phi^{0,a}_{n,p}$, so we should prove only continuity of the operator T_h : $\Phi^{0,a}_{n,p} \to \Phi^{0,a}_{n,p}$. Let us choose some cube $\mathfrak{K}'$ that contains $h(\mathfrak{K})$, where $\mathfrak{K}$ is the fundamental cube with respect to $\mathbf{Z}^n$. Then

$$\| T_h f \|^2_{L_2(\mathfrak{K})} = \int_{\mathfrak{K}} | f(h(x)) |^2\ dx =$$

$$= \int_{h(\mathcal{K})} |f(x)|^2 J_h^{-1}(h^{-1}(x))\, dx \le$$

$$\le C \int_{\mathcal{K}'} |f(x)|^2\, dx = C\|f\|^2_{L_2(\mathcal{K}')} .$$

The same estimate (with the same constant) holds, due to $\mathbb{Z}^n$-periodicity of h, when we replace $\mathcal{K}$ and $\mathcal{K}'$ by $\mathcal{K}+k$ and $\mathcal{K}'+k$ with any $k \in \mathbb{Z}^n$. This means that T_h is continuous in the space $\Phi^{0,a}_{n,p}$, so L_1 is continuous as an operator from $\Phi^{2m,a}_{n,p}$ into $\Phi^{0,a}_{n,p}$. This enables us, as in chapter 3, to regard solutions of the equation $Lu = 0$ with the estimate $|u(x)| \le C \exp a|x|^p$ $(p \ge 1)$ as continuous linear functionals on $\Phi^0_{n,p}$ that annihilate the range of the operator $L_1 : \Phi^s_{n,p} \to \Phi^0_{n,p}$. In order to repeat the considerations of chapter 3, we need only the Fredholm property of the corresponding morphism of Banach bundles. Let us assume that $h_i(x) \equiv x$ for all i such that L_i is of the maximal order $2m$ in $L(x)$. In other words, we allow deviated arguments only in the lower order terms of $L(x,D)$. We assume also that the principal part of $L(x,D)$ (which is now an operator without deviations) is elliptic.

We have to adjust the definition of the bundle $\mathcal{E}^s$ in such way that our nonlocal operators can be regarded as morphisms. The fiber E^s_z over a point $z \in \Omega_n$ is now the subspace of $H^s_{loc}(\mathbb{R}^n)$ consisting of all functions $f(x)$ with the following property: $f(x+k) = z^k f(x)$ for all $k \in \mathbb{Z}^n$. The restriction to the fundamental cube $\mathcal{K}$ obviously provides an identification of the space with the old version of E^s_z (see chapter 2). Defining E^s_z in this way, we may observe that operators L and ${}^tL$ are continuous from E^{2m}_z into E^0_z for every z. After the standard trivialization (see chapter 2), we get the family $L(\lambda) = e^{-\lambda x t} L\, e^{\lambda x}$ $(\lambda \in \mathbb{C}^n)$ of operators on the torus $\mathbb{T}^n$. Since the principal part of $L(\lambda)$ is elliptic, we easily check that $L(\lambda)$ is a Fredholm operator from $H^{2m}(\mathbb{T}^n)$ into $L^2(\mathbb{T}^n)$. Indeed, the principal part corresponds to a Fredholm operator, and the lower order terms (even with deviated arguments) are compact operators (since they can be factored through the compact inclusion $H^{2m}(\mathbb{T}^n) \subset H^{2m-1}(\mathbb{T}^n)$).

Now we may repeat the proof of theorem 3.2.1 to obtain the following result.

Theorem 6.1.1. *Suppose that there is no deviation in arguments of the principal part of the operator L (that is, the corresponding $h_i(x)$ are identically equal to x) and* *the principal part is elliptic. Then the Floquet representation theorem 3.2.1 holds for the equation $Lu=0$.*

If we do have deviations in the principal part, it can destroy the Fredholm property. If, however, this property survives on an appropriate region, we can still obtain the Floquet representation for a more restricted class of solutions. Namely, we can prove the following conditional result.

Theorem 6.1.2. *Let $e^{-\lambda x}L_1(x,D)e^{\lambda x}$ be Fredholm as an operator from $H^{2m}(\mathbb{T}^n)$ into $L^2(\mathbb{T}^n)$ for all $\lambda=(\lambda_1,\dots,\lambda_n)$ such that $|Re\ \lambda_j|<R$. Then the analog of theorem* 3.2.1 *holds for solutions $u(x)$ such that $|u(x)| \le C\ \exp a|x|$ $(a<R)$ provided that Ω_n is replaced by the polyring*

$$\Omega_n(R) = \{z \in \Omega_n \,|\, e^{-R} < |z_j| < e^{R} \text{ for all } j=1, \dots, n\}.$$

The proof is similar to that of theorem 3.2.1. The operators that arise can generate non-Fredholm bundle morphism, but the hypothesis of the theorem ensures that they are Fredholm on the fibers over $\Omega_n(R)$.

In some cases we know conditions that guarantee the Fredholm property for operators with deviated arguments. Then we can verify the conditions of the last theorem. We shall show on some examples how it works in the case of ordinary differential operators.

Consider an ordinary 1-periodic operator L of the form

$$Lu = a(t)u^{(k)}(t) + b(t)u^{(k)}(t+h) + \sum_{i<k} a_i(t)u^{(i)}(h_i(t)) \tag{6.1.4}$$

Here the coefficients a, b and a_i are C^∞-smooth and 1-periodic, h is a constant and h_i are C^∞-diffeomorphisms of $\mathbf{R}$ that commute with integer translations. The operator L_1 has an analogous structure with $a(t)$, $b(t-h)$ instead of $a(t)$, $b(t)$.

Consider now some special cases:

a) *h is rational.* Let $h=p/q$ be its irreducible rational representation. Assume that

$$\Big|\prod_{k=0}^{q-1} a(t+kh)\Big| \ne \Big|\prod_{k=0}^{q-1} b(t+kh)\Big|\,.$$

We define

$$R = p^{-1} \inf \mid ln \mid \Big(\prod_{k=0}^{q-1} a(t+kh)\Big) \,/\, \Big(\prod_{k=0}^{q-1} b(t+kh) \Big| \mid . \tag{6.1.5}$$

Here the infimum is taken over all values of t. (If the denominator is zero and the numerator is not zero, the ratio is assumed to be infinite. If both are zero, the ratio is assumed to be 0.)

b) *h is irrational and $b(t)$, $a(t)$ have no zeros.* Let

$$R = h^{-1} \Big| \int_0^1 ln \Big| \frac{a(t)}{b(t)} \Big| \, dt \Big| . \tag{6.1.6}$$

c) *h is irrational, $a(t)$ has no zeros and either $b(t)$ is zero on a set of non-zero measure or the integral*

$$\int_0^1 ln \,|\, b(t) \,|\, dt$$

is divergent. In this case we let $R = \infty$.

d) *h is irrational, $a(t)$ has no zeros, and* $\int_0^1 ln \,|\, a(t) \,|\, dt > \int_0^1 ln \,|\, b(t) \,|\, dt$. Then we let

$$R = h^{-1} \int_0^1 ln \Big| \frac{a(t)}{b(t)} \Big| \, dt .$$

Theorem 6.1.3. *In all cases a), b), c), d) the conditions of theorem 2 are satisfied for the corresponding values of R. Hence, the corresponding Floquet representations hold.*

Proof consists only of checking the Fredholm conditions for the operators $L(\lambda) = e^{-\lambda t} L_1 e^{\lambda t}$. A simple computation shows that

$$L(\lambda)u(t) = a(t)u^{(k)}(t) + e^{-\lambda h} b(t-h)u^{(k)}(t-h) + \text{lower order terms} =$$

$$= a(t)u^{(k)}(t) + b_\lambda(t)u^{(k)}(t-h) + \ldots$$

a) In this case the results of [16] say that L_λ is Fredholm if

$$\prod_{k=0}^{q-1} a(t+kh) \neq \prod_{k=0}^{q-1} b_\lambda(t+kh) \ (=e^{-p\lambda}\prod_{k=0}^{q-1} b(t+k)).$$

Hence, the condition is

$$e^{-p\lambda} \neq \Big(\prod_{k=0}^{q-1} a(t+kh)\Big) / \Big(\prod_{k=0}^{q-1} b(t+kh)\Big). \tag{6.1.7}$$

It is obvious that the condition $|Re\ \lambda|<R$ for R given in a) is sufficient for (7).

b) In this case the condition of [16] is

$$\int_0^1 ln|a(t)|dt \neq \int_0^1 ln|b_\lambda(t)|dt \ (= \int_0^1 ln|b(t)|dt \ - h\ Re\ \lambda),$$

so the inequality $|Re\ \lambda|<R$ (R given in b)) is sufficient.

c) In this case we see that

$$\int_0^1 ln|b_\lambda(t)|dt$$

is divergent for all λ, so the result of [16] guarantees Fredholm property for all λ.

d) The Fredholm condition of [16] requires in this case the inequality

$$\int_0^1 ln|a(t)|dt > \int_0^1 ln|b_\lambda(t)|dt. \tag{6.1.8}$$

The choice $|Re\ \lambda|<R$ for R given in d) guarantees (8).

6.2. Equations with coefficients that do not depend on some variables

Let $G\subset \mathbb{R}^m$ be a bounded domain with smooth boundary. Consider the solid cylinder $Q=\mathbb{R}^n\times G$. We shall denote by (x,y) points of the product $\mathbb{R}^n\times G$. Let $L(y,D_x,D_y)$ be a linear differential operator of order $2l$, whose coefficients are smooth in $\overline{Q}$ and do not depend on x. We denote by L_0 the principal part of L. Let us also consider boundary operators $B_j(y,D_x,D_y)$ $(j=0, \ldots, l-1)$ of order l_j with coefficients that are smooth in a neighborhood of ∂G and do not depend on x. The additive group of the space $\mathbf{R}^n_x$ acts naturally on $\overline{Q}$ by x-translations. The operator L commutes with the action.

The following definition is a particular case of the one given by M. Atiyah [20] for the more general situation of a general Lie group action.

Definition 6.2.1. The operator L is said to be *transversally elliptic* with respect to the $\mathbf{R}^n$-action if, for every pair of linearly independent vectors $\xi_1, \xi_2 \in \mathbf{R}^m$ and every $y \in G$, the polynomial $L_0(y, 0, \xi_1 + \tau\xi_2)$ of τ has exactly l zeros in both the upper and lower half-planes. (If $n>2$, it is sufficient to assume that $L_0(y, 0, \xi) \neq 0$ for $\xi \in \mathbf{R}^n \backslash 0$).

Let us consider the following boundary value problem in Q:

$$\text{(6.2.1)} \qquad \begin{cases} L(y, D_x, D_y)u = 0 \text{ in } Q \\ B_j(y, D_x, D_y)u|_{\partial Q} = 0, \; j = 0, \ldots, l-1 \,. \end{cases}$$

Consider also the following auxiliary problem in G:

$$\text{(6.2.2)} \qquad \begin{cases} L(y, 0, D_y)\; u(y) = 0 \text{ in } G \\ B_j(y, 0, D_y)\; u(y)|_{\partial Q} = 0, \; j = 0, \ldots, l-1 \,. \end{cases}$$

We will call the problem (1) *transversally elliptic*, if L is transversally elliptic, and (2) is elliptic in the sense of section 5.4.A.

The problem (1) has coefficients that are constant with respect to the x-variable. Our goal is to show that solutions of exponential-polynomial type with respect to x are complete in some classes of solutions of (1). In order to do this, we need to employ some special function spaces.

Let $\alpha, \beta \in (0,1)$, $A>0$, $B>0$ and H be a separable Hilbert space. We define $S^{\beta, B}_{\alpha, A}(\mathbf{R}^n, H)$ as the space of entire H-valued functions $\varphi(z)$ on $\mathbf{C}^n$ such that

$$\|\varphi\|^{B}_{A} = \sup_{z} \|\varphi(z)\| \; exp\,(A|x|^{1/\alpha} - B|y|^{1/(1-\beta)}) < \infty.$$

(Here $z = x + iy$). This is a Banach space with respect to the norm $\|\varphi\|^{B}_{A}$. Now the space $S^{\beta}_{\alpha}(\mathbf{R}^n, H)$ is:

(6.2.3) $$S_{\alpha}^{\beta}(\mathbf{R}^n, H) = \lim_{\substack{A\to\infty \\ B\to 0}} proj\, S_{\alpha,A}^{\beta,B}(\mathbf{R}^n, H).$$

It can be easily shown, that

(6.2.4) $$S_{\alpha}^{\beta}(\mathbf{R}^n, H) = S_{\alpha}^{\beta}(\mathbf{R}^n, \mathbf{C}) \,\widehat{\otimes}\, H$$

(see [341] for the definition of the $\widehat{\otimes}$- product). We shall also need the space

(6.2.5) $$S_{\alpha+}^{\beta+}(\mathbf{R}^n, H) \overset{def}{=} \lim_{\substack{A\to\infty \\ B\to 0}} ind\, S_{\alpha,B}^{\beta,A}(\mathbf{R}^n, H).$$

Lemma 6.2.2. *The Fourier transform is an isomorphism of* $S_{\alpha}^{\beta}(\mathbf{R}^n, H)$ *onto* $S_{\beta+}^{\alpha+}(\mathbf{R}^n, H)$.

In the scalar case (i.e. H=$\mathbf{C}$) this was proved in [321]; see also [210]. In general we can either follow the same proof, or employ (4).

Theorem 6.2.3. *Let us assume that:*

1) The boundary value problem (1) *is transversally elliptic and has no non-zero solutions in* $L_2(Q)$.

2) The Fredholm index of the elliptic problem (2) *is zero.*

Then the set of solutions of (1) *of the form*

(6.2.6) $$u(x,y) = \Big(\sum_{k\in \mathbf{Z}_+^m} x^k g_k(y)\Big)\, exp < \lambda, x >$$

(*where* $\lambda \in \mathbf{C}^n, x^k = \prod_j x_j^{k_j}$ *and the sum is finite*) *is complete in the space of all solutions with the estimate*

(6.2.7) $$||u||_{H^{2l}(G\times\{\mathcal{K}+b\})} \le C_u \exp a_u |b|^{1+\gamma}$$

for some $\gamma < (n-1)^{-1}$. *Here* $\mathcal{K}$ *is a cube in* $\mathbf{R}^n, b \in \mathbf{R}^n$ *and completeness is considered with respect to the weak-star topology of the space*

(6.2.8) $$S_{\alpha}^{\beta}(\mathbf{R}^n, L_2(G))^*$$

for α^{-1}, $(1-\beta)^{-1} \in [1+\gamma, 1+(n-1)^{-1}]$.

Proof. Let us show, first of all, that any solution satisfying (7) corresponds to an element of the space (8). If $f \in S_{\alpha}^{\beta}(\mathbf{R}^n, L_2(G))$, then f can be regarded as a function $f(z,y)$, $z \in \mathbf{C}^n, y \in G$. This allows us to define

$$\langle u, f\rangle = \int_{\mathbf{R}^n} dx \int_G u(x,y) f(x,y) dy.$$

Then,

$$|\langle u, f\rangle| \leq \int_{\mathbf{R}^n} ||u(x,\cdot)||_{L_2(G)} ||f(x,\cdot)||_{L_2(G)} dx \leq$$

$$\leq C \int_{\mathbf{R}^n} ||f||_A^B exp(a_u|x|^{1+\gamma} - A|x|^{1/\alpha}) dx.$$

Since $1+\gamma \leq \alpha^{-1}$, the last integral is convergent for sufficiently large A and we get the estimate

$$|\langle u, f\rangle| \leq C||f||_A^B \quad (B > 0 \text{ is arbitrary}).$$

This means that u belongs to the space (8).

We shall need to introduce an adjoint (with respect to a Green's formula) boundary value problem of the following structure:

$$\text{(6.2.9)} \qquad \begin{cases} L_1 g = f & \text{in } Q \\ C_j g \mid_{\partial Q} = f_j, j = 0, \dots, l-1, \end{cases}$$

where L_1 is the formal transpose operator ${}^tL$ and the problem is transversally elliptic and invariant with respect to $\mathbf{R}^n$-translations. The transversal ellipticity and invariance of L_1 is obvious. We are going to sketch the construction of the appropriate boundary operators C_j (omitting some details that are contained in the proof of theorem 2.1 of chapter 2 in [262]).

Consider a finite covering of ∂Q by relatively open subsets of the form $V_j = \mathbf{R}^n \times U_j$, where $\{U_j\}$ is a finite, relatively open covering of ∂G. It is obvious that we can find a partition of unity subordinate to $\{V_j\}$ that consists of functions depending only on $y \in \mathbf{R}^m$. After a local diffeomorphism, we can assume that $U_j \cap G$ is a semi-ball and V_j is the product of $\mathbf{R}^n$ with this semi-ball. We observe that this transformation acts only in the y-variables. Following the constructions of theorem 2.1, chapter 2 in [262], we can complete the system of operators $\{B_j\}$ on $\partial Q \cap V_j$ to a Dirichlet system by means of operators S_j of the form $D_t^{\mu_j}$. Here D_t is the normal derivative to ∂G. These operators obviously commute with x-translations. Following the proof of the same theorem, and using transversal ellipticity of L, we construct another Dirichlet system $\{C_j, T_j\}$ of boundary operators. After the standard gluing together, we get two Dirichlet systems $\{B_j, S_j\}$, and $\{C_j, T_j\}$ such that problem (9) is transversally elliptic and the following Green's formula holds:

$$(6.2.10) \qquad \int_Q Lu \cdot v dx dy - \int_Q u \cdot L_1 v dx dy = \sum_j \int_{\partial Q} S_j u \cdot C_j v d\sigma - \sum_j \int_{\partial Q} B_j u \cdot T_j v d\sigma.$$

Here functions u and v are assumed to be smooth and at least one of them compactly supported. The construction guarantees that the operators L, L_1, S_j, and B_j, and the sets Q and ∂Q, are invariant with respect to x-translations. We shall also need x-invariance of the operators C_j and T_j. One way to obtain this invariance is to follow the construction closely, but the simplest is the following one. Let us translate C_j and T_j in the x-direction. The translated operators again form a Dirichlet system that satisfies (10). This system, however, is unique [262], so C_j and T_j are invariant with respect to x-translations (i.e., their coefficients do not depend on x).

We shall need the following auxiliary elliptic problem on G:

$$(6.2.11) \qquad \begin{cases} L_1 (y, 0, D_y)\, u = f \\ C_j (y, 0, D_y)\, u|_{\partial G} = f_j. \end{cases}$$

Let us consider the operator

$$\mathcal{L}: S_\alpha^\beta\left(\mathbf{R}^n, H^{2l}(G)\right) \longrightarrow S_\alpha^\beta\left(\mathbf{R}^n, L_2(G) \times \prod_{j=0}^{m-1} H^{2l - l_j - 1/2}(\partial G)\right)$$

defined as

$$\mathcal{L}f = (L_1 f, C_0 f|_{\partial Q}, \ldots, C_{l-1} f|_{\partial Q}).$$

Its continuity is trivial. After the Fourier transform with respect to x, we get the spaces $S^{\alpha+}_{\beta+}(\mathbb{R}^n, H_j)$ instead of $S^{\beta}_{\alpha}(\mathbb{R}^m, H_j)$ $(j=1,2)$, the operator polynomial $L_1(y,z,D_y)$ instead of L_1, and operator polynomials $C_j(y, z, D_y)$ instead of C_j. Here $z \in \mathbb{C}^n$, $H_1 = H^{2l}(G)$ and

$$H_2 = L_2(G) \times \prod_{j=0}^{l-1} H^{2l - l_j - 1/2}(\partial G).$$

Hence, the Fourier image of the operator $\mathcal{L}$ is multiplication by the operator

$$\mathcal{L}(z) = \sum_{\substack{k \in \mathbb{Z}^n \\ |k| \leq 2l}} z^k \mathcal{L}_k, \tag{6.2.12}$$

where $\mathcal{L}_k$ is a continuous operator from $H_1 = H^{2l}(G)$ into

$$H_2^{(k)} = H^{|k|}(G) \times \prod_{j=0}^{l-1} H^{2l - l_j + |k| - 1/2}(\partial G).$$

Continuity follows from the standard embedding theorems (see, for instance, [262, 394]). We conclude (via lemma 1.2.18) that for $k \neq 0$ the operator $\mathcal{L}_k : H_1 \longrightarrow H_2$ belongs to the class $\mathcal{S}_{p/|k|}$ for any $p>n$. When $k=0$, the operator $\mathcal{L}_0$ coincides with $\{L_1(y,0, D_y), C_0(y,0, D_y), \ldots\}$, i.e. with the operator naturally related to problem (11). Due to the ellipticity of (11), we conclude that $\mathcal{L}_0$ is a Fredholm operator (see [262]). The second assumption of the theorem implies that *index* $\mathcal{L}_0 = 0$. Now we deal with the operator of multiplication by the polynomial (12). The operator acts from $S^{\alpha+}_{\beta+}(\mathbb{R}, H_1)$ into $S^{\alpha+}_{\beta+}(\mathbb{R}^m, H_2)$. The leading term of $\mathcal{L}_0$ is Fredholm of index zero; $\mathcal{L}_k \in S_{p/|k|}$ for $p > n$. This means, in particular, that all values of $\mathcal{L}(z)$ are Fredholm operators of index zero. The next step is to show that $\mathcal{L}(z)$ is invertible for some $z \in \mathbb{C}$ (and hence for almost all values of z). Suppose the opposite: $\mathcal{L}(z)$ is nowhere invertible. In other words, $Ker\ \mathcal{L}(z) \neq 0$ for all $z \in \mathbb{C}^n$. Let $p_0 \in \mathbb{R}^n = Re\ \mathbb{C}^n$ be such that

$$\dim Ker\,\mathcal{L}(p_0) = \min_{p\in\mathbf{R}^n} \dim \operatorname{Ker} \mathcal{L}\,(p) \neq 0.$$

Then, in a neighborhood $U \subset \mathbf{R}^n$ of the point p_0, the dimension of $Ker\,\mathcal{L}(p)$ is constant, so

$$\bigcup_{p\in U} Ker\,\mathcal{L}\,(p)$$

and

$$G = \bigcup_{p\in U} Im\,\mathcal{L}(p)$$

are smooth vector bundles over U (see corollary 1.2.14 and theorem 1.4.10). Hence, $G^{\perp} \subset U \times H_2^*$ is also a vector bundle, so we can find a smooth non-zero section $\varphi(p)$ of $G^{\perp}$ which is compactly supported in U. It corresponds to a tuple $(\widehat{\varphi}_0, \widehat{\varphi}_1, ..., \widehat{\varphi}_l)$ of smooth compactly supported functions where $\widehat{\varphi}_0$ has values in $L_2(G)$ and $\widehat{\varphi}_j$ has values in

$$H^{-2l+l_j+1/2}(\partial G) \quad (j{=}1, ..., l).$$

For each p, this tuple satisfies the equation $\mathcal{L}^*(p)\,\widehat{\varphi}\,(p) = 0$, where $\mathcal{L}^*(p)$ is the adjoint operator to $\mathcal{L}(p)$. Due to the assertion 2^0 of the theorem 5.1 of chapter 2 in [262], we conclude that $\widehat{\varphi}_0(p) \in C^\infty(\overline{G})$ and $\widehat{\varphi}_j(p) \in C^\infty(\partial G)$ for every p and j=1, ..., l. Now a routine application of Green's formula (10) shows that $\widehat{\varphi}_0(p,y)$ satisfies the following boundary value problem in G:

(6.2.13)
$$\begin{cases} L(y,p,D_y)\,\widehat{\varphi}_0(p,u){=}0 \text{ in } G \\ B_j\,(y,\,p,\,D_y)\,\widehat{\varphi}_0\,(p,y)|_{\partial G}{=}0,\ j{=}1,\ ...,\ l \end{cases}$$

(see theorem 5.3, chapter 2 of [262]).

Let us apply the inverse Fourier transform to (13). We get a function $\varphi_0(x,y) \in C^\infty(\overline{Q}) \cap L_2(Q)$ that satisfies the boundary problem (1). This contradicts the assumptions of the theorem, so $\mathcal{L}(z)$ is invertible everywhere except on an analytic subset of values of z.

Due to corollary 1.7.2, the necessary and sufficient solvability conditions of the equation $\mathcal{L}(z)x(z) = y(z)$ in holomorphic functions (provided the right hand side is holomorphic) can be written in the form

$$\langle d(\tfrac{\partial}{\partial z})y(z)|_{z=z_0}, h\rangle = 0 \tag{6.2.14}$$

Here $h \in H_2^* = L_2(G) \times \prod_{j=0}^{l-1} H^{-2l+l_j+1/2}(\partial G)$ and $d(\frac{\partial}{\partial z})$ is a linear differential operator with constant coefficients. We have to find out what kind of functionals correspond to the left hand side of (14) under inverse Fourier transform. First of all, we may assume that we deal with the functional

$$y \longrightarrow \langle \frac{\partial^k y}{\partial z^k}|_{z=z_0}, h\rangle.$$

Instead of y, let us substitute the Fourier image of some tuple $(f_0, f_1, \ldots, f_l)$. Then

$$\frac{\partial^k}{\partial z^k}\int_{\mathbf{R}^n} e^{-izx}\Big(\int_G f_0(x,y)h_0(y)dy + \sum_{j=1}^{l}\langle f_j(x,\cdot), h_j\rangle_{\partial G}\Big)\, dx \;\; |_{z=z_0} =$$

$$= \int_Q (-ix)^k e^{-iz_0x} h_0(y) f_0(x,y)dxdy +$$

$$+ \int_{\mathbf{R}^n} \sum_{j=1}^{l} \langle (-ix)^k e^{-iz_0x} h_j, f_j(x,\cdot)\rangle_{\partial G} dx$$

Hence, the functional in (14) can be represented in the form

$$\varphi(f) = \int_Q \Big(\sum_k e^{iz_0x} x^k h_{0,k}(y)\Big) f_0(x,y)dxdy$$

$$+ \sum_{j=1}^{l}\int_{\mathbf{R}^n} \langle \sum_k e^{iz_0x} x^k h_{j,k}(y),\, f_j(x,y)\rangle_{\partial G} dx,$$

where $h_{0,k} \in L_2(G)$ and $h_{j,k} \in H^{-2l+l_j+1/2}(\partial G)$. Let us set

$$v_j(x,y) = \sum_k e^{iz_0 x} x^k h_{j,k}(y),\ j = 0, \dots, l.$$

These functions have the exponential form that we need. The next step is to investigate the smoothness of these functions provided they correspond to a solvability condition (14). We are also going to show that they correspond to an exponential solution of (1).

In order to do this, we have to discuss in more detail the construction of conditions (14). First of all, we show that if the equation $\mathcal{L}(z)x(z) = y(z)$ with a holomorphic right hand side is locally analytically solvable, then it is also globally solvable. Indeed, consider the sheaf morphism $\mathcal{L}^0 : \mathcal{O}^{H_1}_{\mathbb{C}^n} \to \mathcal{O}^{H_2}_{\mathbb{C}^n}$, generated by the operator function $\mathcal{L}$. Denote by $\mathcal{K}$ the kernel and by $\mathcal{R}$ the image of $\mathcal{L}^0$. Then we have the short exact sequence

$$0 \to \mathcal{K} \to \mathcal{O}^{H_1}_{\mathbb{C}^n} \xrightarrow{\mathcal{L}^0} \mathcal{R} \to 0.$$

The corresponding long exact cohomology sequence is:

$$0 \to \Gamma(\mathbb{C}^n, \mathcal{K}) \to \Gamma(\mathbb{C}^n, \mathcal{O}^{H_1}_{\mathbb{C}^n}) \xrightarrow{\mathcal{L}^0} \Gamma(\mathbb{C}^n, \mathcal{R}) \to H^1(\mathbb{C}^n, \mathcal{K}) \to \dots$$

Due to theorem 1.6.14, $H^1(\mathbb{C}^n, \mathcal{K}) = 0$, so the mapping

$$\Gamma(\mathbb{C}^n, \mathcal{O}^{H_1}_{\mathbb{C}^n}) \xrightarrow{\mathcal{L}^0} \Gamma(\mathbb{C}^n, \mathcal{R})$$

is surjective. Assume now the equation $\mathcal{L}(z)x(z) = y(z)$ is locally solvable for some function $y \in \Gamma(\mathbb{C}^n, \mathcal{O}^{H_2})$. Then $y \in \Gamma(\mathbb{C}^n, \mathcal{R})$. This means there exists some $x(z) \in \Gamma(\mathbb{C}^n, \mathcal{O}^{H_1})$ such that $\mathcal{L}(z)x(z) = y(z)$. In other words, the equation is globally solvable. We conclude that solvability conditions (14) can be constructed locally. Let $z_0 \in \mathbb{C}^n$ and $\mathcal{N} = Ker\ \mathcal{L}(z_0)$. Due to ellipticity, $\mathcal{N}$ consists of smooth functions on $\overline{G}$. Let $q_1, \dots, q_r$ be an orthonormal basis of $\mathcal{N}$. Consider the orthogonal complement $\mathcal{M}'$ to $\mathcal{N}$ in $L_2(G)$. Then $\mathcal{M} = \mathcal{M}' \cap H^{2l}(G)$ is a closed complement to $\mathcal{N}$ in $H^{2l}(G)$. The L_2-orthogonal projector Q onto $\mathcal{N}$ is an integral operator with the smooth kernel

$$\sum_{i=1}^{r} q_i(y)\,\overline{q_i(y')}\,.$$

Then operator $P = I - Q$ is a pseudo-differential projector of $H^{2l}(G)$ onto $\mathcal{M}$. Consider the subspace in $H^{2l}(G)$:

$$\mathcal{M}_1(z) = Im\,(\mathcal{L}(z)\,|_{\mathcal{M}}) = Im\,(\,\mathcal{L}(z)\,P)$$

and the subspace $\mathcal{N}_1(z) = \mathcal{M}_1^{\perp}(z)$ in

$$H_2^* = L_2(G) \times \prod_{j=1}^{n} H^{-2l+l_j+1/2}(\partial G).$$

Due to lemma 1.2.11 and theorem 1.6.13, in a neighborhood U of the point z_0 the subspaces $\mathcal{M}_1(z)$ form an analytic bundle. The subspace $\mathcal{N}_1(z)$ in H_2^* is the kernel of the problem transposed (in A. Dikanskii's sense [418]) to L_1. Therefore, due to ellipticity and results of [418], $\mathcal{N}_1(z) \subset C^\infty(\overline{G}) \times \prod C^\infty(\partial G)$. Let $f_1(z), \ldots, f_r(z)$ be a basis of analytic sections of the bundle $\cup \mathcal{N}_1(z)$ over U. We introduce operator functions $\mathcal{R}(z) \in L(H_2, \mathbf{C}^r)$ and $\mathcal{L}'(z) \in L(\mathcal{N}, \mathbf{C}^r)$ as follows:

$$\mathcal{R}(z)\;y = \{(f_1(z),\,y),\,\ldots\,,(f_r(z),\,y)\},\quad \mathcal{L}'(z)\;y = \mathcal{R}(z)\;\mathcal{L}(z)\;y.$$

We claim that holomorphic solvability in U of the equation $\mathcal{L}'(z)\;x(z) = \mathcal{R}(z)\;y(z)$ is equivalent to solvability of the equation $\mathcal{L}(z)x(z) = y(z)$ in U. Indeed, solvability of the second equation obviously implies solvability of the first one. And conversely, let $x_1(z) \in \mathcal{N}$ be an analytic solution of $\mathcal{L}'(z)\;x(z) = \mathcal{R}(z)y(z)$. We denote by $x_2(z)$ the difference $x(z) - x_1(z)$, so $\mathcal{L}(z)x_2(z) = y(z) - \mathcal{L}(z)x_1(z) := y_2(z)$. The function $y_2(z)$ satisfies the condition $\mathcal{R}(z)y_2(z) = \mathcal{R}(z)y(z) - \mathcal{L}'(z)x_1(z) = 0$. Hence, $y_2(z)$ belongs to $\mathcal{M}_1(z)$ for any $z \in U$. Since $\mathcal{L}(z){:}\,\mathcal{M} \to \mathcal{M}_1(z)$ is an isomorphism, there exists an analytic function $x_2(z) \in \mathcal{M}$ such that $\mathcal{L}(z)x_2(z) = y_2(z)$. Now the function $x(z) = x_1(z) + x_2(z)$ provides the analytic solution of the equation $\mathcal{L}(z)x(z) = y(z)$.

Let $x(z) \in \mathcal{N}$ and $x(z) = \sum x_k(z)\;q_k$, where $x_k(z)$ are scalar analytic functions. The equation $\mathcal{L}'(z)\;x(z) = y(z)\; \in \mathcal{N}_1(z)$ can be rewritten as the finite square system of linear equations:

$$\sum <f_i(z),\ L(z)\ q_j>\ x_j(z) = <f_j(z),\ y(z)>.$$

Due to the assumptions of the theorem, the determinant of the system is not identically equal to zero. Therefore, the condition of analytic solvability of the system is: after multiplication by the associate matrix, the right hand side becomes analytically divisible by the determinant. In other words, all solvability conditions can be written in the form:

$$\sum \alpha_{j,k,l} < \frac{\partial^k f_j}{\partial z^k}, \frac{\partial^l y}{\partial z^l} > |_{z=z'} = 0$$

for some constants $\alpha_{j,k,l}$ and some points z'. Comparing with (14) and taking into account the construction of $f_j(z)$, we conclude that all functions $h_{j,k}(y)$ (and hence, $v_j(x,y)$) are smooth.

Due to condition (14), the tuple $\{v_j\}$ satisfies the equality

$$\int_Q v_0(x,y)\ L_1\varphi(x,y)dxdy + \sum_{j=1}^{l} \int_{\partial Q} v_j(x,y)\ C_j\varphi(x,y)d\sigma = 0 \tag{6.2.15}$$

for all smooth compactly supported functions $\varphi \in C_0^\infty(\overline{Q})$. Choosing $\varphi \in C_0^\infty(Q)$ and using (15) together with the Green's formula (10), we get $Lv_0 = 0$ in Q. Applying these formulas again, we get

$$\sum_j \int_{\partial Q} (S_j v_0 - v_j)\ C_j\varphi d\delta - \sum_j \int_{\partial Q} B_j v_0\ T_j\varphi d\delta = 0 \tag{6.2.16}$$

for all $\varphi \in C_0^\infty(\overline{Q})$. Since $\{C_j, T_j\}$ is a Dirichlet family, the set of tuples $\{C_j\varphi, T_j\varphi\}$ for $\varphi \in C_0^\infty(\overline{Q})$ is complete on the boundary ∂Q of Q (see [262]). Hence, equation (16) implies that $B_j v_0 = 0$ and $S_j v_0 = v_j$ for all $j=1, \ldots, l$.

The result of this discussion is that the solvability conditions (14) mean orthogonality of y to all exponential-polynomial (with respect to x) solutions of (1).

We can now start the proof of the theorem's main statement. Suppose that the statement is false. Then, due to the bipolar theorem [342], there exists an element $\varphi \in S_\alpha^\beta(\mathbf{R}^n, H_2)$ that is orthogonal to all exponential-polynomial (with respect to x) solutions of (1), but is not orthogonal to a solution $u(x,y)$ satisfying (7). Let us apply the Fourier transform. The above considerations show that orthogonality of φ to all exponential-polynomial solutions is equivalent to solvability of the equation $L(z)\psi(z)=\widehat{\varphi}(z)$ in entire vector functions. Let $\psi:\mathbf{C}^n \longrightarrow H_1$ be the solution. We shall show that $\psi \in S^{\alpha+}_{\beta+}(\mathbf{R}^n, H_1)$.

In fact, $L(z) = \sum z^k L_k$ where $L_k \in S_{p/|k|}(p>n, |k|>0)$ and $L(z)$ is invertible outside an analytic subset in $\mathbf{C}^n$. Due to the theorem's conditions on α and β, we can chose $p< min\ (\beta^{-1}, (1-\alpha)^{-1})$. According to lemma 1.2.21, we have the representation $L^{-1}(z)=B(z)\Delta^{-1}(z)$, where $B(z)$ is operator valued and $\Delta(z)$ is scalar, both being entire functions of order p and of minimal type. Hence, $\psi(z) = B(z)\widehat{\varphi}(z)/\Delta(z)$. This representation enables us to show that $\psi(z)$ belongs to $S^{\alpha+}_{\beta+}(\mathbf{R}^m, H_1)$. First of all, $B(z)\widehat{\varphi}(z)$ is an entire H_1-valued function that can be estimated as follows

$$||B(z)\widehat{\varphi}(z)|| \le ||B(z)||\ ||\widehat{\varphi}(z)|| \le C\ exp\ A|z|^p exp(-B|x|^{1/\beta} + C|y|^{1/(1-\alpha)}).$$

This leads to the estimate

$$||B(z)\widehat{\varphi}(z)|| \le C\ exp(-\ A_1|z|^{1/\beta} + B_1|y|^{1/(1-\alpha)}),$$

i.e. $B\widehat{\varphi} \in S^{\alpha+}_{\beta+}(\mathbf{R}^m, H_1)$. We show now that division by $\Delta(z)$ leaves this function in the same space. Let us assume that $\Delta(0) \neq 0$ (otherwise we can achieve this by a translation). We may even assume that $\Delta(0) = 1$. We shall estimate $\Delta(z)$ from below, restricting it to each complex line through the origin, so we shall assume that $z \in \mathbf{C}^1$. The estimates will not depend on the choice of complex line. Let us choose a number γ such that

$$0 < \gamma < min\left(\frac{\beta}{1-\alpha}, \frac{1-\alpha}{\beta}\right) < 1.$$

Due to corollary 1.5.7, there exists a system of discs covering $\mathbf{C}^1$ such that every point z belongs to the interior of a disc $B_{R,z}$ with the radius $R \le C(1 + |z|)^\gamma$ and

$$|\Delta(w)| \ge C\ exp\ (-C|z|^p) \text{ for } w \in \partial B_{R,z}.$$

According to the maximum principle,

$$\|\psi(z)\| \le \max_{w \in \partial B_{R_0, z}} \|B(z)\,\hat{\varphi}(z) / \Delta(z)\| \le \tag{6.2.17}$$
$$\le C \; exp\; c|z|^p \max_{w \in \partial B_{R,z}} exp\;(-A|u|^{1/\beta} + B|v|^{1/1-\alpha})$$

(where $u + iv = w$). Consider now the following subsets:

$$U_1 = \{\, z = x + iy \in \mathbf{C}^n \mid |x| \ge 2\,C\,(\,1 + |z|)^\gamma\},$$

$$U_2 = \mathbf{C}^n \setminus U_1.$$

Let z belong to U_1 first (i.e. it belongs to the intersection of U_1 with our complex line). In this case we can estimate for $w \in \partial B_{R_0, z}$:

$$|u| = |x + (u - x)\,| \ge |x| - |u - x| \ge \tfrac{1}{2}\,|x|$$

(because $|x| \ge 2\,C(1+|z|)^\gamma \ge 2|w - z\,| \ge 2\,|u - x|$). Moreover,

$$|v| = |y + (v - y)| \le |y| + c(|x|^\gamma + |y|^\gamma),$$

so

$$\|\psi(z)\| \le C\; exp\left((-A_1|x|^{1/\beta} + B_1|y|^{1/(1-\alpha)} + c_1(|x|^p + |y|^p + |x|^{\gamma/(1-\alpha)} + |y|^{\gamma/(1-\alpha)})\right)$$

Since $p < min\;(1/\alpha,\ 1/(1-\beta))$ and $\gamma/(1-\alpha) < 1/\beta$, $\gamma<1$, we get

$$\|\psi(z)\| \le C_2\; exp\;(-A_2|x|^{1/\beta} + B_2|y|^{1/(1-\alpha)}). \tag{6.2.18}$$

Now let $z \notin U_1$. In this case $|x| \le 2\,C(1+|z|)^\gamma$ and we get (with some new constants):

$$|u| \le C(1 + |z|^\gamma + |y|^\gamma), \tag{6.2.19}$$

$$C^{-1}(1 + |x| + |y|) \leq |v| \leq |y| + C\,(|x|^{\gamma} + |y|^{\gamma}).$$

This leads to the estimate

$$(1.2.20) \qquad ||\psi(z)|| \leq c\ exp\ (B_3\ |y|^{1/(1-\alpha)} + c(|x|^{\gamma/\beta} + |y|^{\gamma/\beta} + |x|^{p} + |y|^{p}) \leq$$

$$c\ exp\ B_4|y|^{1/(1-\alpha)}.$$

We applied here the inequalities $\gamma/\beta < 1/(1-\alpha)$ and $p < 1/(1-\alpha)$. On the other hand, for $x+ iy \in U_2$,

$$exp\ B|y|^{1/(1-\alpha)} = exp\ (-A|x|^{1/\beta} + B|y|^{1/(1-\alpha)} + A|x|^{1/\beta})$$

$$\leq exp\ (-A|x|^{1/\beta} + B|y|^{1/(1-\alpha)} + C|y|^{\gamma/\beta}).$$

Because $\gamma/\beta < 1/(1-\alpha)$, we have

$$exp\ \ B|y|^{1/(1-\alpha)} \leq c\ exp\ \ (-A|x|^{1/\beta} + B'|y|^{1/(1-\alpha)})$$

Together with (20) this leads again to the required estimate. Hence, $\psi \in S^{\alpha\,+}_{\beta\,+}(\mathbb{R}^n, H_1)$. This means that $\hat{\varphi}$ belongs to the image of the operator L on the space $S^{\alpha\,+}_{\beta\,+}(\mathbb{R}^n, H^1)$. After an inverse Fourier transform, we get from $\psi(z)$ a function $q(x,y) \in S^{\beta}_{\alpha}(\mathbb{R}^n, H_1)$ such that $Lq = \varphi$, i.e.

$$(6.2.21) \qquad L_1 q = \varphi_0 \text{ in } Q \text{ and } C_j q|_{\partial Q} = \varphi_i.$$

Let us now consider the expression

$$\langle \varphi, u\rangle = \int_Q \varphi_0\ (x,y)\ u(x,y)\ dxdy + \sum_j \int_{\partial Q} \varphi_j(x,y) S_j u(x,y) d\sigma$$

and substitute (21) into it. We get

$$\langle\varphi,u\rangle = \int_Q (L,q)\; u\; dxdy + \sum_j \int_{\partial Q} c_j q S_j u d\delta$$

The facts that u is a solution of (1) satisfying (7) and that q belongs to $S^{\beta}_{\alpha}(\mathbf{R}^n, H^{2l}(G))$ allow us to use the Green's formula:

$$\langle\varphi,u\rangle = \int_Q q\; Lu\; dxdy + \sum_j \int_{\partial Q} T_j q\; B_j u\; d\sigma.$$

Indeed, let $\kappa(x) \in C_0^{\infty}(\mathbf{R}^n)$, $\kappa(x) = 1$ for $|x| \leq 1$, $\kappa(x) = 0$ for $|x| \geq 2$ and $0 \leq \kappa(x) \leq 1$. Consider the function $U_{\epsilon}(x,y) = \kappa(\varepsilon x)u(x,y)$.

We claim that

$$\text{(6.2.22)} \quad \int_Q L_1 q\; u_{\varepsilon} dxdy - \int_Q qLu_{\varepsilon} dxdy = \sum_j \int_{\partial Q} B_j u_{\varepsilon} T_j q d\sigma - \sum_j \int_{\partial Q} S_j u_{\varepsilon} C_j q d\sigma.$$

This equality definitely holds for $u_{\varepsilon} \in C_0^{\infty}(\overline{Q})$. In our case u_{ε} belongs only to $H^{2l}_{comp}(Q)$. However, the space $\mathfrak{D}(\overline{Q})$ is dense in $H^{2l}_{comp}(Q)$, and all integrals in (22) define continuous functionals of $u_{\varepsilon} \in H^{2l}_{comp}$, so (22) holds. Now we have to justify the limit when $\varepsilon \to 0$ in (21). We show how to do this for the term $\int_Q qLu_{\varepsilon} dxdy$. For the rest of the terms the proof is the same.

$$\left| \int_Q qLu_{\varepsilon} dxdy - \int_Q qLudxdy \right| = \left| \int_Q qL(u_{\varepsilon} - u) dxdy \right| \leq$$

$$\leq C \sum_{\substack{a \in \mathbf{Z}^n \\ |a| > \varepsilon^{-1} - 1}} \|q\|_{L_2(G \times \{\mathcal{K}+a\})} \|u\|_{H^{2l}(G \times \{\mathcal{K}+a\})},$$

where $\mathcal{K}$ is the standard unit cube in $\mathbf{R}^n$. Taking into account the estimates for u and for $q \in S^{\beta}_{\alpha}(\mathbf{R}^n, H^{2l}(G))$, we get:

$$\left| \int_Q qLu_{\varepsilon} dxdy - \int_Q qLudxdy \right| \leq$$

$$\leq C \sum_{a \in \mathbf{Z}^n, |a| > \varepsilon^{-1} - 1} exp(-A|a|^{1/\alpha})\; exp(B|a|^{1+\gamma}) \underset{\varepsilon \to 0}{\longrightarrow} 0$$

Hence,

$$\langle \varphi, u\rangle = \int_Q Ludxdy + \sum_j \int_{\partial Q} T_j q B_j u d\sigma = 0,$$

which contradicts to the assumption that $\langle \varphi, u\rangle \neq 0$. The theorem is proved.

6.3. Invariant differential equations on Riemannian symmetric spaces of non-compact type

The main concepts of analysis on symmetric spaces of semisimple Lie groups can be found in [161, 162, 166, 350]. We shall briefly recall to the reader some of them.

Let G be a noncompact connected semisimple Lie group with finite center and with Lie algebra $\mathfrak{G}$. Let $G=KAN$ be an Iwasawa decomposition of it, $\mathfrak{G}$, $\mathfrak{K}, \mathfrak{A}, \mathfrak{N}$ be the corresponding Lie algebras, and $\langle \cdot, \cdot \rangle$ be the Killing form on $\mathfrak{G}$. We denote by X the Riemannian symmetric space G/K of noncompact type. The action of G on itself by left translations induces an action of G on X. We denote by M the centralizer of A in K, and we let $B = K/M$. We introduce G-invariant measures db and dk on B and K, respectively.

Let $\mathfrak{A}_c$ and $\mathfrak{A}_c^*$ denote the respective complexifications of $\mathfrak{A}$ and of the real dual space $\mathfrak{A}^*$ of $\mathfrak{A}$. The finite Weyl group W acts on $\mathfrak{A}$ and $\mathfrak{A}_c^*$. Let $\rho \in \mathfrak{A}^*$ denote the half-sum (with multiplicities) of the bounded roots that are positive with respect to some Weyl chamber $\mathfrak{A}_+$.

If $g \in G$, then $H(g) \in \mathfrak{A}$ denotes an element such that

$$g \in K \cdot \exp H(g) \cdot N.$$

If $x \in X, b \in B$, and g and k are representatives of x and b, respectively, then $A(x,b) := -H(g^{-1}k) \in \mathfrak{A}$.

Let $D(X)$ denote the algebra of G-invariant linear differential operators on X. It is known [161, 166] that $D(X)$ is isomorphic to a polynomial algebra with finitely many generators. In particular, if $r = \dim_{\mathbf{R}} \mathfrak{A} = 1$, then the Laplace-Beltrami operator Δ on X is a generator in $D(X)$. Moreover, there exists an isomorphism

$$\Gamma: D(X) \to J,$$

where J is the algebra of all W-invariant polynomials on $\mathfrak{a}_c^*$ [161].

If $\lambda \in \mathfrak{a}_c^*$, then we can define the function

$$\varphi_\lambda(x,b) = \exp((i\lambda+\rho)(A(x,b))).$$

Then for $L \in D(X)$,

$$\Gamma\varphi_\lambda = \Gamma_L(\lambda)\varphi_\lambda$$

where Γ_L is the image of L under the isomorphism Γ. In particular,

$$\Gamma_\Delta = -(\langle\lambda,\lambda\rangle + \langle\rho,\rho\rangle)$$

for the Laplace-Beltrami operator Δ.

The symbol δ stands for an irreducible unitary representation of K on a space V^δ that contains a nonzero subspace V_M^δ of M-invariant vectors. Let $d(\delta)$ (respectively, $l(\delta)$) denote $dim\ V^\delta$ (respectively, $dim\ V_M^\delta$). Assume that $e_1,\dots,e_{d(\delta)}$ is a basis of V^δ such that $e_1,\dots,e_{l(\delta)}$ is a basis for V_M^δ. Let $\widehat{\delta}$ denote the representation contragredient to δ. Following [164], we introduce the generalized spherical functions on X according to the formula

$$\Phi_{\lambda,\delta}(x) = \int_K e^{(i\lambda+\rho)(A(x,kM))}\delta(k)dk.$$

We now describe a certain class of matrices introduced by B. Kostant [420]. Let $\mathfrak{B}$ be the orthogonal complement of $\mathfrak{K}$ in $\mathfrak{G}$ with respect to the Killing form. Denote by H the set of K-harmonic polynomials in the symmetric algebra $S(\mathfrak{B}_c)$, and let H^* be the image of H with respect to the canonical symmetrization mapping of $S(\mathfrak{G}_c)$ onto the enveloping algebra $\mathfrak{U}(\mathfrak{G})$. Let $\{\varepsilon_j\}$ be a basis for the space of linear mappings from V^δ to H^* that commute with the action of K. There is a natural decomposition

$$U(\mathfrak{G}) = U(\mathfrak{G})\,\mathfrak{K} + \mathfrak{N}U(\mathfrak{G}) + U(\mathfrak{A}).$$

We now define a mapping $u \to q^u$ from $U(\mathfrak{G})$ to $U(\mathfrak{A})$ by taking the last term in the preceding decomposition. The square matrix $Q^\delta(\lambda)$ of order $l(\delta)$ is then defined by

$$Q^\delta(\lambda)_{j,l} = [\, q^{\varepsilon_j(\varepsilon_l)}(\rho - i\lambda)\,],\ \ 1 \le j,\ l \le l(\delta).$$

Since the elements of $U(\mathfrak{A})$ are identified with polynomials on $\mathfrak{A}^*_c$, $Q^\delta(\lambda)$ is a polynomial matrix.

If V is a real vector space, then $\mathcal{H}(V^*)$ denotes the space of functions on V^* admitting an extension to an entire function $f(\lambda)$ on V^*_c satisfying the condition that there exists a number $R>0$ (depending on the function) such that for any $r>0$,

$$\sup_\lambda |f(\lambda)|(1 + |\lambda|)^r \exp\,(-R\,|Im\ \lambda|) < \infty. \tag{6.3.1}$$

For a particular R this is a Fréchet space in the seminorms on the left-hand side of (1). If we then take the inductive limit as $R \to \infty$, we get a topology on $\mathcal{H}(V^*)$. The usual Paley-Wiener theorem asserts that the Euclidian Fourier transformation carries $\mathfrak{D}(V)$ onto $\mathcal{H}(V^*)$ isomorphically (both linearly and topologically). In view of this, $\mathcal{H}(V^*)$ will be called the *Paley-Wiener space.*

The following spaces are also needed:

$$\mathcal{H}^\delta_0(\mathfrak{A}^*) = \mathcal{H}(\mathfrak{A}^*) \otimes L(V^\delta, V^\delta_M),$$

$$\mathcal{H}^\delta(\mathfrak{A}^*) = \{F \in \mathcal{H}^\delta_0(\mathfrak{A}^*) \mid \text{the function } \widehat{Q^\delta}(\lambda)^{-1}F(\lambda) \text{ is } W\text{-invariant}\},$$

$$\mathcal{H}^\delta_W(\mathfrak{A}^*) = \{F \in \mathcal{H}^\delta_0(\mathfrak{A}^*) \mid \text{the function } F(\lambda) \text{ is } W\text{-invariant}\}.$$

As established in [164], p. 204, the operator of multiplication by $\widehat{Q^\delta}(\lambda)$ gives rise to an isomorphism:

$$\mathcal{H}_W^\delta(\mathfrak{a}^*) \overset{Q^{\hat\delta}}{\to} \mathcal{H}^\delta(\mathfrak{a}^*).$$

Let $\mathfrak{D}_{\hat\delta}(X)$ be the subspace of the space $\mathfrak{D}(X)$ of test functions on X consisting of the functions that transform according to the representation $\hat\delta$ under translations in the argument by elements of the group K. Let $f(x)$ be a function in $\mathfrak{D}_{\hat\delta}(X)$. Define

$$\tilde f(\lambda) = d(\delta) \int_X f(x)\Phi_{\bar\lambda,\delta}(x)^* dx,$$

where * denotes the of adjoint with respect to the inner product in V^δ.

Theorem 6.3.1. (see [164], theorem 7.1). *The mapping $f \to \tilde f$ is a homeomorphism of $\mathfrak{D}_{\hat\delta}(X)$ onto $\mathcal{H}^\delta(\mathfrak{a}^*)$.*

Let $L \in D(X)$, and let L_1 be the operator that is the transpose of L, i.e.,

$$\int Lf \cdot \varphi dx = \int f \cdot L_1 \varphi dx$$

for all $f, \varphi \in \mathfrak{D}(X)$. Then, since the measure dx is invariant, it is easy to see that $L_1 \in D(X)$.

Lemma 6.3.2. $\Gamma_{L_1}(\lambda) \equiv \Gamma_L(-\lambda)$.

Proof. Let $L_2 \in D(X)$ be an operator such that $\Gamma_{L_2}(\lambda) \equiv \Gamma_L(-\lambda)$. It is required to prove that $L_2 = L_1$. For this it suffices, in turn, to verify that

$$\int \varphi \cdot L\psi dx = \int L_2\varphi \cdot \psi dx$$

for all $\varphi, \psi \in \mathfrak{D}(K\backslash G/K)$. After spherical Fourier transformation, the required equality becomes the equality (whose validity is now obvious)

$$\int_{\mathfrak{a}_+} \tilde\varphi(\lambda)\Gamma_L(-\lambda)\tilde\psi(-\lambda)|c(\lambda)|^{-2} d\lambda = \int_{\mathfrak{a}_+} \Gamma_{L_2}(\lambda)\tilde\varphi(\lambda)\tilde\psi(-\lambda)|c(\lambda)|^{-2} d\lambda$$

(here $c(\lambda)$ is Harish-Chandra's c-function [166]). The lemma is proved.

If L is a matrix operator with elements in $\mathcal{D}(X)$, then $\Gamma_L(\lambda)$ is the matrix polynomial obtained by applying Γ to each matrix element of L. Let L_1 be the transpose operator of L with respect to the duality

$$(\varphi,\psi) = \int \sum \varphi_j \psi_j dx.$$

Then the proof of the equality

$$\Gamma_{L_1}(\lambda) \equiv \Gamma_L(-\lambda)^t$$

is similar to that of lemma 2, where A^t is the transpose matrix of A.

Let L be an $r \times s$ matrix differential operator with matrix elements in $\mathcal{D}(X)$, and let L_1 be its transpose. Denote by N the algebraic subset of $\mathfrak{a}_c^*$ consisting of all the vectors $\lambda \in \mathfrak{a}_c^*$ such that the matrix $\Gamma_L(\lambda)$ has nonzero kernel $Ker\ \Gamma_L(\lambda)$.

Theorem 6.3.3. *There exist:*

1) a finite collection of algebraic submanifolds $N_j \subset N$;

2) for each j *and* δ *a matrix linear differential operator* $p_{j,\delta}(\lambda, \partial/\partial\lambda)$ *with coefficients that are polynomials in* $\lambda \in \mathfrak{a}_c^*$ *and are contained in the space*

$$\mathcal{L}\left(L(V^\delta, V^\delta_M),\ L(C^{t_j}, C^s)\right);$$

such that the following conditions are fulfilled:

a) any K-finite solution $u \in [\mathcal{D}'(X)]^s$ *of the equation* $Lu=0$ *can be represented as a finite sum*

$$u(x) = \sum_{j,\delta} \int_{N_j} p_{j,\delta}\left(\lambda, \frac{\partial}{\partial\lambda}\right) \Phi_{\lambda,\delta}(x) d\mu_{j,\delta}(\lambda), \tag{6.3.2}$$

where the measures $\mu_{j,\delta}$ *with values in* C^{t_j} *are such that for any* $a>0$ *there exists an* $R>0$ *with*

$$\int_{N_j} exp\ (a|Im\ \lambda|)\ (1+|\lambda|)^{-R} d|\mu_{j,\delta}(\lambda)| < \infty, \tag{6.3.3}$$

and the integrals in (2) *converge absolutely upon application to any test function* $\varphi \in [\mathfrak{D}(X)]^s$;

b) for any measures $\mu_{j,\delta}$ *satisfying* (3) *the finite sum* (2) *gives a K-finite solution* $u \in [\mathfrak{D}'(X)]^s$ *of the equation* $Lu=0$.

Remarks. 1. The operators $p_{j,\delta}$ have coefficients in the space $L(L(V^\delta, V_M^\delta), L(C^{t_j}, C^s))$. Therefore, after pairing with the measure μ in (2) we get a C^s-valued function.

2. We establish a somewhat weaker assertion, with a) replaced by

a′) for any K-finite solution $u(x)$ *and any* $a>0$ *there exists a collection of measures* $\mu_{j,\delta}$ *such that*

$$(u,\varphi) = \sum_{j,\delta} \int_{N_j} p_{j,\delta}(\Phi_{\lambda,\delta},\varphi) d\mu_{j,\delta}$$

for any function $\varphi \in [\mathfrak{D}(X)]^s$ *supported in the ball of geodesic radius* a *about zero in* X *and*

$$\int_N (1+|\lambda|)^{-R} \, exp \; a|Im\,\lambda| d|\mu(\lambda)| < \infty \tag{6.3.4}$$

for some $R > 0$.

The proof of the theorem in full scope, i.e., the verification that the measures can be chosen independently of the number a in (4), repeats arguments in [325].

Proof. Let $u \in [\mathfrak{D}'(X)]^s$ be a solution of the equation $Lu=0$. This means that $(Lu,\varphi) = 0$ for any function $\varphi \in [\mathfrak{D}(X)]^r$. If, as before, L_1 is the transposed operator of L, then this means that $(u, L_1\varphi) = 0$ for any such function φ. If L_1 is regarded as an operator

$$[\mathfrak{D}(X)]^r \xrightarrow{L_1} [\mathfrak{D}(X)]^s,$$

then u is a functional on $[\mathfrak{D}(X)]^s$ annihilating the range of L_1. It is not hard to see that the converse also holds: any such functional gives rise to a solution of the equation $Lu=0$.

We are interested in K-finite solutions. Each of them is a sum of solutions that transform according to irreducible representations of K. Let δ be an irreducible unitary representation of K and let $\mathfrak{D}_{\hat{\delta}}(X)$ be the space introduced above. Any generalized function $v \in \mathfrak{D}'(X)$ can be restricted to $\mathfrak{D}_{\hat{\delta}}(X)$, and thereby determines an element of $\mathfrak{D}_{\hat{\delta}}(X)^*$. Let $\mathfrak{D}'_{\delta}(X)$ be the subspace of $\mathfrak{D}'(X)$ consisting of all the generalized functions that transform according to the representation δ.

Lemma 6.3.4. *The mapping of restriction to $\mathfrak{D}_{\hat{\delta}}(X)$ of functionals on $\mathfrak{D}(X)$ gives rise to an isomorphism of the spaces $\mathfrak{D}'_{\delta}(X)$ and $\mathfrak{D}_{\hat{\delta}}(X)^*$.*

Proof. Let $d(\delta)$ be the dimension and $\chi_\delta(k)$ the character of the representation δ. For any function $\varphi \in \mathfrak{D}(X)$ the function

$$\varphi_\delta(X) = \frac{1}{d(\delta)}(\chi_\delta * \varphi)(x) = \frac{1}{d(\delta)}\int_K \chi_\delta(k)\varphi(k^{-1}x)dk$$

lies in $\mathfrak{D}_{\hat{\delta}}(X)$. It is elementary to check that $(f,\varphi) = (f,\varphi_\delta)$ for $f \in \mathfrak{D}'_{\delta}(X)$. In view of this, $f=0$ if $f|_{\mathfrak{D}_{\hat{\delta}}} = 0$, i.e. the mapping of restriction to $\mathfrak{D}_{\hat{\delta}}(X)$ is injective. If now $f \in \mathfrak{D}_{\hat{\delta}}(X)^*$, then we introduce $f^\delta \in \mathfrak{D}'(X)$ according to the formula $(f^\delta,\varphi) = (f,\varphi_\delta)$ for any function $\varphi \in \mathfrak{D}(X)$. It is easy to see that f^δ really does lie in the required space, and $f^\delta|_{\mathfrak{D}_{\hat{\delta}}} = f$. Thus surjectivity is also proved. The restriction mapping is continuous, and is a topological isomorphism by the open mapping theorem 1.1.7. The lemma is proved.

As mentioned above, after decomposition into irreducible representations of K, it suffices to consider the solutions $u(x)$ in the spaces $[\mathfrak{D}'_{\delta}]^s$. By lemma 4, such a solution is an element of $[\mathfrak{D}_{\hat{\delta}}(X)^*]^s$ that annihilates the range of the operator

$$L_1\colon [\mathfrak{D}_{\hat{\delta}}]^r \to [\mathfrak{D}_{\hat{\delta}}]^s.$$

(The operator acts in the indicated spaces in view of its G-invariance.)

By theorem 1, after a Fourier transformation we get the operator of multiplication by the matrix-valued function $I \otimes \Gamma_{L_1}(\lambda)$, acting from the space $\mathcal{H}^\delta(\mathfrak{A}^*) \otimes C^r$ to $\mathcal{H}^\delta(\mathfrak{A}^*) \otimes C^s$. Here I is the identity operator on $L(V^\delta, V^s_M)$.

Thus, it is required to describe the cokernel of the operator

$$\mathcal{H}^{\delta}(\mathfrak{A}^*) \otimes C^r \xrightarrow{I \otimes \Gamma_{L_1}} \mathcal{H}^{\delta}(\mathfrak{A}^*) \otimes C^s.$$

Let $\{M_p, d_p(\lambda, \partial/\partial\lambda)\}$ be a collection of algebraic submanifolds of $\mathfrak{A}_c^*$ and normal Noether operators on $\mathfrak{A}_c^*$ corresponding to the polynomial matrix $\Gamma_{L_1}(\lambda)$ (see [324] for the corresponding definitions; in the analytic case analogs of these notions were introduced in the end of section 1.5, starting with definition 1.5.11). Then $\{M_p, I \otimes d_p\}$ is a collection of submanifolds and normal Noether operators for the matrix $I \otimes \Gamma_{L_1}$.

Assume that d_p is a $t_p \times s$ matrix operator. We introduce a function space that is needed in what follows and consists of collections $\{f_p(\lambda)\}$, where the following conditions hold:

1) f_p is an analytic function on M_p with values in $L(V^{\delta}, V^{\delta}_M) \otimes \mathbf{C}^{t_p}$;

2) for any point $\lambda_0 \in \mathfrak{A}_c^*$ there is a function $F(\lambda)$ holomorphic in a neighborhood of λ_0 and with values in $L(V^{\delta}, V^{\delta}_M) \otimes \mathbf{C}^s$ such that for all p

$$\{I \otimes d_p(\lambda, \partial/\partial\lambda)\}\ F(\lambda)|_{M_p} \equiv f_p(\lambda)$$

(for λ close to λ_0);

3) there exists an $a>0$ such that for all $R>0$

$$u_{R,a}(\{f_p\}) = \sup_{p, \lambda \in M_p} |f_p(\lambda)|\ (|\lambda| + 1)^R\ exp\,(-a|\mathrm{Im}\,\lambda|) < \infty.$$

Fixing $a>0$ and introducing a topology by means of the seminorms $u_{R,a}$ with $R>0$, we get a Fréchet space denoted by $\mathcal{H}^{(a)}$. We now define our space:

$$\mathcal{H}_{\delta}(L_1) \overset{def}{=} \varinjlim_{a \to \infty} \mathcal{H}^{(a)}.$$

This space is analogous to the space of p-functions in [324].

We introduce an operator d acting on a function F in $\mathcal{H}_0^{\delta}(\mathfrak{A}^*) \otimes \mathbf{C}^s$ as follows:

$$dF \overset{def}{=} \{(I \otimes d_p)F|_{M_p}\}.$$

This operator acts continuously from $\mathcal{H}_0^{\delta}(\mathfrak{U}^*)\otimes\mathbf{C}^s$ to $\mathcal{H}_\delta(L_1)$.

Lemma 6.3.5. *The sequence*

$$(6.3.5)\qquad \mathcal{H}_0^{\delta}(\mathfrak{U}^*)\otimes\mathbf{C}^r \xrightarrow{I\otimes\Gamma_{L_1}} \mathcal{H}_0^{\delta}(\mathfrak{U}^*)\otimes\mathbf{C}^s \xrightarrow{d} \mathcal{H}_\delta(L_1)\to 0$$

is exact.

The lemma is a particular case of theorem 3 in [324], chapter IV, §5.

Since the function $\Gamma_{L_1}(\lambda)$ is W-invariant, the operator of multiplication by $I\otimes\Gamma_{L_1}$ acts from $\mathcal{H}_W^{\delta}(\mathfrak{U}^*)\otimes\mathbf{C}^r$ to $\mathcal{H}_W^{\delta}(\mathfrak{U}^*)\otimes\mathbf{C}^s$.

We mention that the W-orbit of any point $\lambda\in\mathfrak{U}_c^*$ contains a point at which the matrix $\widehat{Q^\delta}$ is nonsingular (see [164]). Let us now define manifolds N_p and operators ∂_p as follows:

1) in the W-orbit or each manifold M_p choose some M_{p_1} on which $det\, Q^\delta|_{M_{p_1}}$ is not identically equal to zero and let $N_{p_1}=M_{p_1}$ and $\partial_{p_1}=d_{p_1}$;

2) for each of the remaining M_p choose the M_{p_1} in the preceding part lying in the W-orbit of M_p, and let $N_p=M_{p_1}$, while ∂_p is obtained from d_p by an appropriate change of variables, i.e., if k is the change of variables operator in W carrying M_p into M_{p_1}, then $\partial_p f = kd_pk^{-1}f$.

From the collection $\{N_p,\partial_p\}$ we construct the space $\mathcal{H}_\delta^0(L_1)$ and the operator ∂ by the same procedure we used to construct the space $\mathcal{H}_\delta(L_1)$ and the operator d from the collection $\{M_p,d_p\}$.

Lemma 6.3.6. *The sequence*

$$(6.3.6)\qquad \mathcal{H}_W^{\delta}(\mathfrak{U}^*)\otimes\mathbf{C}^r \xrightarrow{I\otimes\Gamma_{L_1}} \mathcal{H}_W^{\delta}(\mathfrak{U}^*)\otimes\mathbf{C}^s \xrightarrow{\partial} \mathcal{H}_\delta^0(L_1)$$

is exact. The operator ∂ in it is a topological homomorphism.

Proof. For a function $f\in\mathcal{H}_W^{\delta}(\mathfrak{U}^*)\otimes\mathbf{C}^s$ the equalities $dF=0$ and $\partial F=0$ are obviously equivalent because of W-invariance. Therefore, if $\partial F=0$, then $F=(I\otimes\Gamma_{L_1})G$, where $G\in\mathcal{H}_\delta^0(\mathfrak{U}^*)\otimes\mathbf{C}^r$, because (5) is exact. However, G is not necessarily W-invariant. Let

$$\widetilde{G}(\lambda) = \frac{1}{|W|} \sum_{k \in W} G(k\lambda).$$

Then $F=(I \otimes \Gamma_{L_1})\widetilde{G}$, by the W-invariance of Γ_{L_1}. This proves that (6) is exact, since it is obvious that (6) is a chain complex.

We now establish that ∂ is a homomorphism. To do this we introduce the mapping σ that carries each component f_p in a collection $\{f_p\} \in \operatorname{Im} \partial$ by W-translations from the set N_p onto all sets M_l such that $N_p=kM_l$ with some $k \in W$. This mapping does not carry $\overset{0}{\mathcal{H}}_\delta(L_1)$ into $\mathcal{H}_\delta(L_1)$, but acts continuously from $Im\ \partial \subset \overset{0}{\mathcal{H}}_\delta(L_1)$ into $\mathcal{H}_\delta(L_1)$.

By lemma 5 and the open mapping theorem 1.1.7, the multivalued mapping $\nu=d^{-1}$ is lower semicontinuous. Let γ be the averaging map with respect to W, which acts from $\overset{\delta}{\mathcal{H}}_0(\mathfrak{A}^*) \otimes \mathbf{C}^s$ onto $\overset{\delta}{\mathcal{H}}_W(\mathfrak{A}^*) \otimes \mathbf{C}^s$.

Consider the lower semicontinuous mapping

$$\gamma\nu\sigma : Im\ \partial \to \overset{\delta}{\mathcal{H}}_W(\mathfrak{A}^*) \otimes \mathbf{C}^s.$$

It is asserted that this mapping is a right inverse for ∂ on $Im\ \partial$. Indeed, let $g=\partial F \in Im\ \partial$, and consider the functions $k=\nu\sigma g$ and $k_\kappa(\lambda) = k(\kappa\lambda)$ $(\kappa \in W)$. Then $dk_\kappa = dk = \sigma g$. Indeed, $dF=\sigma g=dk$, i.e., $k=F+h$, where $h=(I \otimes \Gamma_{L_1})\ \varphi$. But then $k_\kappa=F_\kappa + h_\kappa = F + (I \otimes \Gamma_{L_1})\varphi$, i.e., $dk_\kappa = dF = dk$. This implies that $d\gamma\nu\sigma g = dk = \sigma g$. This means, in turn, that $\partial\gamma\nu\sigma g = g$. Since $\gamma\nu\sigma$ is lower semicontinuous, we get that ∂ is a topological homomorphism. The lemma is proved.

Let $Q_\delta^+(\lambda)$ be the adjoint of $\widehat{Q^\delta}(\lambda)$, i.e., the transpose of the matrix of cofactors, and let $q_\delta(\lambda) = det\ \widehat{Q^\delta}(\lambda)$. Since $(\widehat{Q^\delta})^{-1} = Q_\delta^+ / q_\delta$, it is obvious that for a sufficiently large positive integer α the differential operator

$$\widetilde{\partial} = q_\delta^\alpha \partial \circ \left((\widehat{Q^\delta})^{-1} \otimes i_s\right)$$

with respect to λ has polynomial coefficients (here i_s is the identity operator on $\mathbf{C}^s$, and the sign $\circ$ denotes composition of operators).

We have the commutative diagram

$$\begin{array}{ccc} \mathcal{H}_W^{\delta}(\mathfrak{A}^*)\otimes \mathbf{C}^r & \xrightarrow{I\otimes\Gamma_{L_1}} & \mathcal{H}_W^{\delta}(\mathfrak{A}^*)\otimes \mathbf{C}^s \\ \downarrow\uparrow & & \uparrow\downarrow \\ \mathcal{H}^{\delta}(\mathfrak{A}^*)\otimes \mathbf{C}^r & \xrightarrow{I\otimes\Gamma_{L_1}} & \mathcal{H}^{\delta}(\mathfrak{A}^*)\otimes \mathbf{C}^s \end{array} \tag{6.3.7}$$

where the arrows pointing downward are given by multiplication by the matrix-valued function $Q^{\widehat{\delta}}\otimes I$ (here I is the identity operator in $\mathbf{C}^r$ and $\mathbf{C}^s$), and the arrows pointing upward are given by multiplication by its inverse.

We introduce the space $\mathcal{H}$ of collections $\{f_p\}$, where f_p is a holomorphic function on N_p with values in $L(V^{\delta},V_M^{\delta})\otimes\mathbf{C}^{t_p}$ and estimates analogous to condition 3) in the definition of the space $\mathcal{H}^{\delta}(L_1)$, but without a compatibility condition of the type of 2) in that definition. The topology on $\mathcal{H}$ is introduced in the same way as on $\mathcal{H}_{\delta}(L_1)$. The space $\overset{0}{\mathcal{H}}_{\delta}(L_1)$ introduced earlier is a subspace of this space $\mathcal{H}$.

Lemma 6.3.7. *All the mappings in* (7) *are continuous. The sequence*

$$\mathcal{H}^{\delta}(\mathfrak{A}^*)\otimes\mathbf{C}^r \xrightarrow{I\otimes\Gamma_{L_1}} \mathcal{H}^{\delta}(\mathfrak{A}^*)\otimes\mathbf{C}^s \xrightarrow{\tilde{d}} \mathcal{H} \tag{6.3.8}$$

is exact, and the operator $\tilde{d}$ in it is a topological homomorphism.

Proof. Only the continuity of the arrows pointing upward in (7), i.e., of the operator

$$\mathcal{H}^{\delta}(\mathfrak{A}^*)\otimes\mathbf{C}^r \xrightarrow{(Q^{\widehat{\delta}})^{-1}\otimes i_s} \mathcal{H}_W^{\delta}(\mathfrak{A}^*)\otimes\mathbf{C}^s$$

is not obvious. Recall that $(Q^{\widehat{\delta}})^{-1}=Q_{\delta}^{+}/q_{\delta}$. Since Q_{δ}^{+} is a polynomial matrix, multiplication by it is continuous in $\mathcal{H}_0^{\delta}(\mathfrak{A}^*)$. It remains to show that the mapping of division by the polynomial $q_{\delta}\neq 0$ in $\mathcal{H}(\mathfrak{A}^*)$ is continuous on its domain. But this fact is well known and follows immediately from lower estimates of polynomials (see, for example, [169], proposition 7.6.5).

What has already been proved and the preceding lemma give us that the sequence

$$\mathcal{H}^{\delta}(\mathfrak{A}^*)\otimes \mathbf{C}^r \overset{I\otimes \Gamma_{L_1}}{\rightarrow} \mathcal{H}^{\delta}(\mathfrak{A}^*)\otimes \mathbf{C}^s \overset{\partial\circ((Q^{\hat{\delta}})^{-1}\otimes i_s)}{\longrightarrow} \mathcal{H}^0_{\delta}(L_1)$$

is exact, and the right-hand mapping in it is a topological homomorphism. To prove the lemma it is now sufficient to establish that the operator of division by q_δ in $\mathcal{H}$ is continuous on its domain. In turn, it suffices to solve this problem under the following simplifying assumptions: the polynomial q_δ is irreducible, while the space $\mathcal{H}$ consists of scalar functions $f(z)$ that are analytic on an irreducible algebraic manifold M and satisfy for some a and any $R>0$ the estimate

$$|f(z)| \leq C_R(1+|z|)^{-R}\ exp\ a|Im\ z|.$$

As before, the space $\mathcal{H}$ is endowed with the natural topology of an inductive limit of Fréchet spaces.

If M coincides with $\mathfrak{A}^*_c$, then, as observed earlier, the continuity of the operator of division by q_δ on its domain is established, for example, in [169]. The case $M=\mathfrak{A}^*_c$ can now be reduced to the case $M=\mathfrak{A}^*_c$. Indeed, let S be the set of zeros of the polynomial q_δ in $\mathfrak{A}^*_c$, and consider the algebraic manifold $M\cup S=T$. Let $h\in q_\delta\mathcal{H}$, i.e. $h=q_\delta f$, where $f\in\mathcal{H}$. We introduce the function h_1 on T equal to h on M and zero on S. It is analytic on T. Indeed, let F be a holomorphic extension of the function $q_\delta^{-1}h=f\in\mathcal{H}$ to $\mathfrak{A}^*_c$. Then the function $F_1 = q_\delta F$ is entire in $\mathbf{U}^*_c$ and gives h_1 as the restriction to T.

We introduce the space $\mathcal{H}_1$ in a way analogous to that for $\mathcal{H}$, with M replaced by T. Then the mapping $h \rightarrow h_1$ constructed above is a continuous linear operator γ: $q_\delta\mathcal{H} \rightarrow \mathcal{H}_1$. By theorem 1 in [324], chapter IV, §5, the restriction mapping $\xi:\mathcal{H}(\mathfrak{A}^*_c) \rightarrow \mathcal{H}_1$ is surjective. By the open mapping theorem, the multivalued mapping $\xi^{-1}:\mathcal{H}_1 \rightarrow \mathcal{H}(\mathfrak{A}^*_c)$ is lower semicontinuous. We consider the lower semicontinuous mapping

$$\xi^{-1}\gamma\colon\ q_\delta\mathcal{H} \rightarrow \mathcal{H}(\mathfrak{A}^*_c).$$

Obviously, the range of $\xi^{-1}\gamma$ lies in $q_\delta\mathcal{H}(\mathfrak{A}^*_c)$. Indeed, by construction, $\xi^{-1}\gamma h|_s=0$ for any $h\in q_\delta\mathcal{H}$. This implies that $q_\delta^{-1}\xi^{-1}\gamma h$ is holomorphic (and thus belongs to $\mathcal{H}(\mathfrak{A}^*_c)$). The operator of division by q_δ is continuous on $q_\delta\mathcal{H}(\mathfrak{A}^*_c)$. Thus, the operator $q_\delta^{-1}\xi^{-1}\gamma$: $q_\delta\mathcal{H} \rightarrow \mathcal{H}(\mathfrak{A}^*_c)$ is lower semicontinuous. If now τ_M is the operator of restriction to

M, then we get continuity of the operator

$$q_\delta^{-1} = \tau_M \, q_\delta^{-1} \, \xi^{-1} \gamma : q_\delta \mathcal{H} \to \mathcal{H}.$$

The lemma is proved.

Let us return to the proof of the theorem. As mentioned, our solution $u(x)$ gives rise to an element

$$\widetilde{u} \in (Im(I \otimes \Gamma_{L_1}))^{\perp} \subset (\mathcal{H}^\delta(\mathfrak{A}^*) \otimes \mathbf{C}^s)^*$$

after Fourier transformation. In view of lemma 7, $\widetilde{u}$ then has the form $\langle \widetilde{u}, y \rangle = \langle \varphi, \widetilde{d}\, y \rangle$, where $\varphi \in \mathcal{H}^*$. Therefore,

$$|\langle \varphi, \{f_p\} \rangle| \leq C \sup_{p, \lambda \in N_p} |f_p(\lambda)|(1+|\lambda|)^R \, exp\,(-a|\, Im\, \lambda|) \tag{6.3.9}$$

for some $R>0$ on those $\{f_p\}$ such that the expression on the right-hand side of (9) is finite for all $R>0$. By (9) and the Hahn-Banach and Riesz theorems, the functional φ has the form

$$|\langle \varphi, \{f_p\} \rangle| = \sum_{i,j,p} \int_{N_p} \langle f_p(\lambda)\, e_i, e_j \rangle d\mu_{i,j,p}\,(\lambda)$$

on elements $\{f_p\}$ of the class indicated above, where $\mu_{i,j,p}$ are measures on N_p with the estimates required in the theorem. Thus,

$$\langle \widetilde{u}, y \rangle = \sum_{i,j,p} \int_{N_p} \langle q_\delta^\alpha \partial_p \big((Q^{\widehat{\delta}})^{-1} \otimes i_t\big)\, y e_i, e_j \rangle d\mu_{i,j,p}.$$

If we now substitute the Fourier transform of a function $\varphi \in \mathfrak{D}_{\widehat{\delta}}(X)$ in place of y, we get the representation required in the theorem.

Assertion b) of the theorem is obvious, since the integrands are K-finite solutions of the equation by construction, and the integral converges in the sense indicated because of the estimates on the measures. This proves the theorem.

We shall consider now equations invariant under the Cartan group of motions. Let X_0 be the tangent space of X at "zero" (i.e., in the coset $\{eK\}$, where e is the identity of G). The maximal compact subgroup K acts on X_0 as the stationary subgroup of "zero". Let G_0 denote the group of affine transformations of X_0 (the Cartan group of motions) generated by the transformations in K and the translations of X_0. Let $(\cdot,\cdot)$ be the restriction to X_0 of the Killing form on $\mathfrak{G}$, and let $D(X_0)$ be the algebra of G_0-invariant linear differential operators on X_0.

Following [165], we introduce the generalized Bessel functions on X_0:

$$\psi_{i,j,\delta}(\lambda,x) = \int_K exp\ i(k\lambda,x)\cdot\langle\delta(k^{-1})e_i,e_j\rangle dk,$$

where $\lambda\in\mathfrak{A}_c$ and $x\in X_0$. If $X_0 = \mathbf{R}^2$ and K is the rotation group, these functions are easily shown to be expressible in terms of the usual Bessel functions.

In [165] S.Helgason developed a technique for Fourier transformation on X_0 (in particular, he obtained an analog of the Paley-Wiener theorem) leading to results of the same kind as above. For simplicity we consider the case of a single equation, and not a system. Moreover, we have introduced the generalized Bessel functions not in matrix form (as generalized spherical functions) but in scalar form in order to demonstrate easily a different (coordinate) equivalent way of writing an expansion of a solution.

Theorem 6.3.8. *Let $L\in D(X_0)$. Then there exists a finite collection of algebraic submanifolds $N_\gamma\subset\mathfrak{A}_c$ and linear differential operators $d_{i,j,l,\gamma,\delta}(\lambda,\partial/\partial\lambda)$ with polynomial coefficients on $\mathfrak{A}_c$ such that the following conditions are fulfilled:*

1) any K-finite solution $u\in\mathfrak{D}'(X_0)$ of the equation $Lu=0$ has the form of a finite sum

$$u(x) = \sum_{i,j,l,\gamma,\delta}\int_{N_\gamma} d_{i,j,\gamma,l,\delta}\Big(\lambda,\frac{\partial}{\partial\lambda}\Big)\,\psi_{i,j,\delta}(\lambda,x)\ d\mu_{l,\gamma,\delta}(\lambda), \tag{6.3.10}$$

where the measures satisfy estimates (3) *and convergence of the integral is understood in the same sense as in theorem* 3;

2) for any choice of finitely many measures with the indicated properties the right-hand side of (10) *determines a K-finite solution $u\in\mathfrak{D}'(X_0)$ of the equation $L\ u=0$.*

We do not give the proof. It is similar to that of theorem 3. The reader can find some details in [232].

6.4. Comments and references

There have been many works devoted to the Floquet theory of ordinary differential equations with deviating arguments (see, for instance, [133, 155, 205, 241, 270, 306, 416]). The situation is still not clear enough. Section 6.1 contains the author's results [220, 229] for ordinary and partial differential equations.

There an obvious similarity between constant coefficient and periodic partial differential equations: in both cases we have invariance with respect to a group action. In the first case we are dealing with the natural (transitive) action of the commutative Lie group $\mathbf{R}^n$ on itself. After we apply the Fourier transform in an appropriate space of test functions, we get the operator of multiplication by a polynomial in a space of analytic functions. At this point powerful methods of complex analysis can be applied [42, 99, 273, 274, 324]. In the periodic case the equation is invariant with respect to the natural (nontransitive) action of the commutative discrete group $\mathbf{Z}^n$ on $\mathbf{R}^n$. After an appropriate Fourier transform (F, $\hat{F}$, or $\mathfrak{F}$), we get an operator of multiplication by an operator valued function (see section 2.2). The values of the function are elliptic differential operators on the (compact) orbit space $\mathbf{R}^n/\mathbf{Z}^n$, so they are Fredholm. According to Paley-Wiener type theorems, Fourier images of appropriate spaces of test functions consist of analytic vector functions (see sections 2.1, 2.2). Now we can apply the theory of analytic Fredholm morphisms developed in chapter 1. Then the natural question arises: what conditions on the group, on the action, and on the differential operator are essential for our approach? There are three natural restrictions that seem necessary. The first one is compactness of the orbit space. The second one is transversal ellipticity of the equation. These two conditions together should lead to the Fredholm property of the arising operator function. In order to use the machinery of analytic Fredholm operator functions, we need some kind of Fourier transform on the group which is developed to the level of Paley-Wiener type theorems (including extension into the "complex" domain of non-unitary representations). Under these three assumptions we can hope to develop some kind of generalized exponential expansion; it is reasonable to call it a "Floquet expansion" when the group is discrete. Sections 6.2 and 6.3 provide some simple implementations of this ideology.

In section 6.2 the group is again $\mathbf{R}^n$, but it acts non-transitively on a cylindrical domain of a higher dimension as space. The problem is supposed to be elliptic only in the directions that are transversal to the action (Atiyah's "transversal ellipticity" [20]). The orbit space is a compact manifold (with boundary). In this case we can prove some results about completeness of solutions that are of exponential-polynomial type in the orbit directions (theorem 6.2.3). The results of that section were proved by the author (see [229, 231] for further references). See also some related results in [15, 353].

Theorems 6.3.3 and 6.3.8 of section 6.3 also contain the author's results [226, 229, 231, 232]. Riemannian symmetric spaces have proven to be very good objects for translations from $\mathbf{R}^n$ of results of harmonic analysis and partial differential equation theory [44, 45, 49, 97, 98, 162-166, 191, 226, 228, 231, 232, 251, 327]. Here we consider the natural action of a semi-simple group G on the Riemannian symmetric space of noncompact type $X = G/K$, and a differential equation (or system) that is invariant with respect to the action. In this case we are able to obtain some analog of the exponential expansion. Generalized spherical (or Bessel) functions play here the role of exponents. These results are obtained by a rather simple combination of methods and results developed in [324] and in [164-166] (see also [162, 166, 350] for some surveys of the subject). The history of this problem goes back to F. Karpelevich's work [171], where he described all positive joint eigenfunctions of invariant differential operators on X. S. Helgason described all joint eigenfunctions in the case of spaces of rank one and K-finite joint eigenfunctions (i.e., the ones that have finite Fourier series with respect to K) for arbitrary rank. This task required Paley-Wiener type theorems that were also proved by S. Helgason. The reader can find these results and further references in [162-166, 350]. The corresponding result for spaces of arbitrary rank and without the K-finiteness condition was proved in [419]. This required the technique of hyperfunctions. The author's results in section 6.3 still contain the K-finiteness condition. It would be interesting to get rid of this assumption. Helgason's representation of joint eigenfunctions can be obtained by choosing the $r \times 1$-system that consists of generators of the algebra of invariant differential operators (here r is the rank of the space). As in the case of constant coefficient equations in $\mathbf{R}^n$ (see [42, 45, 48]), the results of section 6.3 can be generalized to some extent to convolution equations [228, 231]; cf. also [44, 45, 49] for some related results. An interesting problem is to develop some kind of Floquet theory for elliptic differential equations on $X = G/K$ that are invariant with respect to a co-compact lattice Γ in G. Such a theory could be important for the spectral theory of automorphic functions [251, 327, 401].

There is another important situation when an analog of Floquet theory is needed for some non-commutative group of periods. In contrast with section 6.3, where we dealt with semi-simple groups, the group is in this case nilpotent. Consider the stationary Schrödinger operator (an analogous consideration works for the Pauli operator) for an electron in periodic (two-dimensional) electric and magnetic fields. The simplest interesting case is that of a constant magnetic field orthogonal to the plane. Assuming that the group of periods is $\mathbb{Z}^2$, we can represent the corresponding Schrödinger operator in the form

$$H = \left(\frac{1}{i}\frac{\partial}{\partial x} - \pi\alpha y\right)^2 + \left(\frac{1}{i}\frac{\partial}{\partial y} + \pi\alpha x\right)^2 + V(x,y) = H_0 + V, \tag{6.4.1}$$

where α is a constant and the electric potential $V(x,y)$ is 1-periodic in every variable. The physical meaning of $2\pi\alpha$ is the flux of the magnetic field through the unit cell of the period's lattice. The interesting circumstance is that in spite of periodicity of the fields, the operator H itself does not commute with $\mathbb{Z}^2$-translations. The reason is that the magnetic field is the *curl* of the magnetic vector potential, so the operator contains linear terms. E. Brown and J. Zak have made the interesting observation [412, 417] that the operator is invariant with respect to an action of a non-commutative discrete group. Consider the operators:

$$\begin{cases} (T_1\phi)(x,y) = \phi(x+1,y)\, e^{i\pi\alpha y} \\ (T_2\phi)(x,y) = \phi(x,y+1)\, e^{-i\pi\alpha x} \end{cases} \tag{6.4.2}$$

It is easy to check that they commute with H:

$$T_j H = H T_j,\ j = 1,2. \tag{6.4.3}$$

On the other hand, they do not commute with each other:

$$T_2 T_1 = e^{2\pi i\alpha}\, T_1 T_2. \tag{6.4.4}$$

We denote by Λ the scalar operator $e^{-2\pi i\alpha}I$. We remind the reader that $\mathbf{R}^3$ can be equipped with the structure of a nilpotent group (the *Heisenberg group*), if we identify this space with the space $\mathbf{H}$ of upper-triangular matrices

$$\begin{bmatrix} 1 & x & z \\ 0 & 1 & y \\ 0 & 0 & 1 \end{bmatrix}$$

with the standard matrix multiplication. The set

$$\Gamma = \left\{ \begin{bmatrix} 1 & l & n \\ 0 & 1 & m \\ 0 & 0 & 1 \end{bmatrix} \;\middle|\; l,\ m,\ n \in \mathbf{Z} \right\}$$

is a co-compact lattice in $\mathbf{H}$. We shall denote the corresponding matrices by $(l,\ m,\ n)$. Consider the representation ρ of the lattice such that $\rho(1,\ 0,\ 0) = T_1$, $\rho(0,\ 1,\ 0) = T_2$, $\rho(0,\ 0,\ 1) = \Lambda$. Equality (4) guarantees that such a representation exists. Equality (3) shows that H is intertwining with ρ. Thus we have some kind of periodicity of H, but the group of periods is not commutative. The spectral theory of the operator H is of significant interest for physics. Much work has been devoted during the last several years to this problem [25, 26, 28, 40, 65, 67, 79, 153, 157-160, 269, 310, 314, 367, 380]. For rational values of α the spectrum still has band structure, but for irrational ones it can be a Cantor set. When α is integer, equality (4) shows that we deal with a representation of the commutative group $\mathbf{Z}^2$, so an analog of Floquet theory can be easily developed. When we deal with a rational number $\alpha = p/q$, then we can find a commutative subgroup of finite index in the symmetry group. Namely, the operators T_j^q $(j = 1,2)$ commute. This enables us to develop "magnetic" Floquet theory in this case. No analogs of Floquet theory have been considered so far for irrational values of α. One wonders whether it can be done. The first step here could be developing an analog of the exponential expansion for solutions of invariant differential equations on the Heisenberg group $\mathbf{H}$ (this would be some type of "fundamental principle"). We expect that this could be done, since harmonic analysis on this group has been developed to a very high level (see, for instance, [121, 129-131, 374, 375]). An analogous problem arises also in [6].

Many problems of the theory of periodic operators can be also posed for liftings of elliptic operators on compact manifolds to non-compact covering manifolds: in the standard periodic case we lift an operator from the torus to the whole Euclidian space. The problems of this sort that have been already discussed include: structure of the spectrum [21, 62, 63, 66, 67, 90, 203, 377-383], Bloch property [203], positive solutions [4, 102, 260], and operators with "magnetic" terms [65]. These problems arise in spectral geometry [62, 63, 90, 377-383] and some other areas [102, 259, 278]. This area has been developing very fast during the last decade, but apparently much work still has to be done.

It is also interesting to note some relations with Floquet theory that arise naturally in the theory of (ordinary) differential and functional-differential equations in complex domains [43, 248, 271].

There is yet another possible generalization of Floquet theory. Consider a periodic ordinary differential system

$$\frac{du}{dt} = A(t)u,$$

where $A(t)$ is a periodic function on $\mathbb{R}$ (or a function on the unit circle) with values in the Lie algebra $\mathfrak{GL}(n, \mathbb{C})$ of all $n \times n$ complex matrices. The corresponding equation for the matrix fundamental solution $\Phi(t) \in GL(n, \mathbb{C})$ is:

$$\frac{d\Phi}{dt}\,\Phi^{-1} = A(t),\ \Phi(0) = I,$$

or

$$d\Phi \circ \Phi^{-1} = A(t)\ dt. \tag{6.4.5}$$

Both sides of (5) can be interpreted as differential forms on the circle with values in the Lie algebra $\mathfrak{GL}(n, \mathbb{C})$. This equation naturally generalizes to functions on manifolds with values in an arbitrary Lie group A. If the manifold is equipped with a linear connection, the notion of a "constant coefficient equation" (5) can be introduced (observe that (5) is not a linear equation). Now the question can be posed as to whether the original equation can be reduced to a constant coefficient one using some natural analog of Floquet-Lyapunov substitution. The reader can find some discussion of this problem in [143, 215, 317]. The theory involves the technique of non-abelian cohomologies [316].

Bibliography

[1] T. Adachi, T. Sunada, and Polly Wee Sy, On the regular representation of a group applied to the spectrum of a tower, preprint 1991.

[2] S. Agmon, Lectures on Exponential Decay of Solutions of Second-Order Elliptic Equations, Princeton Univ. Press 1982

[3] ——, On positivity and decay of second order elliptic equations on Riemannian manifolds, in: Methods of Functional Analysis and Theory of Elliptic Equations, 19-52, Liguori, Naples, 1982.

[4] ——, On positive solutions of elliptic equations with periodic coefficients in $\mathbf{R}^n$, spectral results and extensions to elliptic operators on Riemannian manifolds, Proc. Internat. Conf. on Differential Equations, North-Holland Math. Studies, no.92, 1984, 7-17.

[5] M.S. Agranovich and M.I. Vishik, Elliptic problems with a parameter and parabolic problems of general type, Russian Math. Surveys 19(1964), no.3, 53-157.

[6] M. Agranovsky, C. Berenstein, Der-Chen Chang, and D. Pascuas, The uniqueness of bounded solutions of convolution equations in the Heisenberg group, preprint 1991.

[7] J. Albert, Generic properties of eigenfunctions of elliptic differential operators, Trans. Amer. Math. Soc. 238(1978), 341-354.

[8] G.R. Allan, Ideals of vector-valued functions, Proc. London Math. Soc. 18(1968), 193-216.

[9] ——, Holomorphic vector-valued functions on a domain of holomorphy, J. London Math. Soc. 42(1967), 509-513.

[10] ——, On one-sided inverses in Banach algebras of holomorphic vector-valued functions, J. London Math. Soc. 42(1967), 463-470.

[11] W. Allegretto, On the equivalence of two types of oscillation for elliptic operators, Pacific J. Math. 55(1974), 319-328.

[12] ——, Positive solutions and spectral properties of second order elliptic operators, Pacific J. Math. 92(1981), 15-25.

[13] ——, Criticality and the λ-property for elliptic equations, J. Diff. Equat. 69(1987), 39-45.

[14] G. Alvarez and B. Sundaram, Perturbation theory for Floquet eigenvalues in a zero-range potential, Phys. Rew. A 42(1990), no.1, 452-458.

[15] C. Amick, On the Dirichlet problem for infinite cylinders and equations with transversely varying coefficients, J. Diff. Equat. 30(1978), 248-279.

[16] A.B. Antonevich, Linear Functional Equations. Operator Approach, Univ.Publ., Minsk 1988. (In Russian)

[17] F.M. Arscott, Periodic Differential Equations, Macmillan, NY 1964.

[18] N.W. Ashcroft and N.D.Mermin, Solid State Physics, Holt, Rinehart and Winston, New York-London, 1976.

[19] M.F. Atiyah, K-Theory, Benjamin, New York-Amsterdam 1967.

[20] ——, Elliptic Operators and Compact Groups, Lecture Notes in Math., v. 401, Springer Verlag, Berlin e.a. 1974.

[21] ——, Elliptic operators, discrete groups and von Neumann algebra, Asterisque 32-33(1976), 43-72.

[22] ——, and I.Singer, The index of elliptic operators. III, Ann. of Math. (2) 87 (1968), 546-604.

[23] L. Auslander, I. Gertner and R. Tolimieri, The finite Zak transform and the finite Fourier transform, in: Radar and Sonar, part II, Springer-Verlag, 1991, 21-35.

[24] L. Auslander and R. Tolimieri, Abelian Harmonic Analysis, Theta Functions and Function Algebra on Nilmanifolds, Lect. Notes in Math., v. 436, Springer-Verlag, 1975.

[25] J.E. Avron , I. Herbst and B. Simon, Schrödinger operators with magnetic fields. I. General interactions, Duke Math. J. 45(1978), no.4, 847-883.

[26] ——, ——, ——, Separation of mass in homogeneous magnetic fields, Ann. Phys. 114(1978), 431-451.

[27] —— and B. Simon, Analytic properties of band functions, Ann. Physics 110(1978), 85-101.

[28] ——, ——, Stability of gaps for periodic potentials under variation of a magnetic field, J. Phys. A 18(1985), 2199-2205.

[29] N.S. Bakhvalov and G.P. Panasenko, Homogenization: Averaging Processes in Periodic Media, Kluwer Acad. Publ., Dordrecht-Boston 1989.

[30] S. Banach, Theorie des Operations Lineaires, Warszawa 1932.

[31] H. Bart, Meromorphic Operator Valued Functions, Math. Centre Tracts, v.44, Mathematisch Centrum, Amsterdam, 1973.

[32] ——, Holomorphic relative inverses of operator valued functions, Math. Ann. 208(1974), 179-194.

[33] ——, I. Gohberg and M. Kaashoek, Minimal Factorization of Matrix and Operator Functions, Birkhäuser, 1979.

[34] ——, M.A. Kaashoek and D.C. Lay, Stability properties of finite meromorphic operator functions, Indag. Math. 36(1974), 217-259.

[35] ——, ——, ——, Relative inverses of meromorphic operator functions and associated holomorphic projection functions, Math. Ann. 218(1975), 199-210.

[36] R. G. Bartle and L.M. Graves, Mappings between function spaces, Trans. Amer. Math. Soc. 72(1952), 400-413.

[37] D. Bättig, A toroidal compactification of the two-dimensional Bloch variety, Thesis, ETH-Zürich 1988.

[38] ——, A toroidal compactification of the Fermi surface for the discrete Schrödinger operator, Comment. Math. Helv. 67(1992), 1-16.

[39] ——, H. Knörrer, and E. Trubowitz, A directional compactification of the complex Fermi surface, Compositio Math., 79(1991), no.2, 205-229.

[40] J. Bellisard, C*-algebras in solid state physics - 2D electrons in a uniform magnetic field, in: Operator algebras and applications, vol.2, Cambridge Univ. Press 1988, 49-76.

[41] A. Bensoussan, J.L. Lions, G. Papanicolaou, Asymptotic Analysis of Periodic Structures, North Holland, Amsterdam 1978.

[42] C.A. Berenstein and M.A. Dostal, Analytically Uniform Spaces and their

Applications to Convolution Equations, Lecture Notes in Math., v. 256, Springer-Verlag, Berlin e.a. 1972.

[43] __ and A. Sebbar, On a functional equation of A. Hurwitz, to appear in Adv. Math.

[44] __ and M. Shahshahani, Harmonic analysis and the Pompeiu problem, Amer. J. Math. 105(1983), 1217-1229.

[45] __ and D. Struppa, Complex analysis and mean-periodicity, in: Itogi Nauki i Techniki. Sovremennnye Probl. Math., v.54, VINITI, Moscow 1989, 5-111. (In Russian. English transl. to appear in: Encyclopaedia of Math. Sci., v.54, Springer-Verlag, Berlin e.a. 1993.)

[46] __ and B.A. Taylor, A new look at interpolation theory for entire functions of one variable, Adv. in Math. 33(1979), 109-143.

[47] __, __, Interpolation problems in $\mathbf{C}^n$ with applications to harmonic analysis, J. Analyse Math. 38(1980), 188-254.

[48] __, __, Mean-periodic functions, Internat. J. Math. & Math. Sci. 3(1980), no.2, 199-235.

[49] __ and L.Zalcman, Pompeiu's problem on symmetric spaces, Comment. Math. Helv. 55(1980), 593-621.

[50] F.A. Berezin, M.A. Shubin, The Schrödinger Equation, Kluwer, Dordrecht 1991.

[51] A. Berthier, On the point spectrum of Schrödinger operators, Ann. Scient. Ec. Norm. Super., 4 serie, 15(1982), 1-15.

[52] E. Bishop, Analytic functions with values in a Frechet space, Pacific J. Math. 12(1962), 1177-1192.

[53] D. Bleecher and L. Wilson, Splitting the spectrum of a Riemannian manifold, SIAM J. Math. Anal. 11(1980), no.5, 813-818.

[54] F. Bloch, Über die Quantenmechanik der Elektronen in Kristallgittern, Zeit. für Physik 52(1928), 555-600.

[55] R. Boas, Jr., Entire Functions, Acad. Press, NY 1954.

[56] G. Borg, Eine Umkehrung der Sturm-Liouvillischen Eigenwertaufgabe, Acta Math. 78(1946), 1-96.

[57] A.V. Bratishchev and Yu.F.Korobeinik, A multiple interpolation problem on the semi-axis, Math. USSR-Izvestiya 10(1976), 1049-1074.

[58] ——, ——, A multiple interpolation problem in spaces of entire functions, in: Coll. articles on nat. science, math., geology, geography, and phys., Rostov-on-Don 1976, 10-16. (In Russian)

[59] H.-J. Bremermann, Holomorphic functionals and complex convexity in Banach spaces, Pacific. J. Math. 7(1957), 811-831.

[60] L. Brillouin, Wave Propagation in Periodic Structures, Electric Filters and Crystal Lattices, Dover, NY 1953.

[61] M. D. Bronstein, On the allowable rate of decrease of solutions of differential equations with analytic coefficients in a half-strip, preprint, no. 2962-80, VINITI, Moscow 1960. (In Russian)

[62] R. Brooks, The fundamental groups and the spectrum of the Laplacian, Comment. Math. Helv. 56(1981), 581-598.

[63] ——, The spectral geometry of a tower of coverings, J. Diff. Geom. 23(1986), 97-107.

[64] J. Bros and D. Pesenti, Fredholm theory in complex manifolds with complex parameters: analyticity properties and Landau singularities of the resolvent, J. Math. Pure Appl. 58(1980), 375-400.

[65] J. Brüning and T. Sunada, The Schrödinger operator with a periodic magnetic field, preprint 1992.

[66] ——, ——, On the spectrum of periodic elliptic operators, Nagoya Math. J. 126(1992), 159-171.

[67] ——, ——, On the spectrum of gauge-periodic elliptic operators, preprint 1992.

[68] L. Bungart, Holomorphic functions with values in locally convex spaces and applications to integral formulas, Trans. Amer. Math. Soc. 111(1964), 317-344.

[69] ——, On analytic fiber bundles. I, Holomorphic fiber bundles with infinite-dimensional fibers, Topology 7(1968), 55-68.

[70] V. Buslaev, Semiclassical approximation for equations with periodic coefficients, Russ. Math. Surveys 42(6)(1987), 97-125.

[71] —— and L. Dmitrieva, Bloch electrons in external electric field, in: Schrödinger operators. Standard and non-standard (Dubna, USSR, 1988), World. Sci., 1990, 103-129.

[72] ——, ——, Geometrical aspects of the Bloch electrons theory in external fields, in: Topological Phases in Quantum Theory (Dubna, 1988), World. Sci. Publ., Teanek, NJ 1989, 218-250.

[73] S.B. Chae, Holomorphy and Calculus in Normed Spaces, Marcel Dekker, New York and Basel, 1985.

[74] T.A. Chapman, Locally trivial bundles and microbundles with infinite-

dimensional fibers, Proc. Amer. Math. Soc. 37(1973), 595-602.

[75] Ju. Chuburin, On scattering for the Schrödinger operator in the case of a crystallic film, Teor. Mat. Fiz. 72(1987), no.1, 120-131. (In Russian)

[76] ——, An asymptotic representation of Floquet solutions for the Schrödinger equation in the case of a semi-bounded crystal, Teor. Mat. Fiz. 77(1988), no.3, 472-478. (In Russian)

[77] ——, On scattering for the Schrödinger operator in the case of a semi-bounded crystal, in: Problems of Theory of Periodic Motions, Izhevsk, 1988, no.9, 63-72. (In Russian)

[78] ——, On Scrödinger operator in the case of two joined crystals, in: Problems of Theory of Periodic Motions, Izhevsk, 1990, no.10, 49-55. (In Russian)

[79] H.L. Cycon, R.G. Froese, W. Kirsch, and B. Simon, Schrödinger Operators with Applications to Quantum Mechanics and Global Geometry, Texts and Monographs in Physics, Springer Verlag, Berlin 1987.

[80] G. Da Prato and A. Lunardi, Floquet exponents and stabilizability in time-periodic parabolic systems, Appl. Math. and Optimization 22(1990), 91-113.

[81] B. Dahlberg and E. Trubowitz, A remark on two dimensional potentials, Comment. Math. Helv. 57(1982), 130-134.

[82] Iu.L. Daletskii and M.G. Krein, Stability of Solutions of Differential Equations in Banach Space, Amer. Math. Soc., Providence, RI 1974.

[83] L. Danilov, Zone structure of the spectrum of the Dirac operator with periodic potential, in: Nonlinear oscillations and control theory, Ustinov 1985, 83-90. (In Russian)

[84] ——, On the spectrum of the Dirac operator with periodic potential, Acad. Sci. USSR, Ural Branch, preprint, Sverdlovsk, 1987. (In Russian)

[85] ——, On spectrum of the Dirac operator in $\mathbf{R}^n$ with periodic potential, Theor. Math. Fiz., 85(1990), no.1, 41-53. (In Russian)

[86] ——, The spectrum of the Dirac operator with periodic potential, I, preprint, no.4588-B91, VINITI, Moscow 1991, 36 pp. (In Russian)

[87] V.I. Dergusov, A mathematical investigation of periodical cylindrical wave guides. I, Vestnik Leningr. Univ. 13(1972), 32-40. II, Vestnik Leningr. Univ., no.19, 14-20. (In Russian)

[88] ——, Linear equations with periodic coefficients and their applications to wave guide systems, D.Sc. Dissertation, Leningrad 1975. (In Russian)

[89] ——, On the discreteness of the spectrum of the periodic boundary-value problem associated with the study of periodic wave guides, Siberian Math. J. 21 (1980), 664-672.

[90] H. Donnelly, On the spectrum of towers, Proc. Amer. Math. Soc. 87(1983), 322-329.

[91] B.A. Dubrovin and S.P.Novikov, Ground states in a periodic field. Magnetic Bloch functions and vector bundles, Soviet Math. Dokl. 22(1980), 240-244.

[92] ——, ——,Ground states of a two-dimensional electron in a periodic magnetic field, Soviet Physics JETF 52 (1980), 511-516.

[93] N. Dunford and J. Schwarz, Linear Operators, Spectral Theory, Interscience Publishers, New York-London 1963.

[94] M.S.P. Eastham, The Spectral Theory of Periodic Differential Equations,

Scottish Acad. Press, Edinburgh-London 1973.

[95] J. Eels, Fredholm structures, in "Nonlinear Analysis", Proc. Symposium Pure Math. 18(1970), 62-85.

[96] —— and K.D. Elworthy, on Fredholm manifolds, Internat. Congr. Math., vol.2, Nice 1970, 215-220.

[97] D.V. Efremov , Spectral function asymptotics of second order elliptic operators on Lobachevsky spaces, Vestnik Moscow Univ. Ser. I, Math., Mech., 1988, no.3, 72-74. (In Russian)

[98] —— and M.A. Shubin, Spectrum distribution function and variational principle for authomorphic operators on hyperbolic space, Equations aux derivees partielles, Semianire 1988-1989 Ecole Polytechnique, Expose no. VIII.

[99] L. Ehrenpreis, Fourier Analysis in Several Complex Variables, Wiley-Interscience, New York-London 1970.

[100] K.D. Elworthy, Embeddings, isotopy, and stability of Banach manifolds, Compositio Math. 24(1972), 175-226.

[101] ——, and A.J. Tromba, Fredholm maps and differential structures on Banach manifolds, in: Global Analysis, Proc. Symp. Pure Math. 15(1968), 45-94.

[102] C.L. Epstein, Positive harmonic functions on abelian covers, J. Funct. Anal. 82(1989), 303-315.

[103] G. Eskin, Inverse spectral problem for the Schrödinger operator with periodic magnetic and electric potentials, Equations aux derivees partielles, Seminaire 1988-1989, Ecole Polytechnique, Palaiseau, Expose no.XIII 1989.

[104] ——, Inverse spectral problem for the Schrödinger equation with periodic

vector potential, Commun. Math. Phys 125(1989), 263-300.

[105] __, J. Ralston and E. Trubowitz, On isospectral potentials in $\mathbb{R}^n$, Comm. Pure Appl. Math. 37(1984), 647-676. II, Comm. Pure Appl. Math. 37(1984), 715-753.

[106] L. Euler, De integratione aequationum differentialium altiorum gradum, Miscellanea berol. 7(1743), 193-242.

[107] S. Fassari, On the Schrödinger operator with periodic point interactions in the three-dimensional case, J. Math. Phys. 25(1984), 2910-2917.

[108] H.O. Fattorini, The Cauchy Problem, Addison-Wesley, London e.a. 1983.

[109] J. Feldman, H. Knörrer, and E. Trubowitz, The perturbatively stable spectrum of a periodic Schrödinger operator, Invent. Math. 100(1990), 259-300.

[110] __, __, __, Perturbatively unstable eigenvalues of a periodic Schrödinger operator, ETH-Zürich, preprint 1990.

[111] N.E. Firsova, The trace formula for the perturbed one-dimensional Schrödinger operator with periodic potential, I, in: "Problems of math. Phys.", no. 7, Leningrad, 1974, 162-176. (In Russian)

[112] __, The Riemann surface of a quasimomentum, and scattering theory for a perturbed hill operator, in: Mathematical problems in the theory of wave propagation, vol.7, Zapiski Nauchn. Semin. LOMI 51(1975), 183-196. (In Russian)

[113] __, An inverse scattering problem for the perturbed Hill operator, Mat. Zametki 18(1975), no.6, 831-843. (In Russian)

[114] __, On some spectral identities for the one-dimensional Hill's operator, Teor.

Mat. Fiz. 37(1978), no.2, 281-288. (In Russian)

[115] ——, On the effective mass for the one-dimensional Hill's operator, Vestnik Leningr. Univ., 1979, no.10, 13-18. (In Russian)

[116] ——, On the resonances for the perturbed Hill's operator with an exponentially decreasing impurity potential, Mat. Zametki, 36(1984), no.5, 711-724. (In Russian)

[117] ——, On the Levinson formula for the perturbed Hill's operator, Teor. Mat. Fiz. 62(1985), no.2, 196-209. (In Russian)

[118] ——, Direct and inverse scattering problems for the one-dimensional perturbed Hill's operator, Mat. Sb. 130(1986), no.3, 349-385. (In Russian. English transl. in Soviet Math. Sbornik)

[119] ——, On solution of Cauchy problem for the Korteveg-de Vries equation with initial data that are the sum of periodic and fast decaying functions, Mat. Sbornik 135(1988), no.2, 261-268. (In Russian. English transl. in Soviet Math. Sbornik)

[120] G. Floquet, Sur les equations differentielles lineaires a coefficients periodiques, Ann. Ecole Norm., Ser. 2, 12(1883), 47-89.

[121] G.B. Folland, Harmonic Analysis in Phase Space, Ann. of Math. Study #122, Princeton Univ. Press, Princeton, NJ 1989.

[122] V.N. Fomin, Mathematical Theory of Parametric Resonance in Linear Distributed Systems, Leningr. Univ. Publ., Leningrad 1972.(In Russian)

[123] D. Freed and J. Harvey, Instantons and the spectrum of Bloch electrons in a magnetic field, Phys. Rew. B 41(1990), no.16, 11 328- 11 345.

[124] L. Friedlander, On the spectrum of the periodic problem for the Schrödinger operator, Comm. Partial Diff. Equat. 15(1990), 1631-1647.

[125] M. Fuhrman, Bounded solutions for abstract time-periodic parabolic equations with nonconstant domains, Diff. and Integral. Equations 4(1991), no.3, 493-518.

[126] I.M. Gelfand, Expansion in eigenfunctions of an equation with periodic coefficients, Dokl. Akad. Nauk SSSR 73(1950), 1117-1120. (In Russian)

[127] I.M. Gelfand and V.B. Lidskii, On the structure of the domains of stability of linear canonical systems of differential equations with periodic coefficients, Uspekhi Mat. Nauk 10(1955), 3-40.(In Russian, English transl. in: Amer. Math. Soc. Transl. (2) 8, 143-181).

[128] I.M. Gelfand and G.E. Shilov, Generalized Functions, vol.2. Spaces of Fundamental and Generalized Functions, Academic Press, New York-London 1968.

[129] D. Geller, Fourier analysis on the Heisenberg group, Proc. Nat. Acad. Sci. 74(1977), 1328-1331.

[130] __, Fourier analysis on the Heisenberg group I: Schwartz space, J. Funct. Anal. 36(1980), no.2, 205-254.

[131] __, Analytic Pseudodifferential Operators for the Heisenberg Group and Local Solvability, Princeton Univ. Press, Princeton, NJ 1990.

[132] C. Gerard, Resonance theory for periodic Schrödinger operators, preprint, Ecole Polytechnique Palaiseau, M831.1288, 1988.

[133] O. Germanovich, Linear Periodic Equations of Neutral Type and their Applications, Leningrad State Univ. Publ., Leningrad 1986.

[134] D. Gieseker, H. Knörrer, and E. Trubowitz, An overview of the geometry of algebraic Fermi curves, To appear in Proc. Conf. on Algebraic Geometry, Salt Lake City 1989.

[135] D. Gilbarg and N.S. Trudinger, Elliptic Partial Differential Equations of Second Order, 2nd.ed., Springer Verlag, Berlin 1983.

[136] I.M. Glazman, Direct Methods of Qualitative Spectral Analysis of Singular Differential Operators, Isr. Progr. Sci. Transl., Jerusalem 1965.

[137] A.M. Gleason, The abstract theorem of Cauchy-Weil, Pacific J. Math. 12(1962), 511-526.

[138] I.C. Gohberg, On some topics of spectral theory of finite meromorphic operator functions, Izvestiya Akad. Nauk. Armyan. SSR 5(1971), 160-181.

[139] ——, M. Kaashoek and D. Lay, Equivalence, linearization, and decomposition of holomorphic operator functions, J. Funct. Anal. 28(1978), 102-144.

[140] —— and M.G.Krein, Fundamental aspects of defect numbers, root numbers, and indexes of linear operators, Uspekhi Mat. Nauk 12(1957), no.2, 43-118. (In Russian, English transl. in Amer. Math. Soc. Transl. 13(1964), 185-264.)

[141] ——, ——, Introduction to the Theory of Linear Nonselfadjoint Operators, Amer. Math. Soc., Providence, RI 1969.

[142] —— and E.I. Sigal, An operator generalization of the logarithmic residue theorem and the theorem of Rouché, Math. USSR Sbornik 13(1971), 603-625.

[143] V. Gorbatsevich, Generalized Lyapunov's theory on Maltsev manifolds, Mat. Sbornik 94(1974), 163-177. (In Russian. English transl. in Math. USSR Sbornik.)

[144] B. Gramsch, Meromorphie in der Theorie der Fredholmoperatoren mit Anwendung auf elliptische Differentialoperatoren, Math. Ann. 188(1970), 97-112.

[145] __, Ein Zerlegungsatz für Resolventen elliptischer Operatoren, Math. Z. 133(1973), 219-242.

[146] __, Inversion von Fredholmfunktionen bei stetiger und holomorpher Abhängigkeit von Parametern, Math. Ann. 214(1975), 95-147.

[147] __, Relative Inversion in der Störungstheorie von Operatoren und ψ-Algebren, Math. Ann. 269(1984), 27-71.

[148] __, W. Kaballo, Regularisierung von Fredholmfunktionen, Math. Ann. 232(1978), 151-162.

[149] __, __, Spectral theory for Fredholm functions, in: "Functional Analysis. Surveys and recent results. II", North-Holland Publ., 1980, 319-342.

[150] __, R. Wagner, Eine Bemerkung zur Division von Distributionen durch analytische Operatorfunktionen, Manuscripta Math. 21(1977), 25-42.

[151] H. Grauert, Analytische Faserungen über holomorph-vollständigen Räumen, Math.Ann. 135(1968), 163-173.

[152] A. Grothendieck, Produits Tensoriels Topologiques et Espaces Nucléaires, Mem. Amer. Math. Soc., no.16(1955), 140pp.

[153] V. Guillemin and A. Uribe, Clustering theorems with twisted spectra, Math. Ann. 273(1986), 479-506.

[154] R.C. Gunning and H. Rossi, Analytic Functions of Several Complex Variables, Prentice-Hall, Englewood Cliffs, N.J., 1965.

[155] V.P. Gurarii, N.A. Kulesko, V.I. Matsaev, and Y.A. Palant, Completeness of the system of Floquet solutions of linear first order differential equations with delays that are multiplies of the period of the coefficients, Vestnik Kharkov. Gos. Univ. 67 (1971), 3-10. (In Russian)

[156] P. Hartman, Ordinary Differential Equations, Wiley & Sons, NY 1964.

[157] B. Helffer, Semi-Classical Analysis for the Schrödinger Operator and Applications, Lect. Notes Math. , v.1336, Springer Verlag, Berlin e.a. 1988.

[158] —— , and J. Sjöstrand, Equation de Schrödinger avec champ magnetique et equation de Harper, in : Schrödinger Operators, Lect. Notes Phys., vol. 345, Springer Verlag, Berlin e.a. 1989, 118-197.

[159] ——, ——, On diamagnetism and de Haas-van Alphen effect, Ann. Inst. Henri Poincare, Phys. Theor. 52(1990), no.4, 303-375.

[160] ——, P. Kerdelhue, and J. Sjöstrand, Le papillon de Hofstadter revisite, Suppl. au Bulletin Soc. Math. France, Mem. no.43, 118(1990), fasc. 3.

[161] S. Helgason, Differential Geometry and Symmetric Spaces, Academic Press, New York-London 1962.

[162] ——, Analysis on Lie Groups and Homogeneous Spaces, Regional CBMS conf. series, Amer. Math. Soc., Providence, RI 1972.

[163] ——, The surjectivity of invariant differential operators on symmetric spaces. I, Ann. of Math. (2) 98 (1973), 451-479.

[164] ——, A duality for symmetric spaces with applications to group representations. II. Differential equations and eigen-space representations, Adv. in Math. 22 (1976), 187-219.

[165] __, A duality for symmetric spaces with applications to group representations. III. Tangent space analysis, Adv. in Math. 36(1980), no.3, 297-323.

[166] __, Groups and Geometric Analysis, Acad. Press, New York-London-Toronto 1984.

[167] M. Herve, Several Complex Variables. Local theory, Clarendon Press, Oxford-London 1963

[168] E. Hille and R.S. Phillips, Functional Analysis and Semigroups, Amer. Math. Soc. Colloq. Publ. 31, Amer. Math. Soc., Providence, R.I. 1957.

[169] L.Hörmander, An Introduction to Complex Analysis in Several Variables, Van Nostrand, Toronto-New York-London 1966.

[170] __, Pseudodifferential operators and hypoelliptic equations, Ann. of Math. 2(83) (1966), 129-209.

[171] __, Linear partial differential equations, third edition, Springer-Verlag, Berlin-Heidelberg-New York 1969.

[172] __, On the index of pseudo-differential operators, in: Elliptische Differentialgleichungen, band II, Akademie-Verlag, Berlin 1971, 127-146.

[173] __, Lower bounds at infinity for solutions of differential equations with constant coefficients, Israel J. Math. 16 (1973),103-113.

[174] __, The Analysis of Linear Partial Differential Operators, vol. I-IV, Springer Verlag, Berlin e.a. 1983-1985.

[175] J. Howland, Scattering theory for Hamiltonians periodic in time, Indiana Univ. Math. J. 28(1979), no.3, 471-494.

[176] D. Husemoller, Fibre Bundles, McGraw-Hill, New York-London 1966.

[177] V.A. Iakubovich and V.M. Starzhinski, Linear Differential Equations with Periodic Coefficients, Wiley, NY 1975.

[178] ——, ——, Parametric Resonance in Linear Systems, Nauka, Moscow 1987. (In Russian)

[179] V.I. Iudovich, The linearization method in the stability problem for periodic flows of low viscosity fluids, in: Proceedings of the VI- th winter school on mathematical programming and related problems, Drogobych 1973: Functional analysis and its applications, Moscow 1975, 44-113. (In Russian)

[180] ——, The Linearization Method in Hydrodynamical Stability Theory, Amer. Math. Soc., Providence, RI 1989.

[181] K. Jänich, Vektorraumbündel und der Raum der Fredholmoperatoren, Math. Ann. 161(1965), 129-142.

[182] W. Kaballo, Über holomorphe und meromorphe einseitige Inverse, Manuscripta Math. 13(1974), 1-13.

[183] ——, Projectoren und relative Inversion holomorpher Semi-Fredholmfunktionen, Math. Ann. 219(1976), 85-96.

[184] ——, Holomorphe Systeme von Pseudodifferentialoperatoren, Arch. Math. 29(1977), no.3, 284-299.

[185] ——, Holomorphe Semi-Fredholm Funktionen ohne komplementierte Kerne bzw. Bilder, Math. Nachr. 91(1979), 327-335.

[186] ——, Lifting-Sätze für Vektorfunktionen und (ϵL)-Räume, J. Reine Angew. Math. 309(1979), 55-85.

[187] ——, Lifting-Probleme für H^∞-Funktionen, Arch. Math. 34(1980), 540-549.

[188] ——, Surjektivität und Parameterabhängigkeit bei partiellen Differentialoperatoren, Results in math. 11(1987), 72-82.

[189] ——, G. Thijsse, On holomorphic operator function equations, Integral Equat. and Operator Theory 2(1979), 244-263.

[190] ——, D. Vogt, Lifting-Probleme für Vektorfunktionen und $\otimes$-Sequenzen, Manuscripta Math. 32(1980), 1-27.

[191] F.I. Karpelevich, The geometry of geodesics and the eigenfunctions of the Beltrami-Laplace operator on symmetric spaces, Trans. of the Moscow Math. Soc., 14(1965), Amer. Math. Soc., Providence, RI 1967, 51-199.

[192] Yu.E. Karpeshina, Spectrum and eigenfunctions of Schrödinger operator with point-like potential of the homogeneous lattice type in three-dimensional space, Teor. Math. Phys., 57(1983), no.2, 304-313. (In Russian)

[193] ——, Spectrum and eigenfunctions of the Schrödinger operator in three-dimensional space with point-like potential of the homogeneous two-dimensional lattice type , Teor. Math. Phys., 57(1983), no.3, 414-423. (In Russian)

[194] ——, Geometrical background for the perturbation theory of the polyharmonic operator with periodic potential, in: Topological Phases in Quantum Theory (Dubna, 1988), World Sci. Publ., Teanek, NJ 1989, 251-276.

[195] ——, Analytic perturbation theory for a periodic potential, Izvestiya Akad. Nauk SSSR, Ser. Math., 52(1989), no.1, 45-65. (In Russian. English transl. in Math. USSR-Izvestiya 34(1990))

[196] ——, Perturbation theory for Schrödinger operator with a periodic potential, in: Schrödinger operators. Standard and Non-Standard, World. Sci. Publ. 1990,

131-145.

[197] ——, Perturbation theory for the Schrödinger operator with a periodic potential, Proceedings of the Steklov Inst. of Math., 1991, issue 3, 109-145.

[198] ——, Perturbation formula for the Schrödinger operator with a nonsmooth periodic potential, Math. USSR Sbornik 71(1992), no.1, 101-124.

[199] T. Kato, Perturbation Theory for Linear Operators, Springer Verlag, Berlin e.a., 1980.

[200] V. Katsnelson, On conditions for the basis property of the system of root vectors for a certain class of operators, Funkt. Anal. Prilozh. 1(1967), no.2, 39-51. (In Russian. English transl. in Funct. Anal. Appl. 1(1967).)

[201] M.V. Keldysh, On the eigenvalues and eigenfunctions of certain classes of nonselfadjoint equations, Dokl. Akad. Nauk SSSR 77(1951), 11-14.(In Russian. English transl. in [280].)

[202] H. Knörrer and E. Trubowitz, A directional compactification of the complex Bloch variety, Comment. Math. Helv. 65(1990), 114-149.

[203] T. Kobayashi, K. Ono, T. Sunada, Periodic Schrödinger operator on a manifold, Forum Math. 1(1989), 69-79.

[204] W. Kohn, Analytic properties of Bloch waves and Wannier functions, Phys. Rev. 115(1959), 809-821.

[205] V.B. Kolmanovskii and V.R. Nosov, Stability of Functional-Differential Equations, Acad. Press, London-Orlando 1986.

[206] E.L. Korotyaev, On the spectrum of monodromy operator for the Schrödinger operator with a potential periodic in time, Zapiski Nauch. Semin. LOMI

96(1980), 101-104. (In Russian)

[207] M.A. Krasnosel'skii, Positive solutions of operator equations, Noordhoff, Groningen, 1964.

[208] M.G. Krein and G.Ya. Lyubarskii, On the theory of transmission bands of periodic wave guides, J. Appl. Math. Mech. 25(1961), 29-48. (In Russian)

[209] S.G. Krein, Linear Differential Equations in a Banach Space, Transl. Math. Monographs 29(1971), Amer. Math. Soc., Providence, R.I. 1971.

[210] __ (Ed.), Functional Analysis, Wolters-Noordhoff, Groningen 1972.

[211] __, Linear Equations in a Banach Space, Birkhäuser, Boston 1982.

[212] __ and G.I. Laptev, Boundary value problems for an equation in Hilbert space, Dokl. Akad. Nauk SSSR 146(1962), no.3, 535-538. (In Russian. English transl. in Soviet Math. Dokl.)

[213] __, __, Boundary value problems for second-order differential equations in a Banach space.I, Diff. Uravneniya 2(1966), no.3, 382-390; II, Diff. Uravneniya 2(1966), no.7, 919-926. (In Russian)

[214] __ and V.P. Trofimov, On holomorphic operator functions of several complex variables, Funct. Anal. Appl. 3(1969), 330-331.

[215] __ and N.I. Yatskin, Linear Differential Equations on Manifolds, Voronezh Univ. Publ., Voronezh 1980. (In Russian)

[216] I.M. Krichever, Spectral theory of two-dimensional periodic operators and its applications, Russian Math. Surveys 44(1989), 145-225.

[217] P.A. Kuchment, Banach locally-trivial bundles, Voronezh Gos. Univ. Trudy

Nauchn.-Issled Inst. Mat. 3(1971), 61-77. (In Russian)

[218] ——, On the lifting of functions with values in a quotient space of a Banach space, Mat. Issled., Kishinyov, 10(1975), no.1, , 183-193. (In Russian).

[219] ——, Floquet representation of the solutions of linear hypoelliptic systems of partial differential equations, Functional Anal. Appl. 13 (1979), 59-60.

[220] ——, On the Floquet theory for periodic linear differential equations with deviating arguments, Uspehi Mat. Nauk, 34(1979), no.3, 201-202. (Russian. English translation in Russian Math. Surveys.)

[221] ——, Representation of solutions of linear partial differential equations with constant or periodic coefficients, in: Theory of Operator Equations, Voronezh State Univ. Publ. 1979, 62-69. (In Russian)

[222] ——, An interpolation problem in the spaces of periodic entire functions of a finite order, preprint, no. 4501-80, VINITI, Moscow, 1980, 11 pp. (In Russian).

[223] ——, Bloch solutions of periodic partial differential equations, Functional Anal. Appl. 14 (1980), 52-53.

[224] ——, The absence of L_2-solutions for periodic partial differential equations, Russian Math. Surveys 35(1980), no.2, 247-248.

[225] ——, On Floquet theory for parabolic and elliptic boundary value problems in a cylinder, Soviet Math. Dokl., 23(1981), 522-525.

[226] ——, Representation of solutions of invariant differential equations on certain symmetric spaces, Soviet Math. Dokl., 24(1981), no.1, 104-106.

[227] ——, Hermite's interpolation problem in a class of periodic entire functions of a finite order, in: Linear Operators in Function Spaces,

Voronezh State Univ. Publ., 1981, p. 63-69. (In Russian).

[228] __, On mean-periodic functions on symmetric spaces, Funkt. Anal. i Prilozh., 16(1982), no.3,68-69.(Russian. English translation in Funct. Anal. Appl., v.16(1982), no.3.)

[229] __, Floquet theory for partial differential equations, Russian Math. Surveys 37(1982), no.4, 1-60.

[230] __, Representations of solutions of periodic PDE, Math. USSR Izvestiya, 21(1983), no. 1,93-117.

[231] __, On the spectral synthesis in the spaces of solutions of invariant differential equations, Lecture Notes in Math., v. 1214, Springer Verlag, Berlin e.a. 1986, 85-100.

[232] __, Spherical representations of solutions of invariant differential equations on a Riemannian symmetric space of noncompact type, Math. USSR Izvestiya, 27(1986), no. 3, 535-548.

[233] __, On periodic perturbations which preserve the dispersion of a differential operator, Differencial'nye Uravneniya, 22(1986), no. 9, 1641-1642. (In Russian. English transl. in Differential Equations)

[234] __, To the Floquet theory of periodic difference equations , in : Geometrical and Algebraical Aspects in Several Complex Variables, Cetraro (Italy), June 1989, EditEl, 1991, 203-209.

[235] __, On the behavior of Floquet exponents of some kind of periodic problems, to appear in J. Diff. Equat.

[236] __ and S.Y.Lvin, The absence of non-zero solutions in $L_2(\mathbf{R}^n)$ for periodic partial differential equations, preprint, no. 2738-79, VINITI, Moscow 1979. (In

Russian)

[237] —— and V. Ovchinnikov, On the problem of lifting of holomorphic functions with values in a Banach space, in: Methods of Solving of Operator Equations, Voronezh State Univ. Publ., 1978, p. 94-96. (In Russian).

[238] —— and A. Pankov, Classifying spaces for the equivariant K-theory, Math.USSR Sbornik, v.24, no.1, 1974, p.31-48.

[239] —— and L.B. Zelenko, On the Floquet representations of exponentially increasing solutions of elliptic equations with periodic coefficients, Soviet Math. Dokl., 19(1978), no.2, 506-507.

[240] N. Kuiper, The homotopy type of the unitary group of Hilbert space, Topology 3(1965), 19-30.

[241] N.A. Kulesko and B.Ya. Levin, The completeness of the Floquet solutions for second order differential equations with deviating argument, Sibirsk. Mat. Ž. 18(1977), no.2, 321-326. (In Russian)

[242] A. Kurato and M.P. Kats, Lifting theorems for vector-valued functions, Math. USSR Izvestiya 9(1975), 861-875; 10(1976), 672-674.

[243] O.A. Ladyzhenskaia and N.N. Uraltseva, Linear and Quasilinear Elliptic Equations, Acad. Press, New York 1968.

[244] E. Landau, Vorlesungen über Zahlentheorie, Chelsea Publ. Co., NY 1946-47.

[245] E. M. Landis, Some questions in the qualitative theory of second-order elliptic equations, Russian Math. Surveys 18(1963), no. 1, 1-62

[246] ——, On the behavior of solutions of higher order elliptic equations in unbounded domains, Trans. Moscow. Math. Soc. 31 (1976), 30-54

[247] S. Lang, Introduction to Differential Manifolds, Interscience, New York-London 1962.

[248] H. Langer, R. Mennicken, and M. Möller, On Floquet eigenvalue problems for first order differential systems in the complex domain, J. Reine Angew. Math. 425(1992), 87-121.

[249] P.D. Lax, A stability theorem for solutions of abstract differential equations, and its application to the study of the local behavior of solutions of elliptic equations, Comm. Pure Appl. Math. 9(1956), no.4, 747-766.

[250] __, Periodic solutions of the KdV equation, Comm. Pure Appl. Math. 28(1975), 141-188.

[251] __ and R. Phillips, Scattering Theory for Automorphic Functions, Princeton Univ. Press, Princeton, NJ 1976.

[252] J. Leiterer, Banach coherent analytic Fréchet sheaves, Math. Nachr. 85(1978), 91- 109.

[253] __, Local and global equivalence of meromorphic operator functions, part I, Math. Nachr. 83(1978), 7-29. Part II, Math. Nachr. 84(1978), 145-170.

[254] __, An implicit function theorem in Banach spaces, Ann. Polon. Math. XLVI(1985), 171-175.

[255] P. Lelong and L. Gruman, Entire Functions of Several Complex Variables, Springer Verlag, Berlin e.a. 1986.

[256] B.Ja. Levin, Distribution of Zeros of Entire Functions, Transl. Math. Monogr., vol.5, Amer. Math. Soc., Providence, RI 1964.

[257] V.B. Lidskii, Eigenfunction expansions for equations with periodic coefficients,

Appendix VIII to the Russian (1961) edition of [389].

[258] ——, Summability of series in terms of the principal vectors of non-self-adjoint operators, Trudy Moskow. Mat. Obshch. 11 (1962), 3-35. (In Russian)

[259] V.Ya. Lin, Liouville coverings of complex spaces, and amenable groups, Math. USSR Sbornik, 60(1988), no.1, 197-216.

[260] —— and Pinchover Y., Manifolds with group actions and elliptic operators, preprint, Technion, Haifa, June 1992.

[261] J. Lindenstrauss and L. Tzafriri, On the complemented subspaces problem, Israel J. Math. 9(1971), 263-269.

[262] J.-L. Lions and E.Madgenes, Non-Homogeneous Boundary Value Problems and Applications, vol.1, Springer-Verlag, Berlin-Heidelberg-New York 1972.

[263] A. Lunardi, Bounded solutions of linear periodic abstract parabolic equations, Proc. Roy. Soc. Edinburgh Sect. A 110(1988), 253-279.

[264] ——, Stabilizability of time periodic parabolic equations, Preprint, Scuola Norm. Sup. Pisa 1989.

[265] S.Ya. Lvin, Green's formula and solubility of elliptic problems with boundary conditions of arbitrary order, Voronezh Univ. Trudy Nauchn.-Issled. Inst. Mat. VGU, 1975, no. 17, 49-56. (In Russian)

[266] ——, Green's formula and the solubility of Douglis-Nirenberg elliptic problems with boundary conditions of arbitrary order, preprint,Novosibirsk, VINITI, no.3318-78,1978. (In Russian)

[267] A.M. Lyapunov, General problem of stability, Kharkov 1892; cf. also Collected

Works, vol. 2, Moscow 1956, 7-263. (In Russian)

[268] T. Lyons and D. Sullivan, Function theory, random paths and covering spaces, J. Diff. Geometry 19(1984), 299-323.

[269] A.S. Lyskova, The Schrödinger operator in the bundle of non-zero curvature over the two-dimensional torus, preprint, VINITI, Moscow, 1982, 58 pp.

[270] Y.I. Lyubich and V.A. Tkachenko, On a Floquet theory for equations with retarded argument, Diff. Uravneniya 5 (1969), 490-494.(In Russian. English transl in Diff. Equations)

[271] W. Magnus, Monodromy groups and Hill's equation, Comm. Pure Appl. Math. 29(1976), 701-716.

[272] W. Magnus and S. Winkler, Hill's Equation, Wiley, NY 1966.

[273] B. Malgrange, Existence et approximation des solutions des equations aux derivees partielles et des equations de convolution, Ann. Inst. Fourier 6(1956), 271-355.

[274] __, Sur les systemes differentiels a coefficients constants, Colloq. Internat. Centre Nat. Rech. Scient. 117(1963), 113-122.

[275] F. Mantlik, Partial differential operators depending analytically on a parameter, Ann. Inst. Fourier 41(1991), 577-599.

[276] V.A. Marchenko, Sturm-Liouville Operators and Applications, Birkhäuser, Basel-Boston-Stuttgart 1986.

[277] __ and I.V. Ostrovskii, A characterization of the spectrum of Hill's operator, Mat. Sbornik 97(1975), no.4, 540-606. (In Russian. English transl. in Math. USSR Sbornik)

[278] G.A. Margulis, Positive harmonic functions on nilpotent groups, Soviet Math. Doklady 7(1966), 241-244.

[279] A.S. Markus, On a basis of root vectors of a dissipative operator, Dokl. Akad. Nauk SSSR 132(1960), 524-527. (In Russian. English transl. in Soviet Math. Dokl. 1(1960).)

[280] ——, Introduction to the Spectral Theory of Polynomial Operator Pencils, Transl. Amer. Math. Monogr., v.71, AMS, Providence, RI 1988.

[281] —— and V.I. Matsaev, On the basis property for a certain part of the eigenvectors and associated vectors of a selfadjoint operator pencil, Mat. Sb. 133(1987), 293-313.(In Russian. English transl. in Math USSR Sb. 61(1988))

[282] J. Massera and J. Schäffer, Differential Equations and Function Spaces, Acad. Press, NY 1966.

[283] V.I. Matsaev, On a class of compact operators, Dokl. Akad. Nauk SSSR 139(1961), 548-551. (In Russian. English transl. in Soviet Math. Dokl. 2(1961).)

[284] H.P. McKean and P. van Moerbeke, Hill and Toda curves, Comm. Pure Appl. Math. 32(1979), 1-19.

[285] H.P. McKean and D. Sullivan, Brownian motion and harmonic functions on the class surface of the thrice punctured sphere, Adv. Math. 51(1984), 203-211.

[286] H.P. McKean and E. Trubowitz, Hill's operator and hyperelliptic function theory in the presence of infinitely many branch points, Comm. Pure Appl. Math. 29(1976), 143-226.

[287] ——, ——, Hill's surfaces and their theta functions, Bull. Amer. Math. Soc. 84(1978), no.6, 1042-1085.

[288] V. Meshkov, Exponential decay of solutions for one class of partial differential equations, Dokl. Akad. Nauk SSSR 266(1982), no.5, 1056-1059. (In Russian. English transl. in Soviet Math. Dokl.)

[289] __, A Karleman type inequality and some its applications, Dokl. Akad. Nauk SSSR 288(1986), no.1, 46-49.(In Russian. English transl. in Soviet Math. Dokl.)

[290] __, On possible rate of decay at infinity for solutions of partial differential equations of second order, Mat. Sbornik 182(1991), no.3, 364-383. (In Russian. English transl. in Math. USSR-Sbornik.)

[291] R.E. Meyer and M.C. Shen, On Floquet's theorem for nonseparable partial differential equations, in: Ordinary and partial differential equations, v.III, Proc. of the Eleventh Dundee Conf., 1990, Longman, Wiley&Sons, NY 1991, 146-167.

[292] __, __, On exponential asymptotics for nonseparable wave equations: Approximate spectral bands of periodic potentials in string, SIAM J. Appl. Math. 52(1992), no.3, 730-745.

[293] E. Michael, Continuous selections. I, Ann. of Math. (2) 63(1956), 361-382.

[294] __, Three mapping theorems, Proc. Amer. Math. Soc. 15(1964), 410-415.

[295] W. Miller, Jr., Topics in harmonic analysis with applications to radar and sonar, in: Radar and Sonar, part I, Springer-Verlag, 1991, 66-168.

[296] A.I. Miloslavskii, On the Floquet theory for parabolic equations, Funct. Anal. Appl. 10(1976), 151-153.

[297] __, Floquet theory for abstract parabolic equations with periodic coefficients, Ph.D. Dissertation, Rostov-on-Don 1976. (In Russian)

[298] __, An analog of Floquet representation for solutions of an abstract integro-

differential equation with periodic coefficients, preprint, Kharkov, 1987. (In Russian)

[299] ——, Asymptotics of solutions of abstract integro-differential equations with periodic coefficients, Funkt. Anal. i Prilozh. 22(1988), no.3, 77-78. (In Russian. English transl. in Funct. Anal. Appl. 22(1988).)

[300] R.Mittra and S.W. Lee, Analytic Techniques in the Theory of Guided Waves, Collier-Macmillan, London 1971.

[301] B.S. Mityagin, The homotopy structure of the linear group of a Banach space, Russian Math. Surveys 25(1970), no.5, 59-103.

[302] S. Mizohata, The Theory of Partial Differential Equations, Cambridge Univ. Press., Cambridge-New York 1973

[303] P. van Moerbeke,and D. Mumford, The spectrum of difference operators and algebraic curves, Acta Math. 43(1979), 93-154.

[304] W. Moss and J. Piepenbrink, Positive solutions of elliptic equations, Pacific J. Math. 75(1978),219-226.

[305] M. Murata, Structure of positive solutions to $(-\Delta+V)=0$ in $\mathbf{R}^n$, in: Differential Equations and Mathematical Physics, Lect. Notes Math., v.1285, Springer-Verlag, Berlin e.a. 1987, 342-347.

[306] A. Myshkis, On some problems of the theory of differential equations with deviating arguments, Uspekhi Mat. Nauk 32(1977), no.2, 173-202. (In Russian. English transl. in Russian Math. Surveys 32(1977).)

[307] R. Narasimhan, Introduction to the Theory of Analytic Spaces, Lecture Notes in Math., v. 25, Springer Verlag, Berlin e.a. 1966.

[308] S.A.Nazarov, On the Noether property and asymptotic solutions of elliptic boundary-value problems with periodic coefficients in a cylinder, Russian. Math. Surveys 36(1981), no.1, 256-247.

[309] ——, Elliptic boundary-value problems with periodic coefficients in a cylinder, Math. USSR-Izv. 45 (1981), 89-98.

[310] G. Nenciu, Stability of energy gaps under variation of the magnetic field, Letters in Math. Phys. 11(1986), 127-132.

[311] G. Neubauer, Homotopy properties of semi-Fredholm operators in Banach spaces, Math. Ann. 176(1968), 273-301.

[312] R.G. Newton, Inverse scattering by a local impurity in a periodic potential in one dimension.I, J. Math. Phys. 24(1983), 2152-2162; II, J. Math. Phys. 26(1985), 311-316.

[313] S.P. Novikov, A periodic problem for the Korteweg - de Vries equation. I, Funct. Anal. Appl. 8(1974), 236-246.

[314] ——, Two-dimensional Schrödinger operators in periodic fields, in: Itogi Nauki i Techniki, Contemporary Problems of Math., v.23, VINITI, Moscow 1983,3-32. (In Russian).

[315] F. Odeh, J.B. Keller, Partial differential equations with periodic coefficients and Bloch waves in crystals, J. Math. Phys. 5(1964), 1499-1504.

[316] A. Oniščik, Some notions and applications of non-abelian cohomologies, Trans. Moscow Math. Soc. 17(1967), Amer. Math. Soc., Providence, RI 1969, 49-98.

[317] ——, On completely integrable equations on homogeneous spaces, Mat. Zametki 9(1977), no. 4. (In Russian)

[318] A. Outassourt, Comportment semi-classique pour l'operateur Schroedinger a potential periodique, J. Funct. Anal. 72(1987), 65-93.

[319] R. Ouzilou, Classification homotopique des fibres Banachiques, C.R. Acad. Sci. Paris 268A(1969), 1275-1278.

[320] R. Palais, On the homotopy type of certain groups of operators, Topology 3(1965), 271-279.

[321] V.P. Palamodov, Fourier transforms of infinitely differentiable functions of rapid growth, Trans. Moscow Math. Soc. 11 (1962), 309-350.

[322] ——, Coherent analytic sheaves, in: Modern problems in the theory of analytic functions, Moscow 1966, 246-247. (In Russian)

[323] ——, Differential operators in coherent analytic sheaves, Math. USSR-Sbornik 6(1968), 365-391.

[324] ——, Linear Differential Equations with Constant Coefficients, Springer-Verlag, Berlin e.a. 1970.

[325] ——, A remark on exponential representation of solutions of differential equations with constant coefficients, Mat. Sb. 76(1968), 417-434. (In Russian. English transl. in Math. USSR Sbornik 5(1968).)

[326] ——, Harmonic synthesis of solutions of elliptic equations with periodic coefficients, preprint, 1992, to appear in Ann. Inst. Fourier.

[327] B.S. Pavlov and L.D. Faddeev, Scattering theory and automorphic functions, Zap. Nauchn. Semin. LOMI 27(1972), 161-193. (In Russian)

[328] R. Phelps, Lectures on Choquet's Theorem, Van Nostrand, Princeton, NJ 1966.

[329] J. Piepenbrink, Nonoscillatory elliptic equations, J. Diff. Equat. 15(1974), 541-550.

[330] __, A conjecture of Glazman, J. Diff. Equat. 24(1977), 173-177.

[331] A. Pietsch, Operator Ideals, VEB Verlag Deutscher Wissenschaften, Berlin 1978.

[332] Y. Pinchover, Representation theorems for positive solutions of parabolic equations, Proceedings of Amer. Math. Soc. 104(1988), no.2, 507-515.

[333] __, On positive solutions of elliptic equations with periodic coefficients in unbounded domains, in: Maximum Principles and Eigenvalue Problems in Partial Differential Equations, Pitman Res. Notes in Math. , no. 175, Longman, London 1988, 218-229.

[334] __, Criticality and ground states for second-order elliptic equations, J. Diff. Equat. 80(1989), 237-250.; II, J. Diff. Equat. 87(1990), 353-364.

[335] A. Plis, Non-uniqueness in Cauchy's problem for differential equations of elliptic type, J. Math . Mech. 9(1960), 557-562.

[336] V.N. Popov and M.M. Skriganov, A remark on the structure of spectrum of two-dimensional Schrödinger operator with periodic potential, Zapiski Nauchn. Semin. LOMI 109(1981), Leningrad, 131-133. (In Russian)

[337] J. Pöschel and E. Trubowitz, Inverse Spectral Theory, Acad. Press, NY 1987.

[338] M.H. Protter and H.F. Weinberger, Maximum Principles in Differential Equations, Springer Verlag, New York e.a. 1984.

[339] D. Raikov, Two-sided closed graph theorem for topological linear spaces, Siberian Math.J. 7(1966), no.2, 353-372.(In Russian)

[340] ——, Open mapping and closed graph theorems, Appendix 1 to the Russian translation (Mir, Moscow, 1967) of the book [342].

[341] M. Reed and B. Simon, Methods of Modern Mathematical Physics, v. I-IV, Acad. Press, NY 1972-1978.

[342] A.P. Robertson and W. Robertson, Topological Vector Spaces, Cambridge Univ. Press, 1964.

[343] F.S. Rofe-Beketov, On the spectrum of non-selfadjoint differential operator with periodic coefficients, Doklady Akad. Nauk SSSR, 152(1963), no.6, 1312-1315. (In Russian)

[344] Ya.A. Roitberg and Z.G. Sheftel′, A theorem on homeomorphisms for elliptic systems and its applications , Math. USSR Sbornik 7 (1969), 439-465.

[345] L. I. Ronkin, Introduction to the Theory of Entire Functions of Several Variables, Transl. Math. Monogr., v.44, Amer. Math. Soc., Providence, RI 1974.

[346] ——, Some questions of the distribution of zeros of entire functions of several variables, Math. USSR Sbornik 16 (1972), 363-380.

[347] —— and A. Russakovskii, On extension with bounds of a function that is holomorphic on an algebraic set, Ann. Polon. Math. XLVI(1985), 403-431.

[348] E. Sanchez-Palencia, Non-Homogeneous Media and Vibration Theory, Springer-Verlag, Berlin e.a. 1980.

[349] J.-J. Sansuc, V.A. Tkachenko, Spectral parametrization of non-selfadjoint Hill's operator, preprint, 1992.

[350] H. Schlichtkrull, Hyperfunctions and Harmonic Analysis on Symmetric Spaces, Birkhäuser, Boston e.a. 1984.

[351] G. Segal, Fredholm complexes, Quart. J. Math. Oxford Ser. 21(1970), 385-402.

[352] M.I. Serov, On some properties of non-selfadjoint differential operators of the second order, Doklady Akad. Nauk SSSR 31(1960), no.1, 27-29.(In Russian)

[353] A.S. Shamaev, The behavior of solutions of differential equations in a half-strip, Russian Math. Surveys 35:1 (1981), 231-232.

[354] D. Shenk and M.A. Shubin, Asymptotic expansion of the state density and the spectral density and the spectral function of a Hill operator, Math. USSR Sbornik 56(1987), no.2, 473-490.

[355] I. Shigekawa, Spectral properties of Schrödinger operators with magnetic fields for a spin $\frac{1}{2}$ particle, J. Funct. Anal. 101(1991), 255-285.

[356] E. Shnol′, On the behavior of eigenfunctions of the Schrödinger equation, Mat. Sbornik 42(1957), 273-286. (In Russian)

[357] M.A. Shubin, On holomorphic families of subspaces of a Banach space, Mat. Issled. 5(1970), no.4, 153-165. (In Russian)

[358] ——, Spectral properties and spectrum distribution function of a transversally elliptic operator, Proc. of Petrovskii Seminar, 8(1982), Moscow, 239-258. (In Russian)

[359] ——, Pseudodifferential Operators and Spectral Theory, Springer-Verlag, Berlin e.a. 1987

[360] ——, Weak Bloch property and weight estimates for elliptic operators. Seminaire Equations aux derivees partielles, 1989-1990, Ecole Polytechnique, Palaiseau, Expose n.V., to appear in Asterisque.

[361] ——, Spectral theory of elliptic operators on non-compact manifolds, preprint,

MIT 1991.

[362] A.S. Shvarts, On the homotopic topology of Banach spaces, Soviet Math. Dokl. 5(1964), 57-59.

[363] B. Simon, Trace Ideals, London Math. Soc. Lect. Notes, no.35, Cambridge Univ. Press 1979.

[364] ——, Schrödinger semigroups, Bulletin of the Amer. Math. Soc. 7(1982), no.3, 447-526.

[365] Siu Y.T., Noether-Lasker decomposition of coherent analytic subsheaves, Trans. Amer. Math. Soc. 135(1969), 375-385.

[366] —— and G. Trautman, Gap-sheaves and extensions of coherent analytic subsheaves, Lect. Notes in Math., v. 172, Springer Verlag, Berlin e.a. 1971.

[367] J. Sjöstrand, Microlocal analysis for the periodic magnetic Schrödinger equation and related questions, in: Microlocal Analysis and Applications, Lect. Notes in Phys., v.1495, Springer-Verlag, Berlin e.a. 1991, 237-332.

[368] M.M. Skriganov, Proof of the Bethe-Sommerfeld conjecture in dimension two, Soviet Math. Dokl. 20(1979), 956-959.

[369] ——, On the Bethe-Sommerfeld conjecture, Soviet Math. Dokl. 20(1979), 89-90.

[370] ——, Proof of the Bethe-Sommerfeld conjecture in dimension three, preprint, LOMI, P-6-84, Leningrad 1984. (In Russian)

[371] ——, The spectrum band structure of the three dimensional Schrödinger operator with periodic potential, Invent. Math. 80(1985), 107-121.

[372] ——, Geometric and arithmetic methods in the spectral theory of

multidimensional periodic operators, Proc. of the Steklov Inst. of Math. 171(1985), 1-117.

[373] V.A. Solonnikov, On boundary problems for general linear parabolic systems of differential equations, Proc. of the Steklov Inst. of Math. 83(1965), 1-163. (In Russian)

[374] R. Strichartz, Harmonic analysis as spectral theory of Laplacians, J. Funct. Anal. 87(1989), 51-148.

[375] ——, L^p harmonic analysis and Radon transforms on the Heisenberg group, J. Funct. Anal. 96(1991), 350-406.

[376] D. Sullivan, Related aspects of positivity in Riemannian geometry, J. Diff. Geom. 25 (1987), no.3, 327-351.

[377] T. Sunada, Riemannian coverings and isospectral manifolds, Ann. Math. 121(1985), 169-186.

[378] ——, Fundamental groups and Laplacians, Lect. Notes Math. v.1339, Springer-Verlag, Berlin e.a. 1988, 248-277.

[379] ——, Fundamental groups and Laplacians, Kinokuniya, Tokyo 1988. (In Japanese).

[380] ——, Unitary representations of fundamental groups and the spectrum of twisted Laplacians, Topology 28(1990), no.2, 125-132.

[381] ——, A periodic Schrödinger operator on an abelian cover, J. Faculty of Sci., Univ. Tokyo, Sec. IA, 37(1990), no.3, 575-583.

[382] ——, Group C^*-algebras and the spectrum of a periodic Schrödinger operator on a manifold, Canadian J. of Math. 44(1)(1992), 180-193.

[383] —— and M. Nishio, Trace formulae in spectral geometry, in: Proc. of the International Congress of Mathematicians, Kyoto 1990, Springer-Verlag, Tokyo-Berlin-Heidelberg-New York 1991, 577-585.

[384] V. Sviridov, On completeness of quasi-energy states for the Schrödinger equation with time-periodic potential, preprint, no. 3492-83, VINITI, Moscow 1983. (In Russian)

[385] ——, On completeness of Floquet solutions for hyperbolic equations and systems, preprint, VINITI, no. 2436-85, Moscow 1985. (In Russian)

[386] ——, On completeness of quasi-energy states, Dokl. Akad. Nauk SSSR 274(1984), no.6, 1366-1367. (In Russian. English transl. in Soviet Math. Dokl.)

[387] ——, Some Problems of the Theory of Differential Equations with Periodic Coefficients, Ph.D. Dissertation, Voronezh 1986. (In Russian)

[388] L.E. Thomas, Time dependent approach to scattering from impurities in a crystal, Comm. Math. Phys. 33(1973), 335-343.

[389] E.C. Titchmarsh, Eigenfunction Expansions Associated with Second-Order Differential Equations, Part II, Claredon Press, Oxford 1958.

[390] ——, Theory of Functions, Claredon Press, Oxford-London 1939.

[391] V.A. Tkachenko, Spectral analysis of the one-dimensional Schrödinger operator with a periodic complex-valued potential, Soviet Math. Dokl, 5(1964), 413-415.

[392] ——, Spectral analysis of the non-selfadjoint Hill operator, Doklady Akad. Nauk SSSR, 322(1992), no.2., 248-252.(In Russian. English transl. in Soviet Math. Dokl.)

[393] ——, Discriminants and generic spectra of non-selfadjoint Hill's operators,

preprint, 1992.

[394] H. Triebel, Interpolation Theory, Functional Spaces, Differential Operators, VEB Deutcher-Verlag der Wissenschaften; North Holand Pub. Co., Amsterdam-New York-Oxford 1978.

[395] N. Turbe, Applications of Bloch expansions to periodic elastic and viscoelastic media, Math. Meth. in Appl. Sci., 4(1982), 433-449.

[396] K. Uhlenbeck, Generic properties of eigenfunctions, Amer. J. Math. 98(1976), no.4, 1059-1078.

[397] K.G. Valeev and V.D. Pervak, The study of stability of the solutions of linear partial differential equations with periodic coefficients, Mat. Fiz. (Kiev) 18 (1975), 69-72. (In Russian)

[398] __, __, Justification and application of Hill's method to parabolic equations, in: Physics and Mechanics of Non-Linear Phenomena, Naukova Dumka, Kiev 1975, 5-15. (In Russian)

[399] O.A. Veliev, Asymptotic formulas for the eigenvalues of a multidimensional Schrödinger operator , and periodic differential operators, preprint no.157, Inst. Phys. Acad. Sci. Azerb. SSR, Baku, 1985. (In Russian)

[400] __, Asymptotic formulae for eigenvalues of periodical Schrödinger operator and Bethe-Sommerfeld conjecture, Funct. Anal. i Prilozh. 21(1987), 1-15. (In Russian. English transl. in Funct. Anal. Appl.)

[401] A.B. Venkov, Spectral Theory of Automorphic Functions and its Applications, Kluwer Acad. Publ., Dordrecht 1990.

[402] H. Widom, Perturbing Fredholm operators to obtain invertible operators, J. Funct. Anal. 20(1975), 26-31.

[403] M.I. Weinstein and J.B. Keller, Hill's equation with a large potential, SIAM J. Appl. Math. 45(1985), 200-214.

[404] ——, ——, Asymptotic behavior of stability regions for Hill's equation, SIAM J. Appl. Math. 45(1987), 941-958.

[405] C. Wilcox, Theory of Bloch waves, J. Analyse Math. 33(1978), 146-167.

[406] K. Yajima, Scattering theory for Schrödinger equation with potential periodic in time, J. Math. Soc. Japan 29(1977), 729-743.

[407] ——, Resonances for the AC-Stark effect, Commun. Math. Phys. 78(1982), 331-352.

[408] —— and H. Kitada, Bound states and scattering theory for time periodic Hamiltonians, Annales Inst. H. Poincare, ser.A 39(1983), 145-157.

[409] M.G. Zaidenberg, On Fredholm operators that depend holomorphically on a parameter, Mat. Issled. 6(1971), no.2, 142-149. (In Russian)

[410] ——, On complex-analytic Banach manifolds in connection with problems of operator theory, Voronezh Gos. Univ. Trudy Nauchn.-Issled. Inst. Mat. 3(1971), 35-43. (In Russian)

[411] ——, S.G. Krein, P.A. Kuchment, and A.A. Pankov, Banach bundles and linear operators, Russian Math. Surveys 30(1975), no.5, 101-157.

[412] J. Zak, Magnetic translation group. I, II, Physical Review 134(1964), no. 6A, A1602-A1611.

[413] O. Zariski and P. Samuel, Commutative Algebra, v.I, Van Nostrand, Toronto-New York-London 1962.

[414] J.M. Ziman, Principles of the Theory of Solids, 2nd ed., Cambridge Univ. Press 1972.

[415] Ya.B. Zel'dovich, Scattering and emission of a quantum system in a strong electromagnetic wave, Soviet Physics Uspekhi 16(1973), 427-433.

[416] A.M. Zverkin, On completeness of the system of Floquet solutions for equations with delays, Differencial'nye Uravneniya 4(1968), no.3, 474-478. (In Russian)

[417] E. Brown, Bloch electrons in a uniform magnetic field, Phys. Review A 133(1964), A1038-A1044.

[418] A. Dikanskii, Problems adjoint to elliptic pseudodifferential boundary problems, Dokl. Akad. Nauk SSSR 200(1971), no.5, 1020-1023. (In Russia. English transl. in Soviet Math. Dokl.)

[419] M. Kashiwara, A. Kowata, K. Minemura, K. Okamoto, T. Oshima and M. Tanaka, Eigenfunctions of invariant differential operators on a symmetric space, Ann. Math. 107(1978), 1-39.

[420] B. Kostant, On the existence and irreducibility of certain series of representations, Bull. Amer. Math. Soc. 75(1969), 627-642.

Index of symbols

Index

Titles previously published in the series

OPERATOR THEORY: ADVANCES AND APPLICATIONS
BIRKHÄUSER VERLAG

1. **H. Bart, I. Gohberg, M.A. Kaashoek:** Minimal Factorization of Matrix and Operator Functions, 1979, (3-7643-1139-8)
2. **C. Apostol, R.G. Douglas, B. Sz.-Nagy, D. Voiculescu, Gr. Arsene** (Eds.): Topics in Modern Operator Theroy, 1981, (3-7643-1244-0)
3. **K. Clancey, I. Gohberg:** Factorization of Matrix Functions and Singular Integral Operators, 1981, (3-7643-1297-1)
4. **I. Gohberg** (Ed.): Toeplitz Centennial, 1982, (3-7643-1333-1)
5. **H.G. Kaper, C.G. Lekkerkerker, J. Hejtmanek:** Spectral Methods in Linear Transport Theory, 1982, (3-7643-1372-2)
6. **C. Apostol, R.G. Douglas, B. Sz-Nagy, D. Voiculescu, Gr. Arsene** (Eds.): Invariant Subspaces and Other Topics, 1982, (3-7643-1360-9)
7. **M.G. Krein:** Topics in Differential and Integral Equations and Operator Theory, 1983, (3-7643-1517-2)
8. **I. Gohberg, P. Lancaster, L. Rodman:** Matrices and Indefinite Scalar Products, 1983, (3-7643-1527-X)
9. **H. Baumgärtel, M. Wollenberg:** Mathematical Scattering Theory, 1983, (3-7643-1519-9)
10. **D. Xia:** Spectral Theory of Hyponormal Operators, 1983, (3-7643-1541-5)
11. **C. Apostol, C.M. Pearcy, B. Sz.-Nagy, D. Voiculescu, Gr. Arsene** (Eds.): Dilation Theory, Toeplitz Operators and Other Topics, 1983, (3-7643-1516-4)
12. **H. Dym, I. Gohberg** (Eds.): Topics in Operator Theory Systems and Networks, 1984, (3-7643-1550-4)
13. **G. Heinig, K. Rost:** Algebraic Methods for Toeplitz-like Matrices and Operators, 1984, (3-7643-1643-8)
14. **H. Helson, B. Sz.-Nagy, F.-H. Vasilescu, D.Voiculescu, Gr. Arsene** (Eds.): Spectral Theory of Linear Operators and Related Topics, 1984, (3-7643-1642-X)
15. **H. Baumgärtel:** Analytic Perturbation Theory for Matrices and Operators, 1984, (3-7643-1664-0)
16. **H. König:** Eigenvalue Distribution of Compact Operators, 1986, (3-7643-1755-8)
17. **R.G. Douglas, C.M. Pearcy, B. Sz.-Nagy, F.-H. Vasilescu, D. Voiculescu, Gr. Arsene** (Eds.): Advances in Invariant Subspaces and Other Results of Operator Theory, 1986, (3-7643-1763-9)

18. **I. Gohberg** (Ed.): I. Schur Methods in Operator Theory and Signal Processing, 1986, (3-7643-1776-0)
19. **H. Bart, I. Gohberg, M.A. Kaashoek** (Eds.): Operator Theory and Systems, 1986, (3-7643-1783-3)
20. **D. Amir:** Isometric characterization of Inner Product Spaces, 1986, (3-7643-1774-4)
21. **I. Gohberg, M.A. Kaashoek** (Eds.): Constructive Methods of Wiener-Hopf Factorization, 1986, (3-7643-1826-0)
22. **V.A. Marchenko:** Sturm-Liouville Operators and Applications, 1986, (3-7643-1794-9)
23. **W. Greenberg, C. van der Mee, V. Protopopescu:** Boundary Value Problems in Abstract Kinetic Theory, 1987, (3-7643-1765-5)
24. **H. Helson, B. Sz.-Nagy, F.-H. Vasilescu, D. Voiculescu, Gr. Arsene** (Eds.): Operators in Indefinite Metric Spaces, Scattering Theory and Other Topics, 1987, (3-7643-1843-0)
25. **G.S. Litvinchuk, I.M. Spitkovskii:** Factorization of Measurable Matrix Functions, 1987, (3-7643-1843-X)
26. **N.Y. Krupnik:** Banach Algebras with Symbol and Singular Integral Operators, 1987, (3-7643-1836-8)
27. **A. Bultheel:** Laurent Series and their Pade Approximation, 1987, (3-7643-1940-2)
28. **H. Helson, C.M. Pearcy, F.-H. Vasilescu, D. Voiculescu, Gr. Arsene** (Eds.): Special Classes of Linear Operators and Other Topics, 1988, (3-7643-1970-4)
29. **I. Gohberg** (Ed.): Topics in Operator Theory and Interpolation, 1988, (3-7634-1960-7)
30. **Yu.I. Lyubich:** Introduction to the Theory of Banach Representations of Groups, 1988, (3-7643-2207-1)
31. **E.M. Polishchuk:** Continual Means and Boundary Value Problems in Function Spaces, 1988, (3-7643-2217-9)
32. **I. Gohberg** (Ed.): Topics in Operator Theory. Constantin Apostol Memorial Issue, 1988, (3-7643-2232-2)
33. **I. Gohberg** (Ed.): Topics in Interplation Theory of Rational Matrix-Valued Functions, 1988, (3-7643-2233-0)
34. **I. Gohberg** (Ed.): Orthogonal Matrix-Valued Polynomials and Applications, 1988, (3-7643-2242-X)
35. **I. Gohberg, J.W. Helton, L. Rodman** (Eds.): Contributions to Operator Theory and its Applications, 1988, (3-7643-2221-7)
36. **G.R. Belitskii, Yu.I. Lyubich:** Matrix Norms and their Applications, 1988, (3-7643-2220-9)
37. **K. Schmüdgen:** Unbounded Operator Algebras and Representation Theory, 1990, (3-7643-2321-3)
38. **L. Rodman:** An Introduction to Operator Polynomials, 1989, (3-7643-2324-8)
39. **M. Martin, M. Putinar:** Lectures on Hyponormal Operators, 1989, (3-7643-2329-9)
40. **H. Dym, S. Goldberg, P. Lancaster, M.A. Kaashoek** (Eds.): The Gohberg Anniversary Collection, Volume I, 1989, (3-7643-2307-8)

41. **H. Dym, S. Goldberg, P. Lancaster, M.A. Kaashoek** (Eds.): The Gohberg Anniversary Collection, Volume II, 1989, (3-7643-2308-6)
42. **N.K. Nikolskii** (Ed.): Toeplitz Operators and Spectral Function Theory, 1989, (3-7643-2344-2)
43. **H. Helson, B. Sz.-Nagy, F.-H. Vasilescu, Gr. Arsene** (Eds.): Linear Operators in Function Spaces, 1990, (3-7643-2343-4)
44. **C. Foias, A. Frazho:** The Commutant Lifting Approach to Interpolation Problems, 1990, (3-7643-2461-9)
45. **J.A. Ball, I. Gohberg, L. Rodman:** Interpolation of Rational Matrix Functions, 1990, (3-7643-2476-7)
46. **P. Exner, H. Neidhardt** (Eds.): Order, Disorder and Chaos in Quantum Systems, 1990, (3-7643-2492-9)
47. **I. Gohberg** (Ed.): Extension and Interpolation of Linear Operators and Matrix Functions, 1990, (3-7643-2530-5)
48. **L. de Branges, I. Gohberg, J. Rovnyak** (Eds.): Topics in Operator Theory. Ernst D. Hellinger Memorial Volume, 1990, (3-7643-2532-1)
49. **I. Gohberg, S. Goldberg, M.A. Kaashoek:** Classes of Linear Operators, Volume I, 1990, (3-7643-2531-3)
50. **H. Bart, I. Gohberg, M.A. Kaashoek** (Eds.): Topics in Matrix and Operator Theory, 1991, (3-7643-2570-4)
51. **W. Greenberg, J. Polewczak** (Eds.): Modern Mathematical Methods in Transport Theory, 1991, (3-7643-2571-2)
52. **S. Prössdorf, B. Silbermann:** Numerical Analysis for Integral and Related Operator Equations, 1991, (3-7643-2620-4)
53. **I. Gohberg, N. Krupnik:** One-Dimensional Linear Singular Integral Equations, Volume I, Introduction, 1992, (3-7643-2584-4)
54. **I. Gohberg, N. Krupnik:** One-Dimensional Linear Singular Integral Equations, Volume II, General Theory and Applications, 1992, (3-7643-2796-0)
55. **R.R. Akhmerov, M.I. Kamenskii, A.S. Potapov, A.E. Rodkina, B.N. Sadovskii:** Measures of Noncompactness and Condensing Operators, 1992, (3-7643-2716-2)
56. **I. Gohberg** (Ed.): Time-Variant Systems and Interpolation, 1992, (3-7643-2738-3)
57. **M. Demuth, B. Gramsch, B.W. Schulze** (Eds.): Operator Calculus and Spectral Theory, 1992, (3-7643-2792-8)
58. **I. Gohberg** (Ed.): Continuous and Discrete Fourier Transforms, Extension Problems and Wiener-Hopf Equations, 1992, (ISBN 3-7643-2809-6)
59. **T. Ando, I. Gohberg** (Eds.): Operator Theory and Complex Analysis, 1992, (3-7643-2824-X)
60. **P.A. Kuchment:** Floquet Theory for Partial Differential Equations, 1993, (3-7643-2901-7)

GPSR Compliance

The European Union's (EU) General Product Safety Regulation (GPSR) is a set of rules that requires consumer products to be safe and our obligations to ensure this.

If you have any concerns about our products, you can contact us on ProductSafety@springernature.com

In case Publisher is established outside the EU, the EU authorized representative is:

Springer Nature Customer Service Center GmbH
Europaplatz 3
69115 Heidelberg, Germany

Batch number: 10577878

Printed by Printforce, the Netherlands